# Housekeeping Handbook For Institutions, Business & Industry

(Revised Edition)

by

Edwin B. Feldman, P.E.

*A World of Books That Fill a Need*

Frederick Fell Publishers, Inc. New York, N.Y.

**Library of Congress Cataloging in Publication Data**

Feldman, Edwin B
Housekeeping handbook for institutions, business & industry.

1. Industrial housekeeping. I. Title.
TX955.F4 1978 647 78-15398
ISBN 0-8119-0072-X

New Printing 1982



For information address:

Frederick Fell Publishers, Inc.
386 Park Avenue South
New York, New York 10016

Published simultaneously in Canada by:
Fitzhenry & Whiteside,
Limited, Toronto

MANUFACTURED IN THE UNITED STATES OF AMERICA

1 2 3 4 5 6 7 8 9 0

Parts of this book were previously published under the title: "Industrial Housekeeping"

# Contents

## PART I – MANAGEMENT



(continued)

## PART II – SUPERVISION



## PART III – OPERATIONS

# Foreword

I first understood that there was more to housekeeping than met the eye in 1946 when I obtained a part-time job in a factory which manufactured sanitation supplies.

When I graduated from Georgia Institute of Technology as an Industrial Engineer in 1950, I became a Plant Engineer of that company. Over a period of ten years, and through various job positions, I also spent a good deal of time with the company's sales staff—they asked me questions about time requirements for cleaning, the proper equipment or chemicals to use, procedures, and the like. It was apparent that this information was not available to them or to their customers on a commercial basis without relationship to the sale of products.

In order to satisfy this need, Service Engineering Associates, Inc. was founded in 1960 as the nation's only firm of professional engineers specializing in custodial consulting. (The firm is completely independent, with no ties to contract cleaning or supply firms of any kind.)

This eventually led to the writing of this book, which is based on consulting experiences for clients all over the United States and Canada, as well as exposure to thousands of housekeeping managers and supervisors at seminars which our firm conducts regularly in various cities.

The writing proceeded in stages: First, as some magazine articles in specialty publications, then as a book entitled "Industrial Housekeeping" published by MacMillan; and finally the first edition of this work by Frederick Fell Publishers, Inc., where I received a great deal of help and encouragement from my good friends Frederick Fell and Charles Nurnberg.

This new edition contains a good deal of revised material and a number of totally new chapters, all in an improved arrangement.

So the book has had a great deal of input from various sources, including material from members of the consulting staff of Service Engineering Associates, and seminar participants.

Readers have also made valuable suggestions, which brings us to you: Your criticism and recommendations are eagerly sought!

Edwin B. Feldman, P.E.

# Part I
# MANAGEMENT

# 1 Housekeeping Today

BILLIONS of dollars are spent every year on housekeeping and sanitation. The undertaking involves the efforts of hundreds of thousands of custodial workers. A substantial fraction of the effort is wasted. This enormous waste is recoverable—not in some theoretical sense, but practically.

Our way of life depends upon the efficiency with which we can produce and distribute goods and perform services. Many advances have been made through the economies and efficiencies stemming from mechanization, automation, standardization, invention, better handling of materials, improvement of materials, analysis of methods, and, fortunately, by a number of other means. Both labor and management can take the credit for these improvements; on the other side of the coin, both labor and management, sometimes unwittingly, have also at times been guilty of retarding this process.

Housekeeping, in a sense, has been swept under management's rug. Primary attention is given in industry to the improvement of the unit production cost, and rightfully so. In addition to this, in recent years growing attention has been focused on indirect costs as well. But the fact is, housekeeping has held out as one of the last potential sources of immediate and considerable cost improvement, by resisting or escaping the careful eye and sharp pencil of internal industrial engineers, methods analysts, and housekeeping personnel themselves. The typical industrial housekeeping operation —including those in the plants of our industrial giants—accomplishes at least 25 percent less than it practically should with the funds available, or costs at least 25 percent more than it practically should.

In the typical plant, management has failed to make an investment in a study of the housekeeping function and its subsequent development; yet the performance of a typical housekeeping operation often costs up to five times more today than it did twenty years—less than a generation—ago.

## Cost of Housekeeping

Often management is not aware of the total cost of housekeeping, or aware that the cost may be out of line. That housekeeping and other indirect costs are not minor items is borne out by the fact that the initial cost of many buildings is duplicated in maintenance costs over a period of twenty years or less. An analysis of the total cost of the housekeeping effort is often a good first step in the investigation of a housekeeping situation. Such a cost analysis should include:

Direct wages of workers
Salaries of supervisors
Overtime costs
Cost of cleaning performed on contract
Cost of housekeeping materials
Amortized cost of housekeeping equipment
Costs borne by personnel outside the housekeeping department in cleaning operations

A more detailed picture can be obtained through consideration of such items as:

The space devoted to housekeeping storage and facilities
Overhead costs for housekeeping (utilities, services, etc.)
Purchasing and stock control costs
The value of management time consumed in housekeeping
Fringe benefits

Unfortunately a large segment of management believes it has a more effective or economical housekeeping operation than the evidence warrants. Your housekeeping operation can provide you with a housekeeping level

25 percent higher than you are now receiving, or if you feel your current housekeeping level is adequate, it is costing you 25 percent more than it should. If your operations have been surveyed and programmed by your own methods people, the savings yet available to you may possibly be somewhat less, but they are still considerable. For every typical improvement brought about internally in the housekeeping effort, there are still numerous outmoded practices to be found. It must be remembered that housekeeping is an annually recurring expense (with constantly rising costs) and that any improvement made would tend to be recurrent.

The one category of industry which might be expected to be relatively exempt from outdated housekeeping problems would be the food-processing industry—meat packers, dairy processors, candy makers, canners, bakers, etc. This is not so. Although the processing areas themselves are indeed generally cleaner than those in most other types of industries, the very problem of dealing with perishable products creates a series of difficulties that are superimposed on the range of problems found in other industries. In the non-processing areas of the plant, such as the offices and personnel facilities, the usual problems are found.

Since the fall of 1960, our engineers have been privileged to survey and program the housekeeping operations in over a hundred organizations annually; the work of thousands of housekeeping workers has thus been examined, analyzed, and rescheduled. Conditions found today are often much the same as they must have been decades ago as concerns some aspects of the problem; however, the problem now is much more serious and urgent because of the vastly increased cost involved, primarily due to higher wage standards. And wages account for all but 7 or 8 percent of industry's direct housekeeping costs! Surely, in today's competitive market, *every* road to cost improvement must be explored.

Although good industrial housekeeping provides and sustains a beneficial environment, unfortunately it has been difficult to pin down these benefits in directly measurable terms. But we are learning that these benefits are *real* and *important.*

## Today's Problems

This is the position of many organizations today:

1. Top management is often indifferent, considering housekeeping as a "necessary evil." Management is not conversant with the hazards and costs of poor housekeeping or the advantages of good housekeeping. Demands are made to

"get the place cleaned up" without any conception of how such general results are to be obtained, or their costs.

2. Often, the size of the housekeeping staff is directly proportioned to current production levels, whereas the requirements of housekeeping do not relate directly at all. For example, even though production may have been cut 20 percent, all areas of the plant will still be used and will require cleaning. All rest-room fixtures will be used daily, although not as often. But they will still need daily cleaning. A 20 percent drop in production in a given plant may only decrease the custodial load by 5 percent. Most of the decrease will result from reduced waste collection and removal. The situation changes, of course, when entire floors or buildings are closed.

3. Management has failed to detail the objectives and responsibilities of the housekeeping department and of other personnel, as well, relating to housekeeping.

4. The housekeeping department is not given the same organizational integrity and status as other departments. It may suffer arbitrary budget cuts and yet be given no opportunity or procedure to outline the consequences or alternative possibilities to management. Promising personnel are pulled out of the housekeeping department to be placed in "important" work.

5. Often industrial management does not recognize the housekeeping function as requiring a separate identity, although related functions such as safety and industrial hygiene have been so identified and given special attention. These very services cannot be completely effective without the support of the housekeeping operation. Contributing to this situation is the fact that housekeeping can be dispersed throughout the various departments, in part if not fully. Thus is lost the opportunity for standardization and intensive supervision.

6. The housekeeping department is often given a negative objective, with descriptions and understandings of its function expressed as bad conditions which it should prevent or avoid. It often seems to the custodian that his working life is devoted to keeping things from happening. Rather, housekeeping should be a creative concept, with the housekeeping department considered as providing conditions of benefit.

7. Furthermore, the attitude of the cleaning workers toward themselves is negative and often difficult to cope with. They generally consider themselves "at the bottom of the ladder" with nowhere to go. This attitude is amply reinforced—even justified at times—by existing conditions which require correction.

8. Supervision is spread too thinly. Many times a single housekeeping foreman is responsible for the direct supervision of scores of personnel working on three different shifts.

9. The housekeeping department has to perform nonhousekeeping duties which other departments do not want or which just do not seem to fit anywhere else. This can include such tasks as mail delivery, running errands, and chauffeuring. It must be said, because of the indivisibility of labor, that this situation may sometimes be justified.

10. The housekeeping workers are untrained in their jobs despite the fact that they are custodians of valuable property. In addition, they are dealing with

ever more complicated materials and equipment, and the effect of their work on productivity and quality control is of considerable importance.

11. Suppliers do not give suitable services, for the most part. The average sanitary-supply firm is not organized, or is it financially able, to give comprehensive field service.

12. Little has been done to correct the image of the "janitor" of twenty years ago. He is depicted as shuffling along with a soiled mop and bucket of black water. This contributes markedly to the problems of morale and employee and management attitudes.

13. Labor contracts sometimes create difficult situations for housekeeping, the most prevalent example of which is company-wide seniority. Since it is mistakenly believed that anyone can perform a housekeeping task, the turnover in the housekeeping department becomes significant during periods of changing production levels. In addition, in many plants, the practice of using group leaders (which can offer many advantages) cannot be followed because the best qualified worker may not get the job.

14. Decentralized housekeeping operations are found, which further compounds the problems of training, supervision, status and morale, standardization, etc. Decentralization of certain functions, with a centralized housekeeping core, is desirable in some plants, such as in the textile industry, however.

15. There is very little standardization in housekeeping levels, materials, methods, and equipment. This is often true even in different parts of the same plant.

16. Some organizations perform their basic cleaning operations on a contract basis. A comparison of the objective of industry (obtaining the most cleaning for the housekeeping dollar) with that of the contractor (retaining as much of the contract fee as possible for a profit) indicates the problems inherent in such an arrangement if a firm is not reputable.

17. Housekeeping is performed on a corrective rather than a preventive basis—a matter of "putting out fires." As a result, projects that should be performed on a periodic basis are not adequately programmed and carried out, and some important tasks are *never* performed.

18. Poor appearance, hazardous conditions, and bad odors often indicate housekeeping situations that should not be permitted to exist.

19. Although there are a number of timesaving housekeeping equipment items now on the market, and others being developed, mechanization in the housekeeping department has not kept pace with mechanization in other fields.

One of the fundamental problems, particularly in larger plants, is the attitude of the average employee toward his housekeeping responsibility. Very often, no responsibility is felt at all—and since all housekeeping work, of course, is performed because of human activities, such attitudes seriously affect the work load with which the custodian is faced each day. The situation becomes more serious because the fruits of the custodian's effort deteriorate almost overnight.

In a given heavy manufacturing industry, the proportion of custodians to other personnel may be one to a hundred. Consider the difference between the custodian working *with* the other workers toward better housekeeping conditions and the custodian struggling with the conditions created by a hundred careless workers. The average housewife, at least, is certainly familiar with the work load created by having to clean up after a number of thoughtless visitors!

## The Public Attitude

The same attitudes that workers exhibit while in the plant are also manifested outside the plant and are carried into the home. Although much could be done to improve the methods of household cleaning, the basic problem is that of attitude. Too many of our citizens consider any spot outside their homes or automobiles a suitable depository for litter and waste. For example, Keep America Beautiful, Inc., points out these facts:

> Over fifty million dollars annually are spent to remove litter from our main highways. The figure goes up considerably when one adds the cost of removing litter from parks, beaches, streams, city streets and other public areas. The United States Forest Service spends two and a half million dollars annually to clean up litter from our national forests alone.
>
> Almost unbelievable is the fact that nearly a thousand of us are killed each year, and one hundred thousand seriously injured, because of cars hitting or attempting to avoid objects thrown onto our highways.
>
> Seventy million dollars per year is the tab for fires caused by litter.

Apparently in our "affluent society," the control of waste and litter is considered to be "someone else's job," just as so many other responsibilities of the citizen are passed onto the unknown someone else.

## Determination of Cleanliness

A basic problem with which we are faced is the determination of housekeeping levels, that is, the determination of the standard of cleanliness of a facility and the measurement of the degree of approach to this standard. Certain specific variables, which are subject to direct measurement, are available:

1. Dust count
2. Reflectivity of surfaces
3. Bacteria count
4. Color intensity

***PLEASE HELP US BY CORRECTING THE FOLLOWING CONDITIONS:***

( ) CIGARETTE BUTTS ON FLOOR

( ) COFFEE CUPS ON DESK

( ) UNTIDY DESK

( ) FOOD & LIQUID IN WASTE BASKET

( ) PAPERS ON FLOOR

( ) MISC. MATERIAL ON FLOOR

( ) GENERAL HOUSEKEEPING

( ) OTHER ____________________

____________________

***THANK YOU***

SIGNED ____________________
Foreman
Service & Support

**We Need Your HELP To CLEAN Your Home**

**AWAY FROM HOME (over)**

A "help wanted" card left on the desk of an offender may help attain cooperation for the building service department, if the program is supported by top management.

5. Photometric measurement (light transmission through glass, foot-candle illumination level of lighting fixtures)
6. Thickness of protective coatings
7. Rate of abrasion
8. Coefficient of friction

Usually these are sophisticated determinations and are not practical to use for establishing everyday housekeeping levels. Rather, they are useful for specialized purposes, such as determining conditions in superclean areas or hospital operating rooms, and for product development. For everyday use, an arbitrary standard has to be set at the level desired by management. Achievements are compared with this standard on the basis of estimated percentages, or merely on the basis of whether or not the standard has been achieved, which is more convenient. This problem and its relationship to the frequency of job performance are discussed further in another chapter.

Industry's housekeeping problems have solutions; they are not always perfect solutions or are they easy to obtain. Some specific changes can bring startling results, but the basic and lasting improvements in housekeeping will be achieved by a deliberate and comprehensive approach to the problem, with an investigation of every phase of the housekeeping operation.

## A Housekeeping Check List

Where does your company stand with respect to housekeeping? Perhaps you would like to take a pencil and check your answers to these simple, but revealing, questions. The results should not only prove interesting but may also help you to pinpoint the strengths and weaknesses of your present operations.

### GENERAL

1. Is management satisfied with present levels of housekeeping?
2. Does top management show an active interest in plant housekeeping?
3. Has management voiced its housekeeping objectives?
4. Are you fully aware of, and adhering to, all regulatory statutes affecting housekeeping?
5. Have you ever undertaken an employee awareness program to build up interest in, and cooperation with, better housekeeping?
6. Have reasonable standards been established as goals for the housekeeping program?
7. Has an evaluation ever been made of the general morale of the housekeeping staff?

8. Have any steps been taken to improve morale, or maintain it at existing levels?

9. Is the housekeeping department given the opportunity to advise on surfaces, materials, etc., for new construction or renovation?

### COST CONTROL

10. Have you recently calculated your total annual housekeeping costs, including *all* costs relating to the cleaning function?

11. Have you means of comparing your plant housekeeping costs with those of other plants in your industry?

12. Has plant housekeeping been costed out by department or by job?

13. If so, are these figures adequate to reveal unusual situations?

14. Could some housekeeping jobs, currently performed by outside contractors, be handled at a lower cost by your own housekeeping department?

15. Could certain specialized cleaning jobs, such as window washing, be performed more economically or more safely on contract?

16. Have you studied current housekeeping work methods and procedures to see if savings could be effected?

17. Has full advantage been taken of the economies of mechanizing plant housekeeping?

### WORK STANDARDS

18. Has a time study ever been made of the various housekeeping jobs?

19. Do you have a basis for comparing your frequencies of housekeeping job performance with those found at other well-run establishments in the same industry for the same general types of areas?

20. Do you know whether the man-hours invested in plant housekeeping are consistent with the actual man-hour requirements for the job to be performed?

21. Are both daily and project cleaning performed on a scheduled basis?

22. Have acceptable performance levels been established for housekeeping workers?

23. Do cleaning schedules allow for seasonal variations?

### ORGANIZATION

24. Is plant housekeeping supervised on a centralized basis wherever practical?

25. Have you studied your housekeeping structure recently to determine whether a reorganization might bring about better results or savings?

26. Have definite lines of authority and reporting functions been established for the housekeeping department?

27. Are custodians performing tasks other than housekeeping (serving as messengers or chauffeurs, performing errands, etc.)?

28. Have you ever considered applying the group-leader principle to reduce supervisory costs and improve supervisory efficiency?

29. Have blueprints and floor plans been used with respect to the assignment of housekeeping tasks?

30. Does your housekeeping program permit variations in the size of the staff (due to changing production levels, seasonal variations, etc.) without disturbing the basic scheduling?

31. Can relief be provided for absent housekeeping workers without disrupting cleaning schedules?

32. If a new building or wing is put up, do you have an accurate means of predetermining housekeeping requirements and costs?

#### CUSTODIAL EFFICIENCY

33. Is a formal attempt being made to improve workers' skills and efficiencies through a continual training program in housekeeping methods and procedures?

34. Has the extent of each custodian's responsibility been fully outlined to him?

35. Does each custodian fully understand these responsibilities and their importance?

36. Have you ever formally tested custodians to determine their understanding of their jobs and their duties?

37. Do you adhere to specific qualifications for the hiring of housekeeping personnel?

38. Does your typical housekeeping worker consider himself as one of a team of specialists, responsible to a large degree for the safety of coworkers and the care of plant facilities?

39. Are your workers provided with the latest developments in housekeeping materials?

40. In the purchasing of supplies, is an unwritten code followed to supply "the cheapest and the least"?

41. Are physical examinations given to prospective housekeeping employees?

#### RECORD KEEPING

42. Are adequate records maintained to insure proper protection of surfaces at proper intervals (for example, the resealing of floors).

43. Are records maintained of the suggestions and complaints about plant housekeeping, the cost or amount of time necessary to remedy the situation, and the source?

44. If so, is the complaint record studied periodically to determine how the general situation can be improved?

45. Are regular reports made to enable management to check on housekeeping progress?

46. Are regular reports made to enable management to check on supervisory performance?

47. Are regular reports made to enable supervisors to check on custodial performance?

PLANT OPERATIONS

48. Is your accident record affected to any appreciable extent by faulty housekeeping procedures?

49. Do you have fire hazards which better housekeeping would remove?

50. Do you conduct regularly scheduled inspections of housekeeping and sanitation to avoid safety, health, and fire hazards?

51. Has the safety department been consulted concerning housekeeping practices?

52. Have housekeeping schedules been arranged to permit maximum efficiency along with minimum disruption of production, engineering, and clerical work?

53. Do you have an efficient system for stocking, inventorying, and distributing housekeeping equipment and supplies?

54. Are custodial storage facilities adequate from the standpoints of size, number, location, ventilation, lighting, furnishing, and plumbing?

55. Do employees complain about hand cleaners, odors in rest rooms, and other personnel items?

56. Do you operate your housekeeping system as *preventive* maintenance rather than *corrective* maintenance?

Your answers to these questions may indicate that your housekeeping operation is more fraught with potential improvements than you might at first have suspected.

In accounting, housekeeping falls into the category of costs known as "burden." This is part of the concept that housekeeping personnel are unproductive, second-grade workers who depend on other personnel for their existence. If the housekeeping operation is built on such a limited concept, the effectiveness of this operation will similarly be limited. In order to provide a realistic objective for housekeeping, we have to examine the objectives of the company as a whole—and when we do this we find that the objectives coincide. Housekeeping is an integral part of the whole organization, which is dedicated to making money through the production and distribution of a good product, while providing a good living for its workers. Housekeeping is more than just a service to production departments—it is one of the ways in which the cost of the plant operation is kept to a minimum, and therefore is a means to a greater profit.

When we are able to think of housekeeping in these terms, we are ready to move forward.

# 2

# Up To This Point

LET us take an example: your plant is proud of the product it produces and the position it holds in its industry.

Basically, you have a good team of executives that respects and uses an engineering approach to solving problems and cutting costs. But, despite this, the housekeeping and sanitation operation in your plant is often considered an annoyance and has become a sore subject because you suspect that money can be saved, and efficiency gained, but you hardly know where to start.

What has brought us to this point?

In looking back, it becomes readily clear that sanitation and cleanliness, even from the earliest pages of history, have positive connotations. The advent and advance of great civilizations have always been accompanied by considerable improvements in cleanliness and sanitation. Technology, culture, sociology, enlightenment, and even piety are associated with cleanliness. On the other hand, soil and disarray carry negative associations, such as barbarism, decadence, and ignorance. Your mental image of an Egyptian temple, as it was in its day, or a Greek or Roman building, is likely to be a vision of immaculate surfaces and conditions. But your image of the structures of the Huns or Bushmen is probably one of squalor, filth, and complete disorder.

The concurrent advance of sanitation with progress and civilization has not been accidental, but necessary. The growth of sanitation is completely interwoven and interdependent with the growth of sciences such as medicine, engineering, and city planning.

## Early Sanitation History

The Egyptians furnished us with the earliest recorded uses of sanitation in medicine. Their knowledge and respect of some phases of cleanliness is remarkable. Their embalming practices indicate their knowledge of the antiseptic qualities of certain chemicals. Laws were passed about the responsibility of the individual for cleanliness in his home and city. The priesthood was noted for its attitude toward cleanliness, an attitude that has been perpetuated by clergymen throughout the centuries.

It was clear to the ancient Hebrews that sanitation and health could not be separated. They founded prophylaxis (treatment to prevent disease), and many of their hygienic practices are described in the Bible. The high priests, in addition to their divine calling, also provided the guidance needed in hygiene and sanitation for their times. The laws of Moses contained sanitation principles which, by their fundamental nature, can be considered universally applicable.

Archaeology in Greece has uncovered proof of a considerable respect for, and wide use of, sanitation principles. Many of the practices and devices of the Greeks were not used again until a thousand years later! Most of us know of Hippocrates, called the father of medicine, for his Hippocratic Oath. It outlines the ethical creed of the medical profession and is still widely used and quoted. He should also be remembered for the fact that, in the fourth century B.C., he pointed out the value of hand cleaning, keeping soil excluded from wounds with medical dressings, and using boiling water for cleansing wounds.

In his *History of Medicine,*[1] F. H. Garrison gives us a summary of ancient contributions:

> The Babylonians had some intuition of the nature and prevention of communicable disease; the Jews originated medical jurisprudence and public hygiene and ordained a weekly day of rest; the Chinese introduced anthropometry

[1] Copyright, W. B. Saunders Co.

(measuring the human body), finger prints, massage (osteopathy), and many drugs; and the Hindus demonstrated that skill in operative surgery which has been a permanent possession of the Aryan race ever since.

## Military Sanitation

Even from the earliest times, sanitation has been improved and developed through military necessity. In Deuteronomy, Moses told of the proper hygienic procedures to be followed in the policing of military camps. The Romans were also leaders in military sanitation. Their field hospitals were models of sanitation for their time. Their walled cities were constructed and used, following basically sound sanitary principles, so that health and physical resistance could be maintained for long periods during sieges. Indeed, up into modern times, military medicine has contributed to the advance of hygiene and sanitation. The work of General Walter Reed and Dr. William C. Gorgas, culminating in the completion of the Panama Canal, is well known to all, for example.

Pioneers in modern military hygiene were the Englishmen Sir John Pringle, a pioneer of antisepsis, and James Lind, whose field was naval hygiene.

## Later History

Since civilization and sanitation are interdependent, the conditions of the Dark Ages created a setback in sanitary practices. Sanitation resumed its advance in medieval times. Hospitals were organized and nursing of the sick was undertaken. Nevertheless, the fantastic epidemics of these times directly resulted from poor sanitation practices. In the middle of the fourteenth century, over 60 million people—a fourth of the population—died of bubonic plague, then known as the Black Death.

In 1675, Antony Von Leeuwenhoek provided the tool necessary for the foundation of bacteriology when, through his hobby of grinding lenses, he invented the microscope.

Industrial hygiene and the consideration of occupational diseases began in the eighteenth century. The founder of public hygiene is considered to be the German, Johann Frank. The sanitary engineer of today carries this indispensable work forward.

Tribute for an organized approach to the housekeeping of buildings must go to John Howard who, at about the time of the American Revolution, devised methods for the sanitation of hospitals and prisons. He recommended smooth floors for easy cleaning and set up delousing and isolation systems. He pointed out, for prisons, the relationship between the death rate and floor space per inmate.

Pioneer thinkers about cross-infection and the acquisition of disease through unsanitary conditions were Joseph Lister in England and Ignaz Semmelweis in Austria. Lister's technique was to use a spray of carbolic acid for sterilization. Semmelweis proved that childbed fever was a form of blood poisoning that could be avoided through simple practices of cleanliness. It is interesting that many of the sanitation problems that faced these early pioneers are still troubling our hospitals today. The struggle for sanitary conditions, unfortunately, has been deterred by the development of antibiotics.

The history of sanitation is interesting to read. We find, for example, that sanitary principles have largely guided the development of table manners and social conduct. Unfortunately, the scope of this work does not permit detailed treatment of the subject. Certainly the names of the pioneers and thinkers in this field deserve honor and remembrance, and you may wish to read some of the books on this subject in your library. Based on the dedication and work of a relatively few great men, we now know that sanitation, and the good housekeeping required for sanitation, must be provided if we are to have conditions leading to good health and to other important benefits.

The basic problem confronting us now is the method of providing these conditions. We must face the fact that housekeeping has become a universal problem in American industry.

## Contributors to the Housekeeping Problem

Some members of certain groups have actually intensified the problems which face us, often unknowingly:

1. Architects, who may build structures without regard to the maintenance which must follow. This subject is pursued further in a chapter on "Planning for Better Maintenance."

2. Design engineers and decorators, who may fail to obtain the viewpoint of the housekeeping department and housekeeping consultants on the method and cost of maintaining new or renovated structures.

3. Labor organizations, who may fail to impart to their members a proprietary attitude about sanitation and housekeeping within the industry in which they work.

4. Management, which may fail to consider housekeeping as a part of basic planning, instead of an item after the fact.

5. Our colleges and engineering schools, which may fail because they have

generally ignored housekeeping and sanitation in their curricula. Usually, little is said to the student concerning his responsibilities toward the facilities he is using while in school. The wonderful opportunity of instilling, in the average citizen, a positive attitude toward this fundamental aspect of civilization has been rather generally, and sadly, ignored.

The sanitary-supply industry, unfortunately, must be censured for malfeasance, misfeasance, and nonfeasance of its duites with respect to its customers. Although some firms do provide worthwhile and continuing services to their customers, the largest of which are the industrial users, most often the sanitary-supply industry has attempted to "load" the customer with merchandise and equipment which may or may not be to the customer's advantage.

Much of the poor practice in the sanitary-supply industry results from a fiercely competitive situation with overemphasis on sales and underemphasis on service. Too much attention is given to price and not enough to the problems of use. As management groups become more keenly aware of the magnitude of their investment in housekeeping, and buyers have the opportunity to more clearly relate product to end use, the problem will tend to correct itself.

## Professional Specialization

As in every other aspect of human endeavor, specialization has taken place. The three general classes of professionals engaged in one or more aspects of sanitation, housekeeping, or hygiene are the *sanitarian,* the *sanitary engineer,* and the *industrial engineer.* The various spheres of responsibility and authority for these professions are not yet fully defined and agreed upon.

### THE SANITARIAN

The *sanitarian* is generally concerned with the clean and wholesome nature of perishable products. He also overlaps the field of the sanitary engineer in public hygiene.

The Sanitarian's Joint Council, in November 1956, proposed a uniform definition for sanitarians:

> A Sanitarian—is a person who by education and experience in the physical, biological, and sanitary sciences is qualified to carry out educational, investigational and technical duties in the field of sanitation.

Also in 1956, the *American Journal of Public Health* [2] defined the sanitarian as:

a person whose education and experience in the biological and sanitary sciences qualifies him to engage in the promotion and protection of the public health. He applies technical knowledge to solve problems of a sanitary nature and develops methods and carries out procedures for the control of those factors of man's environment which affect his health, safety, and well-being.

#### THE SANITARY ENGINEER

The *sanitary engineer* is generally concerned with sewage disposal, water purification, industrial diseases, and other such aspects of industrial hygiene.

The American Sanitary Engineering Intersociety Board has adopted the definition of the National Academy of Sciences for the sanitary engineer:

The professional occupational title "Sanitary Engineer" shall apply to a graduate of a full four-year, or longer, course leading to a Bachelor's or higher degree at an educational institution of recognized standing with major study in engineering, who has fitted himself by suitable specialized training, study, and experience (a) to conceive, design, appraise, direct, and manage engineering works and projects developed, as a whole or in part, for the protection and promotion of the public health, particularly as it relates to the improvement of man's environment, and (b) to investigate and correct engineering works and other projects that are capable of injury to the public health by being or becoming faulty in conception, design, direction, or management.

#### THE INDUSTRIAL ENGINEER

The *industrial engineer* is concerned with the productivity of the worker. This involves an interrelationship of variables dealing with machines, materials, and methods, all in a matrix of economic analysis. His job is to investigate, specify, predict, and evaluate results from programs or systems. His tools come from management and the mathematical, mechanical, physical, psychological, and social sciences. The effectiveness of the industrial engineer is often closely related to his ability to communicate with, and inspire confidence and enthusiasm, in others. His stock in trade is *time*.

The industrial engineer becomes intimately familiar with certain adverbs and pronouns since he must ask, and propose answers for, questions such as these:

1. *Who* will do the work and make certain that it is done?
2. *What* work is necessary to be done and can portions of the work be eliminated or combined?

[2] *A.J.P.H.*, **46** (September 1956), pp. 1149–1156.

3. *When* is the work to be performed?
4. *Where* will the work be done?
5. *How* will the work be done—can the sequence be changed or the operation simplified?
6. *Why* is the work being done—do all persons involved have an understanding of the objectives?

The industrial engineer can work in concert with the sanitarian in a food-processing plant or the sanitary engineer in a major industry, or both, to produce the best results. One does not eliminate the need for the other.

## Titles

No doubt, a suitable title should be found for the individual specializing in the approach to housekeeping as described in this work—that is, the consideration of housekeeping as an interacting problem in both technology *and* technique. Or, if you prefer, a problem in both science *and* art. Consider the title *service engineer* for a person who is responsible to apply the skills of industrial engineering to plant service functions such as housekeeping.

The whole question of titles is confusing and requires clarification and standardization. A study of the names applied to housekeeping supervisors and to the persons performing the work themselves is eloquent testimony to the lack of interindustry coordination in housekeeping planning. For supervisory personnel, we find these titles being used, among others:

Building and grounds manager
Building service director
Custodian
Director of buildings
Director of sanitation
Executive housekeeper
Head custodian
Head janitor
Housekeeping department manager
Housekeeping foreman
Housekeeping superintendent
Housekeeping supervisor
Manager of sanitation
Sanitarian
Service foreman
Service manager
Utility foreman

The person actually performing the work may be called:

Aide
Attendant
Cleaner
Custodian
Housekeeper
Janitor
Laborer
Mop boy
Operator
Orderly
Porter
Sanitationist
Sanitor
Serviceman
Utility man
Worker

The lists are not at all complete.

## Early Systems

Some serious attempts have been made to formalize the programming of housekeeping operations. One of these, which became a matter of interest in the early 1950's, was called the "Gilbert Formula." The method was described by F. L. Gilbert, at that time the manager of the Lincoln-Liberty Building in Philadelphia for the Jackson-Cross Company. He had worked out the system for use in his own building. The objective was to provide an equitable distribution of the work load among the cleaning personnel while eliminating wasted time. The principle involved a computing of units of work to be performed rather than the former approach, which was the assigning of numbers of square feet. This required a complete inventory of all items to be cleaned, assignment of standard units or fractions of units based on cleaning times for each item, and the apportionment of the total numbers of units among the personnel. It was estimated that a woman, for example, could perform approximately 200 such work units per hour.

Another approach consisted of variations on a unit method, in which the cleaning time estimated or calculated for the most typical type of area or room in a structure was considered the standard unit, and the time required to clean other areas was prorated as a multiple or fraction of this unit.

Various types of standards systems have been attempted, some following the lines of the "Gilbert Formula," but, rather than using work units, the figuring was done in seconds or minutes per task.

One of the most prevalent approaches has been the assignment of personnel on the basis of the gross number of square feet involved, there being, of course, different standards set up for various types of areas.

All of these systems are inadequate because the standards set up for specific situations do not have general applicability, and, vice versa, general or typical standards do not have specific applicability. These are piecemeal approaches to the problem and often exclude, by design or accident, such vital aspects of housekeeping as organization, methodology, specialization, team cleaning, mechanization, etc. Of utmost importance is the fact that optimum housekeeping cannot be obtained from the standpoint of the numerical approach—often treated almost as numerology—alone.

## Units of Measure

Some very misleading units of measure have come into general use. They are derived from averages taken from all types of cleaning operations, including those in industries, hospitals, office buildings, and schools. Often they can lead us to believe that a more effective housekeeping program is being utilized than may actually be the case. Or, the reverse may be true. Decisions based on averages and statistics, as is well known, can be faulty and expensive without a knowledge of the basis of these figures and their limitations. If the average automobile fatality were to occur at 60 miles per hour, this would not mean that a person is in no danger at 50 miles per hour. We should be wary of judging the housekeeping situation on the basis of annual costs per square foot, annual hours per square foot, numbers of square feet per worker, cost or hours per unit of production, etc.

Management has generally determined the extent of housekeeping by an empirical system analogous to destructive testing. Repeated reductions are made in housekeeping expenditures until a barely tolerable level is reached. Actually, the process as described is an oversimplification because the drop in the housekeeping level often goes so far that rectifying measures are taken, and perhaps later reversed. Many organizations have seen a cyclic building and cutting of the housekeeping staff. This is a trial-and-error toying with dollars that management would not countenance for a moment in other aspects of the industrial operation—and it wastes dollars that could be saved through a consistent and reasonable attitude which regards housekeeping as a constructive and continuing function.

# 3 Where Is Housekeeping Going?

WE are in the middle of a struggle for profits. It is a constant struggle against rising costs through greater efficiency and productivity. The battle is at times fascinating, exasperating, enervating, triumphant, annoying, and exhilarating. It is a constructive and meaningful pursuit, the very foundation of our nation's socio-political system.

Housekeeping, like all other indirect costs (not contributing directly to production), is involved in this struggle whether it wants to be or not.

Management is beginning to scrutinize its indirect costs, and most particularly housekeeping, carefully because of the growing realization of several important factors. Why are so many fingers being pointed, and so many eyes cast, at housekeeping?

First, pressure has built up because of the tremendous increases in cost. Almost in the last dozen years alone, labor costs have risen about one-half, while materials costs have risen about one-fourth. These increases tend to compound and magnify indirect industrial costs, which are now over 50 billion dollars annually.

Second, there is the growing realization that housekeeping is indeed a management concern. Unlike most support functions, there is the overriding consideration of *degree* of performance. Housekeeping cuts across every department and affects every employee, and management must become involved because of authorizations for capital equipment, union considerations, and the determination of status and organizational structure for the housekeeping department.

In addition, the investment per maintenance worker has increased to the point where the individual custodian is charged, on the average, with the upkeep of more than the $100,000 in capital investment. Very often, he maintains an area valued in the *millions* of dollars.

Finally, management is becoming aware that savings which can result from better control of indirect services can be surprisingly high and obtained in a reasonable period of time.

To a certain extent, management's changing attitude toward the housekeeping operation is the result of being victimized by failure to act previously. Whereas management had let housekeeping go unnoticed, sometimes labor had not. This has had the result that some of the advances in sanitation, which have benefitted both management and labor, have arisen from the demands of organized labor. Thus, improved working conditions through better housekeeping have been considered as a point of bargaining.

Management has found that the empirical determination of the housekeeping force creates conditions which are unacceptable. When the work force is curtailed beyond the abilities of the workers to perform their jobs, it creates a situation where only corrective maintenance—"putting out fires"—can be utilized. Under such conditions, the housekeeping force will normally utilize all its time for repetitive daily work and will be unable to accomplish such intermittent projects as rewaxing, stripping, sealing, wall washing, light and vent cleaning, general vacuuming, etc.

The effects of understaffing in the housekeeping department are not always immediately apparent, but after a time these effects may become drastic and quite expensive. From the standpoint of surface protection, for example, unprotected composition flooring will wear thin and have to be replaced at considerable cost or hard floors may be stained irreparably. From a sanitation standpoint, soap film on lavatories, scale in toilet bowls and urinals, and excessive dust on air vents and surfaces may all carry and transmit disease. From the standpoint of appearance, floors and baseboards will become darkened due to wax build-up, rest rooms will become offensive in appearance and odor, woodwork will darken due to excessive dust retention, and illumination intensity may be reduced by 25 percent or more because of dust on light fixtures. Carpeting is damaged by the fibers breaking or loosening when particles of soil or sand are not removed frequently through vacuuming and dry cleaning. The expense required to live with and later rectify such conditions is quite out of proportion to the cost of preventing their occurrence.

Industry has suffered long enough from having the responsibility for organizing housekeeping placed in the hands of the worker himself. Because they lack qualified technical guidance, in their attempt to solve their problems and ease the work load, the custodian and his supervisor have used a trial-and-error system in their choice of methods, materials, equipment, and other aspects for the cleaning operation. Unquestionably, some of the methods developed in this way have become useful and are desirable, but, on the other hand, many of them are inefficient and needlessly expensive. Unfortunately, the "do-it-yourself" approach is actually recommended by a few sources. One booklet suggests that the custodian list all the duties he is to perform daily, weekly, monthly, etc., and, "when all these steps have been taken, a custodian can prepare his work schedule"! Nothing could be more damaging to the progress of housekeeping and sanitation in industry. We cannot leave the determination of frequencies, schedules, or standards to the untrained custodian. Despite the suggestion that "the custodian's recommendations should always be given top priority in the purchase of operational and maintenance supplies," this does not lead to sound purchasing and material-use practices. To the contrary, it leads to some of the unsound practices from which the sanitary-supply industry suffers. Housekeeping is everybody's job, but it must begin at the management level.

## Awareness of Benefit

Management is now fixing its eye on the target of improving indirect costs. Housekeeping is the bull's-eye in that target.

Good housekeeping is being accepted as good economy by serious businessmen. Most successful plants are clean plants, not because the fruits of success have permitted good housekeeping to follow, but rather because the viewpoint of management that provided a clean plant in the first place led to success. This growing realization brings many changes.

Management now turns to the industrial engineer, whose tools and techniques have been most directly and immediately responsible for the advances that have been made in the organization of the housekeeping effort. The industrial engineer, or service engineer, while applying his techniques to the identification and scheduling of work performed by the custodian, can also assist management in bringing other tools to bear on the problem. Through the application of these forces, housekeeping will cease to hold its unenviable position in American industry of being the outstanding exception to the application of standards and measurements.

## The Internal Approach

An approach by internal personnel to housekeeping may bring only limited results for several reasons. First, housekeeping is not an exact science—it is "loaded" with inexpressible variables. An accurate evaluation can only be provided through extensive research and experience with a variety of housekeeping situations. This point is made positively when we examine the many variables we must consider in order to determine the time required to maintain a particular floor area, for example:

1. Consider the variables of the man doing the work:

   Ability to work with others
   Age
   Attitude toward change
   Experience
   Fatigue factors
   Health
   Intelligence and literacy
   Manual dexterity
   Morale
   Physical strength
   Quality of supervision
   Intensity of supervision
   Training
   Motivation and incentive

2. Next are the variables of the surface to be cleaned:

   Condition (smoothness, porosity, damage)
   Degree of soiling
   Type of soil
   Total area involved
   Traffic
   When and how last cleaned

3. And finally, the variables of work conditions:

   Amount of obstruction
   Interference by other employees
   Comfort level
   Desired quality of result
   Distance to custodial facilities
   Evaporation rate
   Illumination intensity
   Method of performing the work
   Quality of materials being used
   Type, size, and condition of equipment being used

Second, an internal approach is difficult because engineered housekeeping is a very recent development in industry; therefore reference material is difficult to locate and is very limited and general in content.

Then, where is the time to do this? The greatest pressures come in other fields, such as equipment maintenance or construction, because of the greater total savings possible. The individual in charge of housekeeping often has more persons to supervise than he can possibly handle, and, in addition, he may have been saddled with responsibility for various functions which just do not seem to fit into other departments or are not welcomed by them.

A further point is that internal personnel have limited housekeeping experience. A housekeeping foreman in a typical plant may be justly proud of twenty years' service in his department. But insofar as this service relates to the exchange of ideas and information on housekeeping, and the application of engineering principles to it, such a work history might well be equivalent to six months of service repeated forty times.

Finally, the communications problem, particularly in the larger plants, presents a limiting factor because of housekeeping's permeation into all facets of the operation.

## The Smaller Plant

We might ask how large a housekeeping operation must be before a comprehensive study and program become economically feasible. In the smaller housekeeping operations, those having the equivalent of less than a dozen persons, unquestionably there should be sufficient schedules provided to inform the worker as to what work he is to do, when it is to be done, and how it is to be accomplished. Although such schedules do not usually involve the complexity and comprehensive implementation phases of the larger programs, nevertheless the results of good organization and planning can certainly bring worthwhile benefits.

Thus, although the larger operations should receive a comprehensive analysis and programming, the smaller ones can often most benefit by a limited use of the consultant. In addition to the general improvements in organization and scheduling which this can bring, surprising savings can sometimes be made on specific aspects of the housekeeping operation. Here are some examples:

1. In a chemical plant in the Carolinas, turnover was a big problem. It was possible to save two workers by using four permanent men who were able to give direction, as group leaders, and letting them give leadership by example to the newer men under them. Savings, in current dollars even after a 15 percent wage increase to the group leaders, were $15,000 per year.

2. In an engine plant in the Midwest, a concentrated cleaner-disinfectant dispensed through an automatic proportioner eliminated waste, used one product instead of two, avoided transportation of small containers, and was 45 percent

of the former cost *in use dilution* (although more expensive per gallon of product). Savings were $6,200 per year.

3. In a textile plant in the South, the proposed addition of $6,400 in labor costs indicated a cancellation of $16,600 in annual rest-room maintenance contracts. Savings: $10,200 per year.

4. The purchase of $4,450 in new maintenance equipment (while deliberately taking out of service some old machines even though still operable) permitted a Midwestern operation to assume the housekeeping functions of its new wing with two less persons than budgeted. Savings with a five-year amortization of equipment: $14,000.

The techniques and tools utilized in the small operation are different from those used in the larger plant, yet the same type of results in cost improvement can be obtained, although not generally to the same degree.

## Moving Ahead

Since housekeeping is in competition with other departments for the limited amount of money at management's disposal, it is the engineer's place to convince management of the legitimate values of good sanitation. In the size of the housekeeping staff, for example, the counselor has a basic point to make: there is an optimum size of the housekeeping staff, and money is being wasted at a point below or above this figure. This can be interestingly pointed out in a *reductio ad absurdum* exposition. On the one hand, if we assume a reduction of the housekeeping staff to zero personnel, we find that the cleaning frequencies must be extended to an infinite period. In such a situation, it is easy to visualize that within a relatively short period of time the plant would become so choked up with waste, soil, rubbish, and filth that the entire operation would grind to a standstill. On the other hand, at the other ridiculous extreme, if there were one housekeeping worker for each production worker, cigarette ends could be caught in midair before they hit the floor, and cleaning tasks might be performed daily, whereas perhaps monthly would be sufficient. As the figures are changed between these extremes—increased from zero and decreased from some very large number—there lies a point at which any further increase would provide unnecessarily frequent cleaning operations, or any further decrease would create conditions which were detrimental and uneconomical. This concept can be visualized as a graph, where the line of increasing housekeeping personnel is tangent with a line of economic benefit at the optimum point.

In other words, the size of the housekeeping staff should be at that point where the additional cost of another custodian would not provide the company with a suitable return on the investment in him. Or, it can also be stated, where the decrease of one custodian will cause a dollar loss considerably beyond the total cost of maintaining him in his job.

## A Reevaluation of Housekeeping

Management is reevaluating its opinion of the organizational status of the housekeeping department. It is tending to centralize its housekeeping operations to gain the benefits of standardization for all aspects of the work and also to gain a closer control. Housekeeping is at least being considered for a place in the organizational structure on a level with other plant functions. Situations where the housekeeping effort is combined with that of another department are being abandoned because when an immediate supervisor is responsible for two basic functions one of them must suffer to the detriment of the entire organization. Historically, housekeeping has always been the loser in such an arrangement. It is natural for the supervisor to favor the job which he feels carries more prestige, requires more skill, has better public acceptance in terms of status, and which is not likely to receive numerous complaints. There are many unhappy supervisors who have been saddled with such situations. Housekeeping must have its spokesman and champion if its objectives are to be met.

Management is beginning to consider housekeeping at the planning stage. Much more is said about this important subject later.

Housekeeping is rightfully being considered as a *preventive* maintenance function rather than a *corrective* maintenance function—the same change in emphasis that has occurred in the other maintenance fields.

Now it is worthwhile to mention that the housekeeping level of a facility is limited by its condition of physical maintenance. This is true from the standpoint of sanitation as well as appearance. Good maintenance does not guarantee good housekeeping, but it does support good housekeeping.

## Cleaning Materials

Archaic methods are being abandoned, and proper materials and equipment provided. Rather than satisfying the whims of the custodian in product preference, for example, management is providing those materials which will provide the highest level of efficiency.

All too slowly, the consideration of products is being dropped as the item of greatest concern in the housekeeping operation. In time consumption, for example, many housekeeping department managers spend whole days each week talking with suppliers, despite the fact that product suppliers, necessarily, are within a few percentage points of each other in price at the various performance levels. Housekeeping supplies represent only about 6 to 8 percent of the total cost of housekeeping for which the manager or supervisor is responsible. In this respect, the basic business principle of investing time in proportion to the potential dollar yield is being violated regularly. In relation to overall achievement, product alone is of relatively minor importance in the housekeeping operation (assuming it is of reasonable quality, which is to be expected of major manufacturers). After all, a drum of wax, for example, is of absolutely no value at *any* cost unless the wax is applied properly, at the correct time and location, and with a regard for the economies of labor and materials consumption.

Your purchasing agent will probably state that if he would spend his *full time* attempting to cut costs in the purchasing of housekeeping and sanitation materials of reasonable quality, the *most* that he could save would be around 10 percent of the materials cost, or less than 1 percent of the total housekeeping expense. Such a saving, incidentally, would probably involve buying each item from a different supplier, thereby losing certain administrative benefits. It would certainly preclude providing supplier service of any significance, since low volume and a narrow profit margin would necessarily limit such activity.

Buying housekeeping materials on price alone is a trap into which many have fallen. Obviously, performance of the materials must suffer under such a buying practice. Nothing can lead to rising costs faster than this because labor costs must rise to provide more effort on application, more frequent cleaning operations, etc. Some time ago, a presentation on the engineering aspects of housekeeping was given to the Building and Grounds Maintenance Council of the City of New York. An interesting discussion period followed, particularly since this was the first outside presentation before the impressive and knowledgeable body of maintenance managers. A point very emphatically brought out was the difficulty of providing a sound housekeeping operation when products are purchased on the basis of price alone. In government, of course, purchase by bids must be the rule, but the problems can be alleviated by careful wording and policing of product specifications. Industry realizes that it must purchase products on the basis of an analysis of the *total* value of the offering in terms of final cost, performance, and service rendered. Most purchasing personnel are now moving in this direction in the purchase of housekeeping materials.

The unions unquestionably must be encouraged to continue their great work for the betterment of the American worker, but they must also realize that this includes improved productivity for each person. There is a growing awareness and determination that the limitations on general employee efficiency and effectiveness imposed by organized labor require correction. Unless this picture changes, it may not be necessary to look across the seas for our most dangerous and insidious problem—it will be among us.

## Contract Cleaning

Many companies have concluded, often after a series of unhappy experiences, that their objectives cannot be satisfied through the use of outside cleaning contractors. This is strikingly demonstrated by the fact that among the users of general cleaning contractors, about half have returned to an internal system, and those now using general cleaning contractors have switched contractors an average of a full half-dozen times.

Although the basic reason for the failure of contract cleaning to provide worthwhile results is the fundamental disparity in objectives, there are also specific reasons worthy of study.[1]

1. It is inevitable that management will lose most of its control over housekeeping operations.
2. It weakens management's authority over the quality of the personnel working in its plant on this function as well as their performance and the actual products being used. W. E. Fore, of North American Aviation, points out that "contract maintenance lacks a proprietary sense of responsibility on the part of the individual maintenance employee."
3. The costs are too high for the services rendered. Phillip A. Rapino, of Lebanon Steel Foundry, says, "What can be done by an outside maintenance contractor to achieve optimum results can be had just as well by an equally competent home-owned contractor working for his salary, his self-respect, his continued renown, and, of course, for his company."
4. The contractor's employees, many of whom are temporary or part-time employees, will not exhibit loyalty and incentive to the client management. "It will be difficult to build up a cost-conscious maintenance organization with a crew made up of men who are not employees of the parent organization," says John B. Whitlock, of Armco Steel Corporation.
5. One of the most annoying problems is the lowered level of housekeeping achievable.
6. There is the natural tendency for the contractor to use cheaper materials

[1] Quotations in Items 2, 3, 4, and 11 reprinted by special permission of FACTORY, July 1959. Copyright by McGraw-Hill Publishing Co., Inc.

of lower quality. This can result in actual physical damage to the surfaces being cleaned.

7. Cleaning personnel are at times poorly trained and may cause accidents to themselves or to others and damage to furniture and fixtures by poor handling of equipment, etc.

8. Supervision often undergoes a progressive deterioration, and responsibility is difficult to fix or arbitrate.

9. Flexibility for handling emergencies or special jobs, or to permit correction of faults and provide for last-minute changes, cannot be provided.

10. Security is weakened because of turnover, supervision, and selection problems. Security is manifested both in personal items as well as classified or general business items.

11. The general attitude of contract maintenance employees toward the conditions created by client management is poor, owing to lack of management control and the inability to instill a company spirit in outsiders. Albert Chapman of Ford Motor Company says, "In most of our plants, a certain *esprit de corps* exists among our maintenance personnel that we believe would be impossible to attain through contract maintenance."

12. Management is at the mercy of the contractor upon renegotiation for the cleaning service because management is unable to form a new housekeeping department on short notice.

Too often the farming out of basic industrial housekeeping is a default to expediency, inevitably leading to sacrifice of quality and control, or, at the least, of cost.

Of course, the advantages to contract cleaning should not be overlooked:

1. The contractor may have specialized equipment which would be expensive to obtain for a specific type of cleaning, such as overhead work.

2. The hazardous nature of the work may lead to the use of a contractor whose personnel are properly trained and equipped for such a job. An example is window cleaning.

3. Very infrequent cleaning jobs do not require adjusting the size of the housekeeping work force when contractors are used.

4. There may be a wide disparity in the wage rates of contract cleaning personnel and internal workers, bearing on the economic analysis.

5. Contractors may be called for emergency cleaning or cleaning of a temporary structure.

6. A cleaning contract can give a fixed-dollar figure as a basis for budgeting.

7. There may be advantages from the standpoint of labor organization.

For those areas where contract services do provide legitimate values, precautions must be taken:

1. The careful wording of specifications is of utmost importance. The areas of responsibility and the frequency of cleaning should be precisely spelled out.

Many contracts generously allow tremendous latitudes of freedom with expressions such as "when needed" and "as required."

2. The experience of the contractor, both in type and size of accounts and longevity of contracts in force, should be checked carefully.

3. The contractor should specify the organization of the housekeeping staff to be used for the work as well as the extent and method of their supervision.

4. References should be checked, both for financial soundness and job performance integrity.

5. Qualifications which the contractor uses for hiring and screening his personnel should be investigated.

6. The contractor should furnish a schedule of costs on a unit basis for any additional services that may be required.

7. The effect of using the contractor with respect to your own union situation should be carefully considered.

8. The quality—even the product name at times—of the materials and equipment to be used in the work should be specified.

9. The cost, method of payment, and notice of termination of the agreement should be precisely defined.

Although each case must be carefully evaluated on its own merits, the signing of a contract for a general housekeeping service in industry may be viewed as a possible act of desperation or resignation.

Henry Ford once pointed out that his first step after acquiring a business would be to see that the plant was put in a clean and orderly condition and then to see that it was kept that way. He recognized, therefore, the benefits of good housekeeping. These benefits are many, both direct and indirect, and their aggregate result is to make of housekeeping a profitable undertaking.

The *Production Handbook*[2] certainly is correct when it indicates improved productivity through good housekeeping. It says, "The production rate is increased because of the orderly, businesslike conditions and removal of obstacles to production."

## Benefits of a Housekeeping Program

A sound housekeeping program will provide these meaningful benefits, all of which, in the final analysis, must manifest themselves in improved cost:

1. Better hygienic conditions are provided, leading to improved health and decreased absenteeism.
2. Safety is improved.
3. Fire hazards are removed or lessened.

[2] Copyright , The Ronald Press Co.

4. Materials are protected from damage through soiling or contamination.
5. Morale is improved through better overall appearance and general conditions.
6. Abnormal wear, and resulting premature repair and replacement of surfaces, is avoided.
7. Promotional benefits to the organization are provided by public acceptance and aesthetic appreciation.
8. Quality control is enhanced and inspection time reduced.
9. Training and orientation time for new or transferred employees is reduced.
10. Greater flexibility is provided to handle emergencies, special assignments, changing weather conditions, etc.
11. Turnover and absenteeism in the housekeeping department is reduced through improved status and morale.
12. Better internal communications are provided between the housekeeping department and other phases of the operation.
13. Budgets may be accurately prognosticated and substantiated.
14. Other types of indirect services are facilitated.
15. The effectiveness and efficiency of the housekeeping worker is improved.

The future undoubtedly will bring surprising developments in the technology of housekeeping equipment. It is likely that we will see devices which combine ultrasonic with electrostatic cleaning, so that a device placed in the center of a room would loosen all soil from every surface in that area and collect it on or within the machine for disposal. Probably radioisotopes will become a regular tool of housekeeping specialists.

Those who prophesy the future as an era of ideal sanitation and cleanliness do so not because such conditions are "nice," but rather because they are necessary to improved technology and social advancement.

George W. Jernstedt, manager of manufacturing planning for Westinghouse Electric Corporation, recently said,[3] in looking only four or five years ahead,

> Aggressive manufacturing plants—those capable of survival in an era of intense competition—will be employing between twenty-five per cent and fifty per cent new methods and new equipment presently not in production. Each of us would do well to reflect on this situation in light of the way we are currently working.

Housekeeping must not fall behind in our rapidly changing industrial scene. We must be prepared to use new materials and equipment and new ideas and methods, as they are developed.

[3] Quoted with special permission from *Plant Management and Engineering*. Copyright Hitchcock Publishing Company.

In the meantime, there is a more urgent problem, that is, to take advantage of the tools already at hand and to apply existing knowledge and experience, so as to obtain the yet unrealized returns from our housekeeping operations.

# 4

# The Next Step

THE typical industrial operation benefits from efficiencies gained from a carefully planned analysis of the complete operation. Performance standards are established and systems and controls set up to eliminate discrepancies in output, method of performance, and materials. These studies use fairly generally accepted steps in such an analysis:

1. The problem must be defined.
2. Relevant facts must be obtained.
3. These facts should be critically examined and tested.
4. Methods and courses of action must be proposed.
5. Decisions should be made about the course to follow.
6. These decisions should be acted upon and the methods installed.
7. Periodic controls should be used to maintain a standard accepted practice.
8. Periodically, further improvement should be made where possible.

Should not the same techniques utilized to provide efficiencies in direct operations be utilized in housekeeping, along with other indirect functions?

At times it appears that management has an impossible number of functions to perform which are fundamental to the operation of the enterprise. But can management abdicate or eschew its responsibilities to the housekeeping department on the basis that there are other functions to perform? Management must concern itself with *all* functions that are basic to the operation of a sound enterprise, and housekeeping is assuredly one of these.

The assignment of a specific number of square feet to each worker based on an arbitrary schedule devised by guess, experience, and taking averages may have served its purpose to some extent during the time before more sophisticated techniques became available. When wage rates were lower and the struggle to earn profits was possibly less exacting than now, there were less pressures to improve the situation than exist today.

We must avoid the situation existing in many plants today, that is, housekeeping supervision feels that more workers are needed in order to accomplish the housekeeping tasks and objectives assigned to it, whereas, on the other hand, management feels that housekeeping is not operating efficiently enough and should be able to use fewer workers. Certainly arbitrary determinations of the size of the housekeeping department can only prolong this opposition, to the organization's detriment.

Roger M. Blough, Chairman of the Board of the United States Steel Corporation, recently said[1] that "the application of effective change . . . throughout all operating activities is the only solution to maintaining our position in a competitive world market." We now see that the realm of housekeeping, because of generally low levels of productivity, is one of the prime target areas for this "application of effective change."

The general and specific improvement of industrial housekeeping requires both long-range and short-range activities. Let us first consider the long-range objectives which will bring improvements to industry and our society as a whole.

## Long-Range Objectives

Industrial housekeeping will be able to move forward more rapidly and on a broader scale if standards can be adopted covering these aspects of the housekeeping operation:

Job titles and definitions
Frequencies of job performance
Materials specifications (This can include use dilutions, container sizes, and possible color and odor coding for identification.)
Standard times

[1] Quoted with special permission from *Plant Management and Engineering*. Copyright Hitchcock Publishing Company.

Employee selection qualifications, including physical examination and education requirements

Definition of housekeeping department responsibilities (With particular consideration of such items as overhead cleaning, lamp cleaning, seal application, window washing, pest control, grounds maintenance, chip sorting, lubricant recovery, etc.)

The proposal and promotion of such standards for industry indicates the need for a national association devoting its interest *exclusively* to sanitation and housekeeping in industry. No doubt such an association would publish its own periodical for the exchange of ideas and information relating to industrial housekeeping and would claim as its readers plant engineers, housekeeping department supervisory personnel, methods and standards personnel, safety engineers, buyers, etc. Such an organization could further its aims through the establishment of a comprehensive library, the provision of lectures, seminars, and social events, and the preparation and showing of films concerning housekeeping.

The recently formed British Institute of Cleaning Science is certainly a step in this direction, although it is not directed to industry as such. The commendable objectives of this organization are worth quoting in full:

To promote and develop the science of cleaning.

To develop liaison with other institutes and co-operate fully with other organizations in the world having similar objects.

To hold meetings in London and the Provinces.

To provide facilities for courses of instruction leading to a diploma of the Institute.

To form a library and to publish proceedings and reports.

To take such steps as may be decided to ensure that all members of the Institute are qualified as such.

The proposed American society would certainly work in liaison with its British counterpart.

In his *Understanding Modern Art,*[2] Leo Katz, as early as 1936, described our educational system as requiring a "sugar-coating." He says:

> Even science and art have to be presented in such a way that they could be sold to the public. The usual procedure was to expect a completely passive audience who were willing to expose themselves to more or less charming methods of persuasion and salesmanship. The teacher or lecturer had to make all of the effort to keep his audience interested, and, in most cases, the pupils were satisfied merely to decide whether the performance was good or bad.

[2] Copyright 1936, The Delphian Society.

Perhaps subjects dealing with housekeeping have been omitted from our curricula because of the inability of such courses to conform to such outrageous standards. It is difficult to put a sugar-coating on the removal of soil and waste, no matter how fundamental and important is this subject.

Many industrial engineering, industrial management, and business management graduates face the problems of housekeeping today with no preparation whatever. Certainly a survey course should be provided to acquaint the student with the problems and opportunities in this field.

Professional training on the college level must come for future managers of our industrial housekeeping operations. In industry today, we find many able housekeeping department managers, but it is likely that many would be more effective in their work if they had received proper training in this direction. These persons lead housekeeping departments which may have an annual budget in the hundreds of thousands of dollars per year.

I would recommend that the diploma designate the graduate as "Bachelor of Service Management." The course might be taught at liberal arts colleges, technical institutes (where it might be an engineering course), or perhaps in special schools set up for the purpose.

The curriculum should include these subjects:

- Personnel management
- Industrial organization and management
- Basic time study and motion economy
- Industrial safety
- The chemistry of maintenance materials
- Mechanizing the housekeeping effort
- Industrial hygiene
- Industrial psychology
- Housekeeping standards
- Custodial facilities and inventory control
- Training the custodian

In addition to the basic course of study, as well as the required subjects given above, electives such as the following could be offered:

- Bacteriology
- Basic drafting
- Fluid mechanics
- Food-plant sanitation
- Graphical representation
- Mechanics
- Preparation of reports
- Public speaking
- Radiation safety
- Speed reading
- Thermodynamics
- Typing

These latter courses provide the basis for growth into the general maintenance field, and then into management, as well as broadening potential effectiveness in housekeeping.

The University of Washington, through its School of Home Economics, has made a major step forward through its executive housekeeping course. This is directed toward hospitals and other institutions, however, and does not provide suitable training for industry.

In summary, our long-range goals should be in the direction of professionalizing the housekeeping field.

## Short-Range Objectives

The only possible short-term objective than can yield worthwhile results is a comprehensive approach to housekeeping and cleaning from an industrial management and industrial engineering standpoint. Naturally, this will never be done unless management is convinced of the value of such an undertaking. Where this value is in question, it may be desirable to obtain the services of a consultant to help "sell" the program. Under certain conditions, the formation of a special housekeeping committee for the purpose of investigating the problems and evaluating potential benefits from a program would be wise. Such a committee might contain personnel from the production, maintenance, quality control, industrial engineering, industrial relations, safety, purchasing, and housekeeping departments.

The decision to undertake such a program should include a decision on how the work is to be done, whether by internal personnel, through the use of a consultant, or a combination of both.

## The Consultant

Actually, a consultant can be of considerable value even where an internal program has already been launched. He can assist in the development of the general program framework, while internal engineering personnel complete the various details. Or, he can be retained to analyze and assist in the implementation of an internally prepared program.

Consultant services on a *per diem* basis can sometimes bring quick and considerable savings or improvements, but, of course, they can never bring the optimum benefits of a comprehensive approach. *Per diem* services should be considered for such cases as these:

1. Where an internal program has been completed, which would benefit from a review by an experienced specialist.
2. Where an objective opinion is required on a point of grievance, perhaps involving apportionment of work loads, materials selection, or standards.

3. Where specific improvements are required for a situation that will be short-lived, such as the cleaning of a plant that is scheduled for relocation. Here the advantages of planning for better maintenance should be remembered.

4. Where fiscal limitations permit only limited expenditures.

5. Where the client wants a first evaluation of the approach, and then samples of the work, of a particular consulting organization.

The use of an outside consultant may provide several advantages over using internal engineers. This is his *raison d'être,* after all, and each dollar invested in such consultative services should bring several dollars in return.

First, there is the question of time consumption. It has been estimated that a time study man in a new field takes up to three times as long to perform his work as one experienced in that special field. The fact that the quality of the work will also differ should be added.

Second, although each organization is different (and furthermore, continually changing), yet in a number of useful ways each relates to the other. The consulting engineer can act as a vehicle to permit plant engineers, methods personnel, and housekeeping department heads to exchange ideas and avoid repetition of errors made in other plants.

Third, the consultant has established an engineering organization for the sole purpose of undertaking an organized approach to housekeeping. It has no other functions to perform and no other masters to serve than its clients. The libraries and case history files which such an organization can build are valuable sources of information.

Finally, the objective status of the consultant permits the avoidance of political situations within the organization, lends weight to budgetary recommendations, and, with management's support, can cut across departmental lines to obtain organization-wide support and cooperation where needed.

In the food-processing industries, the service engineer works in harmony with the sanitarian, whose principal function is to make sure that the product appears clean, free from contamination, is fresh, and packaged under sanitary conditions. The typical sanitarian, while providing his indispensable function of controlling the condition of food products, is not technically qualified, through training or experience, to engineer the basic operation of the housekeeping function.

Yes, there is a strong movement toward the use of professional industrial engineering consultants specializing in this field—the organization and standardization of housekeeping programs—not to take over the housekeeping operation, but to provide advice and ideas.

The selection of the consultant should not be taken lightly. Neither should the advice of a consultant be followed blindly. Orin F. Nolting, as executive director of the International City Managers' Association, pointed out in *The American City* that one successful city manager of long service said that he considered that one of his important duties was "protecting his municipality from the consultants his council had to retain!"

Investigate the consultant carefully. Be sure suitable answers are provided to these questions:

1. What is the firm's experience in the investigation and programming of housekeeping operations?
2. What are the qualifications of the personnel who will be responsible for the work? Especially consider the principal of the firm, whose registration as a professional engineer in your state (or a verified professional record in the National Council of State Boards of Engineering Examiners) should be a requirement. It is the principal of the firm, just as it is the president or general manager of a business, who establishes the character of the organization. Check on such factors as education, experience, capital, and community standing.
3. How soon can the work be started, and what is the target date for completion? It is necessary to be careful not to force a rush completion date, as this can limit the results.
4. Can he show the forms to be used in the survey, as well as samples of work performed for others?
5. Will he provide a list of reference accounts?
6. What is the exact fee to be paid, and its method of payment? Avoid the selection of a consultant on a "bid" basis. Such services as these cannot be provided on a specification basis, and certainly no two consultants are equally qualified for the specific work at hand. Various court decisions have upheld the viewpoint that it is not necessary to select consultants on the basis of the fee.

The greatest injustices a consultant can perpetrate on management are to agree to perform services for which there will be no reasonable compensation, to guarantee completion times which cannot be practically achieved, or to exaggerate potential benefits. Beware of organizations making claims of *typical* savings of 30 or 40 percent. A person must also be cautious in taking action to bring apparent or paper savings which might result in long- or even short-term losses in housekeeping or other areas.

The use of a consultant is not a panacea. A successful program is normally the result of a close and effective working relationship between management and the consultant, and it must be borne in mind that the responsibility and authority for decision making must always rest with management, who must make decisions on recommendations in terms of acceptance, rejection, modification, or further reconsideration. A company's best people should be assigned to work with the consultant in order to obtain the greatest possible benefit for its organization.

In addition to the most frequent desire of management to improve productivity of the housekeeping worker, resulting in direct dollar savings or improved levels of housekeeping, or both, there may be other reasons why consultants may be called in to advise on housekeeping:

1. In some indefinable sense, management may not be satisfied with housekeeping conditions.
2. Management wishes to obtain an objective buffer against labor complaints or to instigate new action to avoid such problems, such as the application of standards or work measurement in housekeeping.
3. There may be specific quality control problems, such as the setting up of a cleaning program for a "white room," where too many rejects may be due to a high dust count.
4. Management may desire an objective evaluation of a supervisory problem.

In any case, no matter why the consultant was called in originally, he should perform a meaningful and legitimate service, and certainly within the ethics of the profession.

## The Supplier Program

A very interesting occurrence is taking place in the sanitary-supply industry, leading to the offering of engineering services through a product consumption arrangement. Market studies have shown that the typical sanitary-supply customer spends only a small percentage of his housekeeping dollars for housekeeping materials and equipment and is much more interested in getting the total housekeeping job done rather than in merely buying, for example, a drum of wax. The typical buyer or user of housekeeping materials actually has only a limited amount of information about the proper use of housekeeping materials and, beyond this, knows even less about the industrial engineering and industrial management principles necessary to the development of a sound housekeeping program.

The typical sanitary-chemical supplier provides field services limited to demonstrations and discussions of the use of specific products, supplemented with preprinted data concerning equipment and general methods.

The future will see larger suppliers going beyond this point, with the provision of programming services

In considering such a service, we should make certain that:

1. The actual engineering is performed by qualified and registered personnel. Preferably, the engineering work should be purchased by the supplier from others, since personnel within the corporation may be given other duties to perform, with conflicting objectives, which may detract from their work in housekeeping.
2. The organization is beyond the experimental stage and able to apply a fund of experience to your situation. The best check of this is a review of specimen programs and contact with large reference accounts.
3. It must not be a "gimmick." If the supplier cannot furnish proof of the technical competency of the engineers performing the work, its "program" may be more of a sales device than a true service.

Although the supplier program provides a value for the purchaser of housekeeping materials, it is not without certain limitations both for the supplier and the customer:

1. Arrangements for the program are often made below the higher management levels, thus limiting the effectiveness of the program through lack of top management support and recognition.
2. The money invested by the supplier on engineering services, the time invested on field implementation services, and, therefore, the comprehensiveness of the program must be related to the potential dollar volume forthcoming from the account. Only a fixed percentage of sales, or potential sales, can be devoted to service if the account is to remain profitable.
3. Without management's support, the supplier is faced with a constant pressure from purchasing personnel, on the basis of unit prices alone, even if there is no regard for overall value analysis. It is naturally impossible for large suppliers to be competitive with every product sold by local jobbers, particularly when use-dilution ratios, solids content, and other indicators of product quality are not taken into consideration.
4. At times, the supplier becomes subject to the vagaries of the custodian himself, who, considering the program a threat, may strike back through a complaint about the product, or may condemn specific products on the basis of choice of color, odor, or viscosity.
5. The normal sale of chemical products results in an immediate profit on every transaction. The supplier providing a true engineered program, on the other hand, must make a considerable initial investment and can only begin to realize a profit after a number of months, or sometimes even years, of continuous volume. Substantial losses can occur if an account curtails its purchases or discontinues the program entirely, because of transfer of the person with whom the arrangement was made, general company directives which may affect the arrangement, or for other reasons.

The decision to use internal engineers, consultants, or a program offered by a supplier must be made on the basis of an analysis of the extent of the problems with which a person is faced and the potential gains and limitations from each arrangement, or combination of arrangements.

## A Comparison of Systems

Let us consider, for a moment, a tale of two cities. The only comparison between these two cities and the two cities of Charles Dickens' immortal story is the fact that a revolution is involved. In Dickens' story it was the French Revolution; in our story it is a revolution in approach to housekeeping and sanitation. Certainly, the housekeeping theme cannot compare to the French Revolution in drama or impact—yet, its timeliness is of importance to all. Housekeeping can represent a real benefit or danger to each of us, depending on how it is performed.

Our two cities each contain a division of a large industrial concern. We compare, on the one hand, an organization which has operated housekeeping on the basis of a loose arrangement that gives no real scheduling or training in methods, materials, or machines with an organization that has, on the other hand, developed and is using a standardized and organized approach to housekeeping. Let us, therefore, compare "O. K. City" with "Superba" and see what benefits standardization and organization can bring.

In O. K. City Joe, the janitor, is going about his work. Joe was hired a few days ago as a replacement for a man who had quit because he felt he was being treated as a necessary evil and no one cared what he was doing or cooperated to help make his work easier. When he was hired, Joe was given a list of rules and regulations and then told he would have to keep the north end of the first floor of his building clean. He was to work with Jack, another janitor, for a few days so he could learn the ropes. Jack had been there a long time, and Joe learned from him most of the fine points of how housekeeping was done twenty years ago.

Joe, the janitor, is now at his storage closet. The dim lighting, lack of ventilation, and disarray and litter in the closet put Joe in a frame of mind that will necessarily be expressed in his daily work. There are three detergents stored in the closet, and Jack told him that the red one was best. Joe mixes his cleaning solution by putting one cup of detergent into the pail of water. He adds another cup for good measure.

Let us follow Joe into a rest room, his first cleaning job of the day. Since he has had no specific training in rest-room cleaning and has no written instructions to follow, Joe decides to clean only those items which appear to need cleaning. He has been told that the proper way to clean a rest-room fixture is to swab it with detergent solution, then to pour in some pine oil disinfectant. He has already discovered, however, that the use of the acid-type descaler, which he has been told to use whenever scale forms, does a much quicker job, so he is now using that on a daily basis. About half the insoluble powdered cleanser which he pours into the lavatories goes straight down the drain without performing any work at all (this will provide the plumbers with a trap cleaning problem at a later date). Joe next mops the floor, having mixed in some pine oil disinfectant with his detergent. He has not washed his mop since the previous day's work and is using a single bucket.

Everyone who uses this rest room is annoyed with it because it always smells bad. Although part of the odor is caused by the choice of cleaning materials and disinfectant, most of it is a permanent odor which is locked into the scale under the rims of the fixtures or trapped in the cement grout between the ceramic tiles of the floors. In order to overcome these odors, a rather expensive drip-fluid service is utilized, which covers the permanent odors and the pine oil smell, which has a still stronger odor.

When Joe has finished the rest room in O. K. City, he goes about his other work. We can find him during the day sweeping a dusty corridor with a push broom or, perhaps, a dust mop that is incorrectly treated, and thereby at one time creating a cloud of dust or, at another time, leaving streaks on the floor and softening the tile. He handles the dust mop as if it were a wet mop, swinging it from side to side, exercising every muscle in his body and creating more dust than he is collecting. Later, Joe can be seen scrubbing an aisle with a 12-inch floor machine, taking four hours to do a job that could be done in one hour with a 24-inch machine.

When he is finally ready to return to his home in O. K. City, he is tired and unhappy. He has worked hard but seems to have achieved little, and he has heard many complaints and threats. No one seems to understand his problems. He is thinking about looking for another job.

But now, let us look into the City of Superba. Here we see custodian Don Prince who has been on the job just one week. When Don was hired, he was first given a thorough medical examination for his own protection and to make sure he was not *creating* health problems rather than helping to *eliminate* them. During his first day on the job, he attended an orientation meeting where he was introduced to his group leader and foreman. Later, he would meet the housekeeping department manager. At that meeting he was issued his uniform. With custodians from other buildings, he watched a film entitled *The Importance of Sanitation* and heard a consultant, the guest speaker, talk about the meaning of Don's new job, not only to himself and to the people working around him, but also to everyone who might walk into the plant. The foreman gave Don a little package containing a personnel guide, safety precautions, a description of his work area, and detailed step-by-step instructions on the method of performing each job which he was to do. He was urged to ask questions freely and was told that he would be given ample time to learn all the details involved. The meeting then continued with a demonstration of the various materials that were to be used, how they were to be measured, and some demonstrations of the use of equipment. After the meeting, on-the-job training was given by the group leader who did less of the work himself as Don learned more. Don knew that if he ran into difficulty his group leader was available for questioning and that at the next scheduled training class he would have an opportunity to ask further questions and make suggestions.

Don Prince approaches his cleaning closet with confidence and purpose. He knows he has been given the proper materials and equipment with which to do his job, and he knows that job is important. If he gets into difficulty, he can easily get help. He understands exactly what is expected of him in terms of a reasonable day's work and is sure that as long as he gives it he will be shown respect and understanding.

The condition of the custodial storage closet is a reflection of the Superba Division's and Don's attitude toward the work to be done. The closet has been properly stocked, according to a supply list posted on the door, for the entire week's needs. All the materials are properly organized on the shelves and the clean dust mops and wet mops are placed in tool holders on the wall. The closet is well ventilated and well lighted. The utility sink is fitted with a proportioning mechanism that introduces the exact quantity of detergent needed to prepare an efficient cleaning solution. For other materials which have to be measured, Don has graduated measuring cups. Only a half-dozen chemical materials are stored in this closet as this is all that is required for the daily repetitive work. Don will keep his equipment clean and orderly, not only because he knows he can do a better job that way, but also because he remembers that his foreman often makes unannounced inspections and usually looks into the closets first.

Don places his work materials on his cart, which saves a lot of time walking back and forth between his work area and the closet. He moves his cart to the rest room which he must clean first, props open the door, and puts out his "caution" sign.

Don checks his rest-room cleaning schedule to see in what order the jobs must be done, what materials and equipment he will need, and what methods are to be employed to get the most amount of work done with the least consumption of time and energy. His cleaning material is a combination detergent-disinfectant-deodorant, which means he will only have one job to do rather than two or three. His calendar indicates that descaling of the fixtures will not be required until the following Tuesday. He uses a small mirror from time to time to make certain that the undersides of the fixtures are properly cleaned. The Superba Division's standardized approach to controlling rest-room odors is to eliminate permanent odors by properly sealing the floors (which also simplifies their cleaning) and by proper cleaning and disinfecting techniques for all surfaces within the room, and then providing proper ventilation for removal of transient odors.

Custodian Don Prince's work throughout the balance of the day follows standardized methods and uses standardized materials and equipment, which permit him to accomplish a great amount of work with only a reasonable effort. It depends on a well-developed employee cooperation program to keep the housekeeping staff at a minimum.

In order to create such a different working climate and to provide a higher level of housekeeping and sanitation, how does the Superba Division differ from the O. K. City Division? Is it because the people involved in the administration and performance of housekeeping are smarter? No, we realize that there are usually no such great differences from one group to another. Does it mean that they have been given a bigger housekeeping budget? No, to the contrary, although there were some initial expenditures for equipment and facilities, Superba's costs run below those of O. K. City year after year.

Then how did the Superba Division get where it is?

First, management had to realize that the housekeeping operation offered an immediate and positive opportunity for improving profits and that the savings could be realized with a reasonable investment in time and money. It then determined that it would provide a comprehensive housekeeping program involving an analysis, determination of standards, organizational appraisal, work assignment, and including such diverse aspects of implementation as training courses, publicity campaigns, inspections and grading, interdepartmental liaison, etc.

Such a fundamental decision was founded on the acceptance of the principle that good housekeeping is good business and that the setting up of a sound housekeeping program is a legitimate business investment. Without such a broad concept, it would be difficult indeed to realize worthwhile advantages from the housekeeping operation, and there would be the tendency to make penny-wise and pound-foolish decisions concerning housekeeping.

Let us agree that it requires courage, judgment, and initiative to break with tradition in order to take full advantage of the opportunity before us. The results are worth the time.

# 5

# The Survey

THE first step in the preparation of a housekeeping program is to take a survey. Other names may be given to this activity, but the purpose is simply to define the housekeeping problem. The carrying out of this simple overall objective often is in the form of voluminous forms, notes, plans, drawings, printed material, etc., and the quantity of the material collected appears quite formidable and confusing to the novice. The actual program document, the subject of the following chapter, is prepared on the basis of an analysis of this survey material.

## Medical Analogies

The survey of an industrial plant by an engineer to establish the nature of the cleaning problem is analogous, in certain respects, to a medical examination of a patient by a physician to investigate the condition of the mind and body. Let us study some of these analogies:

1. Withholding of pertinent data can lead to faulty diagnosis and treatment.
2. Specialization by the professional brings benefits to the client.
3. "Do it yourself" diagnosis and treatment has limited value, or even brings negative results.

4. Partial examinations create problems. If the physician is limited to an examination only of the left side of the body, for example, it is clear that the results from such a service might be a considerable hazard because of failure to discover the source of a serious problem. If the engineer, similarly, is asked to investigate only one portion of a facility, certain fundamental aspects of housekeeping must remain unresolved, such as organization, management support, employee awareness and cooperation, standardization, etc.

5. The earlier the treatment is given, the more beneficial it is to the client.

6. The client benefits most when choosing an experienced professional of good reputation and when providing him with all possible cooperation.

## Arranging the Survey

When arrangements are made for a housekeeping survey, it is sometimes after many months, or even years, of moving in this direction through conversations, correspondence, management meetings, etc. By the time the surveyor arrives, both management and the housekeeping department are very anxious to learn as quickly as possible the nature of the basic problems and the opportunities for improvement. Under such conditions, the consultant should not make the mistake of giving premature pronouncements or judgments which he may later find it necessary to correct, on the basis of a more careful analysis of the survey data. The survey period should be restricted to a gathering of information and data and making personal contacts which lay the groundwork for the program to follow.

It is difficult to determine in advance, with any degree of accuracy, the time that will be required for a survey, except that it is likely to fall within certain limits learned through experience. Naturally, the time spent on the survey will depend on the actual size of the physical facility, the detail with which various subjects will be pursued, the support given to the survey by management, and other factors. Since the program will be founded on the survey data, its effectiveness will be limited by the completeness and accuracy of the survey. A good survey, therefore, is absolutely essential to the preparation of a sound housekeeping program.

The conduct of the survey should be very carefully planned in advance in order to make certain that all aspects of the inquiry are covered and the work consumes the least amount of management and personnel time. Obviously, the survey agenda must be flexible enough to permit adjustments because of unexpected conditions, such as the calling of meetings, or sickness. Proper planning of the survey will set the tone for the program to follow and will assure the greatest possible benefits from the time invested.

The surveyor's role must be to obtain as much useful information as possible. Actually, he must obtain *all* of the information necessary to the preparation of the program. The surveying engineer must therefore be adept at working with people at all levels of the organizational pyramid in order to evoke a positive response. A worthwhile approach in this circumstance is to use Will Rogers' wonderful attitude "I never met a man I didn't like." The program must not be limited by the development of negative feelings or even active resistance brought out during the survey by expressions of surprise or distaste at unorthodox methods, voiced or implied criticism, impatience, haughtiness, or condescension.

## Advance Data

Certain data can be prepared by the subject organization in advance of the actual survey, thus permitting the time spent during the formal survey to be of the greatest possible value. Such data should be limited to counts, listings, and measurements not subject to. interpretation, such as square footage determinations, fixture counts, the collection and arrangement of plans, etc. This approach is of particular value if a consultant is retained, or internal personnel are costed, on a *per diem* basis. The consultant should furnish a list of such preliminary data required or desirable, together with an indication of the method of arranging, and possibly even collecting, the information.

It is desirable to make an announcement of a forthcoming survey to department heads and to all housekeeping department personnel so that they will understand what is taking place. Furthermore at the beginning of the survey, it is well to have a meeting of the housekeeping department to indicate the nature and extent of the survey and, primarily, to point out to them that one of the basic purposes of the program will be to benefit them as individuals. A representative of management should introduce the survey engineer, explaining the purpose of his work in positive terms. Housekeeping personnel must be reassured about the beneficial nature of the study, and, if this can be agreed upon with management, the fact that there will be no direct lay-off as a result of the program.

In some plants the security situation will hamper the work of the surveyor. This is true not only of defense industries, but also of other types of industries which may have secret processes or products under development. As many steps as possible should be taken to simplify the security arrangements in advance of the actual survey.

## Survey Forms

The consultant, in knowing what facts are needed, and in knowing the emphasis to place on each field of inquiry, is able to conduct an efficient survey which saves the time of all persons involved. Providing detailed and meaningful survey forms, toward this end, is an important contribution of the consultant. Omission of a single basic question may lead to the wrong approach in attempting to solve a problem. The survey forms must be general enough to permit use in a wide variety of situations and yet complete enough so that only a limited amount of special data need be recorded apart from these forms. Much time and, in some cases, possible errors can be saved if the forms are provided with boxes to check, selected words to circle or cross out, and other devices to avoid handwriting. Specialized forms may be used for certain situations having unusual problems or arrangements.

Naturally, the forms must relate to the program document to follow. They must provide all the data needed for the preparation of the program, and, for economy of time, should exclude all extraneous material. One logical arrangement of survey forms is based on the engineer's six "M" variables. An outline of some of the subjects that each of these forms should cover follows:

MANAGEMENT

Objectives
Limitations
Target completion dates
Distribution of information
Attitudes
Contemplated changes and activities
Planned construction or renovation
Basic statistics
Scope of the study
General observations

MEN

Departmental organization
Size of the housekeeping staff
Work assignments
Relief arrangements
Emergency scheduling
Individual limitations
Titles
Hiring qualifications
Specific personnel problems

MONEY

Budgetary limitations
Custodial wages
Indirect costs
Cost of equipment and supplies
Contract cleaning costs

METHODS

Job performance frequency
Method of performing each cleaning function
Housekeeping jobs performed by nonhousekeeping personnel
Nonhousekeeping jobs performed by housekeeping personnel
When the work is performed

MATERIALS

Central storage facilities
Custodial closets and utility sink location
Purchasing methods
Distribution methods
Product specifications and uses

MACHINES

Application
Size
Location
Condition

## Survey Equipment

For a comprehensive survey, the engineer would require the following equipment:

1. Survey forms
2. Graph paper
3. Analysis paper
4. Legal pads for notes
5. Slide rule
6. Architect's scale
7. Photometer
8. Litmus paper
9. Steel tape
10. Colored pencils
11. Portable dictating machine, for some cases

The conducting of a survey by a team rather than an individual provides advantages in certain situations, such as where the work must be completed within a given time limit. In such cases, the work must be especially well arranged to prevent duplication and confusion, and to insure completeness. When two persons are taking the survey, a logical division of the work requires one man to perform the basic physical survey while the other man performs the basic personnel survey.

Personnel from housekeeping, methods, purchasing, safety, and other departments should be encouraged to participate in the survey. This will

SERVICE ENGINEERING
ASSOCIATES

# CUSTODIAL FACILITIES AND INVENTORY

FOR ____________________

DATE ____________________

RESPONDENT ____________________

1. Describe the central storage area:
   Size ____________________
   Location ____________________
   Notes: ____________________

2. Is the central storage area properly equipped? (Check off if OK)
   Water ____________ Tool Holders ____________
   Drain ____________ Ventilation ____________
   Lighting ____________ Shelving ____________
   Racks ____________ Identification Signs ____________

3. Who is responsible for issuing materials from central storage? ____________________

4. Are materials issued in original containers? (Explain) ____________________

5. Where are custodial closets and cabinets located (indicate on plans or describe here, and show which contain utlity sinks).

6. Are closets and cabinets properly equipped? (Check off if OK)
   Water ____________ Tool Holders ____________
   Drain ____________ Ventilation ____________
   Lighting ____________ Shelving ____________
   Racks ____________ Identification Signs ____________

7. Describe the requisitioning and purchasing procedure (see below) ____________________

8. What records are maintained with reference to:
   a. Inventory Control ____________________
   b. Requisitions ____________________
   c. Purchases ____________________
   d. Stock distribution ____________________

9. What types of measuring and dispensing devices are in use? ____________________

   What types are needed? ____________________

10. Other notes: ____________________

11. IMPORTANT: Complete Materials and Equipment Inventory on the other side of this form.

## MATERIALS INVENTORY

| MATERIAL | BRAND NAME | PRICE | USE DILUTION | ANNUAL CONSUMPTION | AMT. ON HAND | MINIMUM | ORDER QUANTITY | USES |
|---|---|---|---|---|---|---|---|---|
| Floor Cleaner | | | | | | | | |
| | | | | | | | | |
| Cleaner - Disinf. | | | | | | | | |
| Wax Remover | | | | | | | | |
| Rug Cleaner | | | | | | | | |
| Floor Wax | | | | | | | | |
| | | | | | | | | |
| Mop Treatment | | | | | | | | |
| Floor Seal | | | | | | | | |
| | | | | | | | | |
| Disinfectant | | | | | | | | |
| Furniture Polish | | | | | | | | |
| Glass Cleaner | | | | | | | | |
| Cleanser | | | | | | | | |
| Metal Cleaner | | | | | | | | |
| Hand Soap, Liq. | | | | | | | | |
| Hand Soap, Pdr. | | | | | | | | |
| Deodorant | | | | | | | | |
| Bowl Cleaner | | | | | | | | |
| Insecticide | | | | | | | | |
| | | | | | | | | |
| | | | | | | | | |

## EQUIPMENT INVENTORY

| EQUIPMENT | NUMBER ON HAND | MAKE | SIZE | CONDITION | STORAGE LOCATION | USE AREAS | USES |
|---|---|---|---|---|---|---|---|
| Auto Scrubbers | | | | | | | |
| Power Sweepers | | | | | | | |
| Floor Machines | | | | | | | |
| Wet & Dry Vacs | | | | | | | |
| Dry Vacs | | | | | | | |
| Pack Vacs | | | | | | | |
| Applicators | | | | | | | |
| Brooms | | | | | | | |
| Brushes | | | | | | | |
| Buckets | | | | | | | |
| Dispensers | | | | | | | |
| Dust Mops | | | | | | | |
| Maid Carts | | | | | | | |
| Matting | | | | | | | |
| Mopping Outfits | | | | | | | |
| Receptacles | | | | | | | |
| Sprayers | | | | | | | |
| Squeegees | | | | | | | |
| Waste Carts | | | | | | | |
| Wet Mops | | | | | | | |
| Wringers | | | | | | | |

give them a proprietary attitude concerning the program, which is important because, as business leader Robert C. Hood put it, "People support what they help create." In addition, they will be likely to learn a good deal about their own housekeeping operation, and possibly their own general business operation, in the process.

Management determines the need for a housekeeping survey, and the program to follow, on the basis of recognition of this method as a means to achieve specific objectives. These management desires and goals are fundamental and must be carefully recorded and agreed on between management and the engineer. This may be done prior to the survey, or as one of the first steps during the survey.

## Communications

In the survey we begin to encounter problems in communications which are to confront us throughout the development and implementation of the housekeeping program. The information must be collected carefully, from well-selected sources. A check of the information should be performed in some cases. On a given question, two different persons within an organization will often provide conflicting information, which may arise from ignorance, reliance on obsolete data, misinterpretation of data, or clerical error. As a matter of fact, an individual may provide self-contradictory survey information on occasion.

Oftentimes wishful thinking enters into responses to survey questions. This is particularly true on questions dealing with frequency of job performance. These answers often indicate what the personnel should be doing, or would like to be doing, rather than what they are actually doing.

And, of course, sometimes information is distorted through fear. The person giving the information may feel that his position is in jeopardy because of some failure to perform properly, and therefore he colors all of his answers accordingly. Or, occasionally, fear will prompt an attempt to discredit the surveyor by having him make his recommendations on the basis of faulty information.

In making a survey, it is good practice to identify the source of the information. When people realize that they will be personally identified with data which are requested of them, they are particularly careful to provide accurate and relevant material. Thus, as in most human contacts, personal recognition has its rewards.

Only through personal contact can a "feel" for the situation be realized.

Thus, the surveyor learns what areas of change are possible and can be implemented due to individual or company personalities.

A rapid familiarization tour should take place very early in the survey, so that the data later collected can be related to the physical situation. Very often we find that supervisory personnel are impatient to conduct a tour of their plant. This healthy pride in the company and its facilities is a response which always kindles optimism for the housekeeping program to follow. The quick walk-through tour, naturally, must be followed later by a detailed physical survey.

## Floor Plans

The fundamental requirement of the physical survey is an accurate and complete floor plan of the physical plant. We see that this is obviously necessary where a program is being provided for a building still under construction, where the amount of written survey data will be quite limited. The helpfulness of the plans is increased materially when they are properly marked to show information such as:

- Color coding for floor types
- Traffic load
- Rest-room fixture counts
- Area names
- Area uses
- Work routes
- Square footages

Of course, the bulk of the cleaning problems are caused by the people working within the buildings, rather than by the buildings themselves. Therefore, the plans alone cannot constitute the entire survey.

For multibuilding facilities, a plot plan is required to show the spatial relationships of the various structures and the actual distances between them. Insurance drawings, such as prepared by Factory Insurance Association or Factory Mutuals, are useful for this purpose. Additional information or understanding can be obtained from aerial photographs in some cases.

## Other Data

The analysis of the personnel within the housekeeping department, their supervision, and various lines of authority within the housekeeping department and upward to management suggest the need for a carefully prepared organization chart. The organization chart should also show other functions performed by housekeeping department supervisors, staff relationships, cases of overlapping supervision, problems with titles, etc.

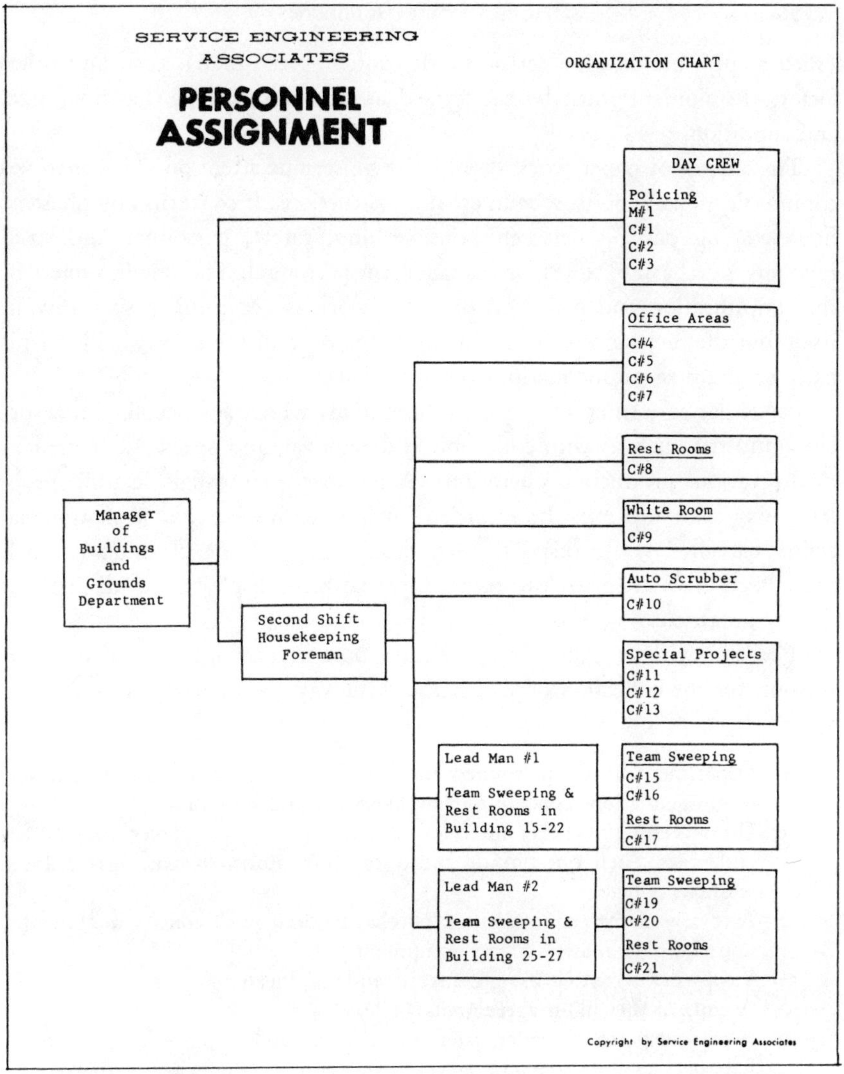


The determination of current housekeeping costs and existing or planned budgets should be carefully made. This is very often an elusive subject, since housekeeping may bear the cost of functions not related to cleaning, while other departments may bear certain housekeeping costs. The figures obtained will not be meaningful unless all costs dealing with cleaning are included, such as overtime, contract costs (for dust mop treatment, laundering, venetian blind cleaning, window washing and rest-room service), etc.

Housekeeping materials must be inventoried and surveyed as to type, amount on hand, ordering quantities and minimums, product characteristics (such as pH, use-dilution ratios, solids content, free alkali), cost, and other factors. Equipment must be catalogued as to number, type, location, size, and condition.

The subject of paper work should receive serious attention. This involves communications, reports, schedules, and instructions. It concerns any phase of housekeeping passing between housekeeping, safety, personnel, industrial relations, purchasing, stores, or management personnel. The whole subject of the simplification and provision of paper work is becoming a speciality in itself, but the inclusion of forms in use at the time of the survey will permit basic program recommendations on this subject.

Particular care must be given to those items which are peculiar to a specific industry, such as chip collection and removal, and special soils created in the various production operations. A number of unusual cleaning problems may arise and must be recorded, such as with the removal of unusual stains, classified waste disposal, bird droppings, and the cleaning of such specialized areas as superclean rooms, gymnasiums, display areas, dispensary, dietary areas, etc.

The surveying engineer may require other types of information, permission for the collection of which should always be cleared through management:

1. Department heads questioned about their specific problems, interest, and recommendations concerning housekeeping and sanitation.
2. Housekeeping workers interviewed relative to their problems, attitudes, and ideas. Such questioning must strictly conform to existing employee relations policies.
3. A review of the requisitioning, purchasing, and stock control methods for housekeeping materials and equipment.
4. A copy of current cleaning contracts and specifications.
5. A copy of the union agreement.
6. A copy of personnel rules, safety precautions, and employee's handbook.
7. The personnel department surveyed concerning its policies for the recruiting, selection, and training of housekeeping personnel.
8. The industrial relations department contacted to determine the availability of means to encourage greater employee awareness and cooperation about housekeeping.
9. The safety department consulted regarding the safety aspects of housekeeping.
10. The medical department interviewed regarding problems of industrial hygiene and cross-infection.
11. An attempt made to determine plant-wide employee attitudes about housekeeping and sanitation.
12. A study made of the method of negotiating cleaning contracts.

13. Current methods of work performance recorded for later comparative analysis.

At the time of the survey, management may find it desirable to question other department heads or personnel about housekeeping in an attempt to determine company-wide attitudes. In this case, the questions should give an opportunity for the persons involved to indicate their opinions concerning the housekeeping services which they like, those which they dislike, where housekeeping might be improved, where they feel their departments could benefit from an improved overall housekeeping effort, and such other information as may seem desirable. Surveys of this type have brought out worthwhile and interesting suggestions, while at the same time focusing attention on the company-wide aspect of good housekeeping.

Certain situations may indicate the need for some of the more sophisticated tools of the industrial engineer. Stop-clock, or time-lapse, photography may be used to analyze the situation at a central housekeeping storage area, for example. Work measurement techniques may be used to gauge worker productivity, perhaps as a prelude to the development of a complete or partial incentive system. Work sampling may be applied to indicate the percentage of consumed time applied to actual housekeeping work.

On the occasion of the survey, perhaps at its completion, a recorded inspection should be made to provide a benchmark against which future performance may be measured. Any one of several measuring or grading systems, which are discussed in a later chapter, can be used for this purpose. Sometimes, it will be more desirable to conduct the first inspection during the installation phase of the program.

## Report of Survey

The progress of the program must be suitably documented in order to keep management appraised, to help organize the work ahead, and to keep interest in the project alive. The proper close of the well-defined survey phase of the program is a report, which should include:

1. Acknowledgement of expressions of appreciation for assistance, guidance, and cooperation.
2. A review of survey activities, indicating the nature of each activity (meeting, tour, completion of questionnaire), the names of the persons present, and the nature of the information gained, along with the date, time, and place of the occurrence.
3. Questions of basic importance should be repeated, together with the answers provided and the identification of the individual supplying the information.

4. Any plans, schedules, charts or tables should be properly identified together with their source.

5. The report of survey is a legitimate vehicle for initial observations on which action can be taken before formal programming.

6. Finally, the report of survey can provide a tentative calendar for future activities.

### The Intermediate Report

If the housekeeping investigation is being performed wholly or in part by consultants, it may be desirable to provide an intermediate formal evaluation before the preparation of the program document is begun. Such a report goes further than the previously discussed report of survey and eliminates its need.

The report would include:

1. Acknowledgements
2. Review of survey activities
3. Verification of basic objectives and limitations
4. Identification of source material
5. Evaluation of current housekeeping operations
6. Specific recommendations for both short-term and long-term improvements, leading to reduced cost or improved quality, including:
   a. The cost of implementing the change
   b. The time required to make the change
   c. The estimated annual cost saving or percentage of quality improvement
7. Tentative calendar for additional activities

### The Housekeeping Audit

Where less than a complete study is desired by management, a good deal of useful information can still be obtained with an audit. This consists of a general review of indicators of performance, followed by a report giving this information:

1. The current status of custodial operations
2. Opportunities for cost improvement
3. Opportunities for quality improvement
4. Considerations concerning contract cleaning
5. Recommended additional studies

An audit can also provide recommendations on organization and staffing where example portions of the facility are measured and time allowances applied. Naturally, such information cannot be as accurate as full-scale analysis.

# 6

# Preparing the Program Document

THE primary objective of your housekeeping program is to extend the effectiveness of each person performing cleaning operations. This will provide a higher housekeeping level, create a situation requiring less workers, thereby saving money, or, in some cases, yield a combination of the two benefits.

In order to accomplish this fundamental purpose, we quickly see that it cannot be done with a new brand of wax, a new "miracle" type of equipment, by hiring an assistant foreman, or any other one-shot approach. There is no simple panacea which will solve the typical set of housekeeping problems in an optimum fashion. What is required is a comprehensive initial approach to the problem, followed by a continuous effort to put that approach into practice.

We can define organization as that optimum arrangement of all variables which will provide maximum results. The program document, therefore, provides that organization. Through this organization we are able to obtain control, which is the positive handling of a situation through planning. The basic elements of control are: (1) the written plan, (2) the measurement of performance against the plan, (3) the evaluation of results, and (4) the application of corrective action.

It is primarily through such control, designed to yield increasing worker productivity, that we are able to improve the housekeeping operation.

## Technology and Technique

Housekeeping is an art as well as a science. A strict numerical approach to the problem, wherein we deal primarily with standard times and work loads, brings only limited results. It is the interdependency of the six "M" variables (management, men, money, methods, materials, machines) that creates a situation that cannot be reduced to mathematical formula. If there were only ten possible choices for each of the basic types of variables, such a small selection alone would provide a million possible combinations! In certain phases of the program, therefore, we must depart from technology and begin to use technique. This, therefore, is the realm of the consultant, and one of the reasons that he can quickly provide a worthwhile return on the investment in his time.

There must obviously be a considerable difference in the effectiveness of a single program prepared with a limited backlog of experience in this complex field, as compared with a program built on a broad and deep range of consulting experience. In designing mechanical devices, we do not go back to the invention of the wheel as the basis of each design; proven standards and methods can be used as great timesavers, although we must use care to be sure they are the best available. The building of knowledge, as if it were blocks being stacked one on the other, is the fundamental reason for our advancing technology. The same approach can be used in the housekeeping program, where the experienced consultant can save time-consuming and expensive steps in the determination of frequencies, methods, work loads, and time-study information.

The written program cannot be a substitute for management, but, rather, it must depend upon management for its full utilization and benefit. No consultant should attempt to exercise authority or undertake responsibilities rightfully belonging to you and your management. An understanding of these points should be written into a statement as early as possible, if necessary.

The written program is often referred to as the "formal" document. The word is good as long as it refers to the comprehensive and detailed nature of the program, rather than to a program perhaps which might be pedantic, obscure, or stuffy.

For a complex situation, it may be desirable to deliberately prepare the program document in two drafts. The first draft should be designed for discussion purposes only. This is especially useful where a number of basic changes or departures for your organization are recommended and require a great deal of study and consideration. Once various decisions have been made about these more fundamental recommendations, the program can be redrafted to include only those items which have been accepted, modified, or will be considered further. Those items which the organization finds impractical to implement should be completely deleted. You can actually try out some of the recommended procedures in a first-draft document before decision is made on its final acceptance. Another approach to this situation is to submit an interim proposal, rather than an entire program document, containing only the more fundamental recommendations.

It must be made clear that in accepting a housekeeping program, management is not providing agreement in advance to use every recommendation which the program specifies or implies. In a sense, the program can be considered a book of ideas. Those ideas which the client feels are not applicable to his given situation, which he feels are contrary to his company's operating principles, or which appear incapable of implementation because of personality or other reasons, may be merely deleted or at least modified. The interdependence of recommendations must be considered from this point of view, since situations may occur where the deletion of a suggestion may require the adjustment of one which was accepted. It is the responsibility of the consultant to point out these relationships, as well as to point out the relative importance of the various recommendations for overall program effectiveness.

As we shall see in the installation and implementation phases of the work, the program should be considered a target to be approached carefully and methodically over a measured period of time, rather than a plan for a revolution which may disrupt the entire operation.

Consider all recommendations as positive opportunities for improvement to advance the status of the custodian, rather than as criticisms. This includes both those recommendations specifically enumerated and those implied as frequencies, methodology, etc., within the various instructions of the program. The program will propose the solution of problems that are rather typically found in industry and is in no way a criticism of the housekeeping department, its supervision, or the management of the organization.

All problems are not capable of solution at any given moment in time. For practical purposes we must often concern ourselves with a short-term approach to the problem (at least within the life of the corporation involved).

Sometimes we may have to be satisfied with a partial alleviation of a basic problem. Problems of this type are usually corollaries of general problems such as those associated with labor movements, psychology, sociology, etc.

Many interesting and worthwhile concepts and viewpoints can be derived from the study of military history. Captain B. H. Liddell Hart, in describing the qualities of the great captains of history, points out that an important requirement is "a sense of what is possible." The most detailed and imposing housekeeping program conceivable is of little value if it is beyond the capability of your organization to implement. The practical applicability of recommendations, either for short-term or long-term use, must be continually borne in mind.

## Program Format

The format for the program document, including the various types and names for the forms or sections to be used, is a matter of personal choice insofar as the details are concerned. But, fundamentally, they must organize the information to accomplish, as simply and as logically as possible, the basic purposes of the document. Some of these purposes are to:

Indicate lines of authority
Provide job descriptions
Set up the recommended method of job performance
Indicate frequencies of job performance
Assign workers to the areas for which they will be responsible
Provide standards for materials, equipment, and methods
Convey recommendations to management
Provide reference data on various aspects of housekeeping
Record time standards
Provide a professional authority for the actions to be taken
Outline the implementation phases to follow

One logical arrangement of the program document is to base it on the six "M" variables. This has the further advantage of correlating the program to the survey, if it was also set up on this basis. The section dealing with management would indicate the proposed supervisory structure of the housekeeping organization and the climate which has been created for its operation, for example. You can reproduce certain pages of the program, if carefully prepared, in quantity to be used directly by the custodians and maids themselves as instruction sheets.

In addition to the material described above, which must become part

## INDEX

COMPOSITE CLIENT



-continued next page-

SERVICE ENGINEERING ASSOCIATES/INC.

## INDEX

COMPOSITE CLIENT



SERVICE ENGINEERING ASSOCIATES/INC.

A custodial manual or program must provide sufficient detail for effective performance, fiscal justification and continuity in case of turnover (see next three pages).

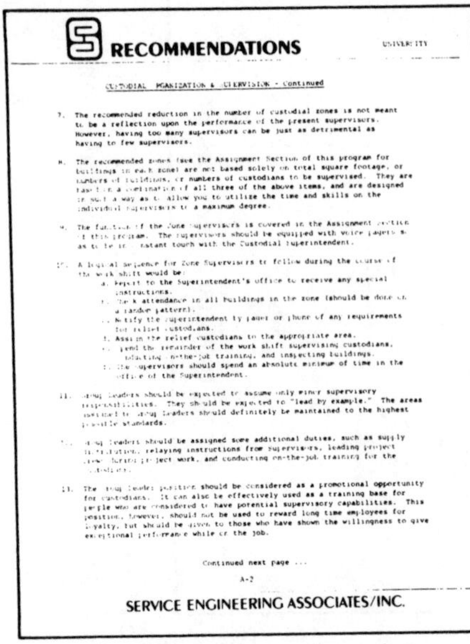

RECOMMENDATIONS UNIVERSITY

CUSTODIAL ORGANIZATION & SUPERVISION - Continued

7. The recommended reduction in the number of custodial zones is not meant to be a reflection upon the performance of the present supervisors. However, having too many supervisors can be just as detrimental as having to few supervisors.

8. The recommended zones (see the Assignment Section of this program for buildings in each zone) are not based solely on total square footage, or numbers of buildings, or numbers of custodians to be supervised. They are based on a combination of all three of the above items, and are designed in such a way as to allow you to utilize the time and skills on the individual supervisors to a maximum degree.

9. The function of the Zone Supervisors is covered in the Assignment section of this program. The supervisors should be equipped with voice pagers so as to be in constant touch with the Custodial Superintendent.

10. A logical sequence for Zone Supervisors to follow during the course of the work shift would be:
   a. Report to the Superintendent's office to receive any special instructions.
   b. Check attendance in all buildings in the zone (should be done on a random pattern).
   c. Notify the Superintendent by pager or phone of any requirements for relief custodians.
   d. Assign the relief custodians to the appropriate area.
   e. Spend the remainder of the work shift supervising custodians, conducting on-the-job training, and inspecting buildings.
   f. The supervisors should spend an absolute minimum of time in the office of the Superintendent.

11. Group Leaders should be expected to assume only minor supervisory responsibilities. They should be expected to "lead by example." The areas assigned to Group Leaders should definitely be maintained to the highest possible standards.

12. Group Leaders should be assigned some additional duties, such as supply distribution, relaying instructions from Supervisors, leading project crews during project work, and conducting on-the-job training for the custodians.

13. The Group Leader position should be considered as a promotional opportunity for custodians. It can also be effectively used as a training base for people who are considered to have potential supervisory capabilities. This position, however, should not be used to reward long time employees for loyalty, but should be given to those who have shown the willingness to give exceptional performance while on the job.

Continued next page ...

A-2

SERVICE ENGINEERING ASSOCIATES/INC.

The Recommendations section of the program provides both a status report and a suggested course of action to enable management to achieve immediate and long range housekeeping objectives.

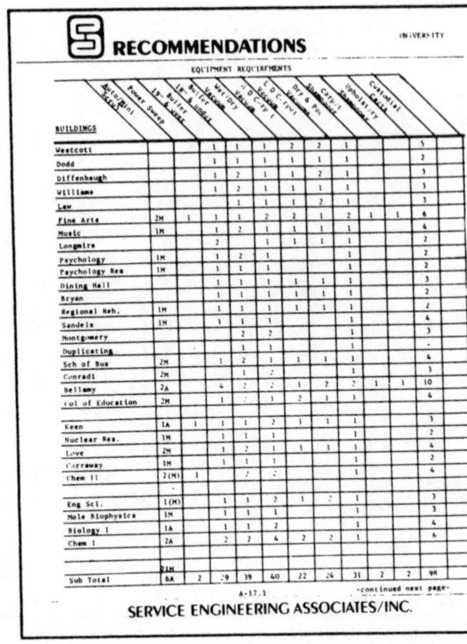

RECOMMENDATIONS UNIVERSITY

EQUIPMENT REQUIREMENTS

| BUILDINGS | Auto/Mini Scrub | Power Sweep | 19" Buffer & over | 18" Buffer & under | Wet/Dry Vacuum | ? D Carpet Vacuum | L D Carpet Vacuum | Dry & Pac Vacuum | Carpet Shampooer | Upholstery Shampooer | Custodial Carts |
|---|---|---|---|---|---|---|---|---|---|---|---|
| Westcott | | | 1 | 1 | 1 | 2 | 2 | 1 | | | 3 |
| Dodd | | | 1 | 1 | 1 | 1 | 1 | 1 | | | 2 |
| Diffenbaugh | | | 1 | 2 | 1 | 1 | 2 | 1 | | | 3 |
| Williams | | | 1 | 2 | 1 | 1 | 1 | 1 | | | 3 |
| Law | | | | 1 | 1 | 1 | 2 | 1 | | | 3 |
| Fine Arts | 2M | 1 | 1 | 1 | 2 | 2 | 1 | 2 | 1 | 1 | 6 |
| Music | 1M | | 1 | 2 | 1 | 1 | 1 | 1 | | | 4 |
| Longmire | | | 2 | | 1 | 1 | 1 | 1 | | | 2 |
| Psychology | 1M | | 1 | 2 | 1 | | | 1 | | | 2 |
| Psychology Res | 1M | | 1 | 1 | 1 | | | 1 | | | 2 |
| Dining Hall | | | 1 | 1 | 1 | 1 | 1 | 1 | | | 3 |
| Bryan | | | 1 | 1 | 1 | 1 | 1 | 1 | | | 2 |
| Regional Reh. | 1M | | 1 | 1 | 1 | 1 | 1 | 1 | | | 2 |
| Sandels | 1M | | 1 | 1 | 1 | | | 1 | | | 4 |
| Montgomery | | | | 2 | 2 | | | 1 | | | 3 |
| Duplicating | | | | 1 | 1 | | | 1 | | | - |
| Sch of Bus | 2M | | 1 | 2 | 1 | 1 | 1 | 1 | | | 4 |
| Conradi | 2M | | | 1 | 2 | | | 1 | | | 3 |
| Bellamy | 2A | | 4 | 2 | 2 | 1 | 2 | 2 | 1 | 1 | 10 |
| Sch of Education | 2M | | 1 | 2 | 1 | 2 | 1 | 1 | | | 4 |
| Keen | 1A | 1 | 1 | 1 | 2 | 1 | 1 | 1 | | | 3 |
| Nuclear Res. | 1M | | 1 | 1 | 1 | | | 1 | | | 2 |
| Love | 2M | | 1 | 2 | 1 | 1 | 1 | 1 | | | 4 |
| Carraway | 1M | | 1 | 1 | 1 | | | 1 | | | 2 |
| Chem II | 2(M) | 1 | | 2 | 2 | | | 1 | | | 4 |
| Eng Sci. | 1(M) | | 1 | 1 | 2 | 1 | 2 | 1 | | | 3 |
| Mole Biophysics | 1M | | 1 | 1 | 1 | | | 1 | | | 3 |
| Biology I | 1A | | 1 | 1 | 2 | | | 1 | | | 4 |
| Chem I | 2A | | 2 | 2 | 4 | 2 | 2 | 1 | | | 6 |
| Sub Total | 21M 6A | 2 | 29 | 39 | 40 | 22 | 24 | 31 | 2 | 2 | 98 |

A-17.1 -continued next page-

SERVICE ENGINEERING ASSOCIATES/INC.

Recommendations include programmed equipment requirements indicating the type, number, and location of the suggested equipment, based on individual work assignments.

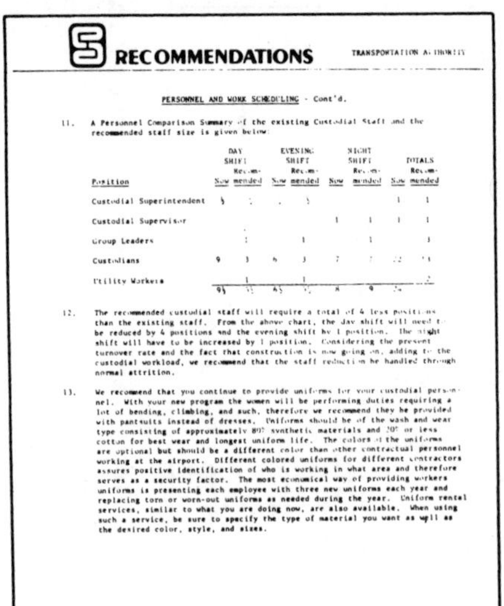

RECOMMENDATIONS TRANSPORTATION AUTHORITY

PERSONNEL AND WORK SCHEDULING - Cont'd.

11. A Personnel Comparison Summary of the existing Custodial Staff and the recommended staff size is given below:

| Position | Day Shift Now | Day Shift Recommended | Evening Shift Now | Evening Shift Recommended | Night Shift Now | Night Shift Recommended | Totals Now | Totals Recommended |
|---|---|---|---|---|---|---|---|---|
| Custodial Superintendent | ½ | [illegible] | [illegible] | ½ | | | 1 | 1 |
| Custodial Supervisor | | | | | 1 | 1 | 1 | 1 |
| Group Leaders | | 1 | | 1 | | 1 | | 3 |
| Custodians | 9 | 3 | 6 | 3 | 7 | 7 | 22 | [illegible] |
| Utility Workers | | 1 | | 1 | | | | 2 |
| | 9½ | [illegible] | 6½ | [illegible] | 8 | 9 | 24 | |

12. The recommended custodial staff will require a total of 4 less positions than the existing staff. From the above chart, the day shift will need to be reduced by 4 positions and the evening shift by 1 position. The night shift will have to be increased by 1 position. Considering the present turnover rate and the fact that construction is now going on, adding to the custodial workload, we recommend that the staff reduction be handled through normal attrition.

13. We recommend that you continue to provide uniforms for your custodial personnel. With your new program the women will be performing duties requiring a lot of bending, climbing, and such, therefore we recommend they be provided with pantsuits instead of dresses. Uniforms should be of the wash and wear type consisting of approximately 80% synthetic materials and 20% or less cotton for best wear and longest uniform life. The colors of the uniforms are optional but should be a different color than other contractual personnel working at the airport. Different colored uniforms for different contractors assures positive identification of who is working in what area and therefore serves as a security factor. The most economical way of providing workers uniforms is presenting each employee with three new uniforms each year and replacing torn or worn-out uniforms as needed during the year. Uniform rental services, similar to what you are doing now, are also available. When using such a service, be sure to specify the type of material you want as well as the desired color, style, and sizes.

A-5

SERVICE ENGINEERING ASSOCIATES/INC.

A personnel summary page in the Recommendations section may show the comparison between existing personnel in various job categories, where significant change is indicated.

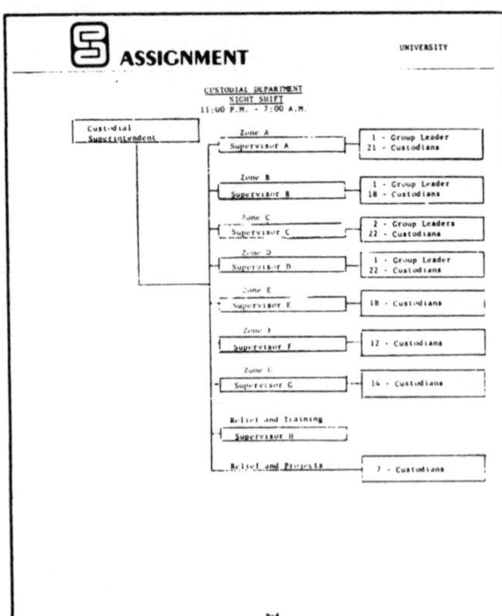

ASSIGNMENT UNIVERSITY

CUSTODIAL DEPARTMENT
NIGHT SHIFT
11:00 P.M. - 7:00 A.M.

- Custodial Superintendent
  - Zone A — Supervisor A — 1 - Group Leader, 21 - Custodians
  - Zone B — Supervisor B — 1 - Group Leader, 18 - Custodians
  - Zone C — Supervisor C — 2 - Group Leaders, 22 - Custodians
  - Zone D — Supervisor D — 1 - Group Leader, 22 - Custodians
  - Zone E — Supervisor E — 18 - Custodians
  - Zone F — Supervisor F — 12 - Custodians
  - Zone G — Supervisor G — 14 - Custodians
  - Relief and Training — Supervisor H
  - Relief and Projects — 7 - Custodians

B-4

SERVICE ENGINEERING ASSOCIATES/INC.

The Assignment section typically begins with an Organization Chart for the entire housekeeping department, supplemented by individual organization charts for each supervisor in larger organizations.

## ASSIGNMENT

UNIVERSITY HOUSING

AREA "B"

McMahon Hall

**Group Leader B-1:** PROCEDURES: 1,2,3,4,5,6,8,10

Perform the routine cleaning of the "B", "C", and "D" levels of the Central area, including all elevators. In addition, clean the public rest rooms and stairwells on the first floor of the Central area. Perform other duties as assigned.

BASIC EQUIPMENT: one set of Group D equipment; one set of Group C equipment for the Central area; and one small tank vacuum.

**Custodian B-2:** PROCEDURES: 1,4,5

Perform the routine cleaning of the student "cluster areas" in the North Wing. (Must clean two floors per day.)

BASIC EQUIPMENT: one set of Group D equipment; one upright carpet vacuum.

**Custodian B-3:** PROCEDURES: 1,2,5,7

Perform the routine cleaning of the North Wing eleventh - second floors, except for the "cluster areas."

BASIC EQUIPMENT: one set each of Group C and D equipment; one wet/dry vacuum; one pack-type vacuum.

**Custodian B-4:** PROCEDURES: 1,4,5

Perform the routine cleaning of the student "cluster areas" in the South Wing (must clean two floors per day).

BASIC EQUIPMENT: one set of Group D equipment; one upright carpet vacuum.

**Custodian B-5:** PROCEDURES: 1,2,5,7

Perform the routine cleaning of the South Wing eleventh - second floors, except for the "cluster areas."

BASIC EQUIPMENT: one set each of Group C and D equipment; one wet/dry vacuum; one pack-type vacuum.

**Custodian B-6:** PROCEDURES: 1,5,8

Perform the routine cleaning of the first floor Central area, except for stairwells and public rest rooms.

BASIC EQUIPMENT: one set each of Group A and B equipment; one wet/dry vacuum; one upright carpet vacuum.

B-14

SERVICE ENGINEERING ASSOCIATES/INC.

Individual job assignments represent measured, reasonable work loads for each position. They indicate the area to be cleaned, equipment to be used, and the procedures to be followed.

## PROCEDURE

FEDERAL RESERVE BANK

Procedure #1

VAULTS

DAILY: Empty Waste from waste receptacle into waste bag on the cart. Damp wipe soiled waste receptacles. Replace plastic liners when used in waste receptacles. Empty ash trays and urns.

Dust all ledges and furniture tops using a treated cloth, dusting mitts, or dust mop with short handles.

Clean smudges and soil from glass in displays, partitions and doors, using glass cleaner in spray-bottle and a clean cloth.

Vacuum any carpets, or mats, and check carpets for spot-cleaning.

Dust mop floors with a wide treated dust mop, keeping the dust mop head on the floor at all times. Pick up soil from floor with dust pan. Vacuum mop head periodically to remove dust, when a vacuum cleaner is available, or shake out into empty plastic bag, with mop head in the bag.

Spot-clean walls, door facings and doors. Use a detergent solution in a spray bottle and a cloth or a sponge. Rinse with sponge and clear water in plastic spray bottle as needed.

NOTE: For floor care procedures involving the use of an auto-scrubber, see Procedure #2.

TWICE WEEKLY: Using a floor machine equipped with a spray pad, spray-buff floors which are coated with floor finish. (Spray-buff floor finish.) After buffing, dust mop the floor as necessary.

WEEKLY: Using a cloth, dusting mitts or small dust mop with short handle, dust vertical furniture surfaces, wall vents and vertical wall trim.

Clean any metal trim with polish.

C-1

SERVICE ENGINEERING ASSOCIATES/INC.

A standard procedure is developed for each type of area. Stressing efficiency and safety, they are based on optimum frequencies, techniques, equipment, and materials. They form the foundation for a training program.

## PROJECT CALENDAR

MAIN PLANT--N. CHICAGO

Building: ______

Floor or Area: ______

PROJECTS TO BE PERFORMED BY SCHEDULED PROJECTS PERSONNEL

| PROJECT | AREA | FREQUENCY | JAN | FEB | MAR | APR | MAY | JUN | JUL | AUG | SEP | OCT | NOV | DEC |
|---|---|---|---|---|---|---|---|---|---|---|---|---|---|---|
| Machine Scrub Floors in Showers | Shower Areas | Monthly | | | | | | | | | | | | |
| Strip & Rewax Elevator Floors | Elevators | Monthly | | | | | | | | | | | | |
| Vacuum Vents | All Areas | Monthly | | | | | | | | | | | | |
| Vacuum and Damp Wipe Blinds | All Areas | Monthly | | | | | | | | | | | | |
| Wash Waste Receptacles | All Areas | Monthly | | | | | | | | | | | | |
| Vacuum Clean Walls | All Areas | Monthly | | | | | | | | | | | | |
| Wash Vents | All Areas | Quarterly | | | | | | | | | | | | |
| Vacuum or Dust High Overhead Pipes, Light Covers, Fixtures, etc. | All Areas | Quarterly | | | | | | | | | | | | |
| Strip & Refinish Heavily Trafficked Composition Floors | All Areas | Each 4 Months | | | | | | | | | | | | |
| Machine Scrub Floors in Rest Rooms, Shops, Glass Washing Areas, Shaker Rooms, and Sterilizer Rooms | Areas Shown in Column to Left | Semiannually | | | | | | | | | | | | |
| Wash Inside Windows & Glass Partitions | All Areas | Semiannually | | | | | | | | | | | | |
| Wash Outside Windows (Contract) | All Areas | Semiannually | | | | | | | | | | | | |
| Strip and Refinish Composition Floors | All Areas | Annually | | | | | | | | | | | | |
| Vacuum Clean Ceilings | All Areas | Annually | | | | | | | | | | | | |
| Wash Light Diffusors (In Washematic) | All Areas | Annually | | | | | | | | | | | | |
| Shampoo Carpets | Hallways | Annually or As Needed | | | | | | | | | | | | |
| Wash, Rinse, & Dry Venetian Blinds | All Areas | Each 2 Years | | | | | | | | | | | | |
| Shampoo Carpets | Offices | Each 2 Years or As Needed | | | | | | | | | | | | |
| Strip Discolored Floors | All Areas | As Needed | | | | | | | | | | | | |
| Relamp Fixtures | All Areas | As Needed | | | | | | | | | | | | |
| Wash Walls | All Areas | As Needed | | | | | | | | | | | | |
| Clean Production Line | Bldg. C-11 | As Needed | | | | | | | | | | | | |
| Machine Scrub Floors in Sterile Or Restricted Areas Where Required | Sterile or Restricted | When Required | | | | | | | | | | | | |

D-2

SERVICE ENGINEERING ASSOCIATES/INC.

Although the Procedures section covers routine activities, it is also necessary to develop a Project Calendar for scheduling and controlling non-routine activities, which are often overlooked.

## PROJECT METHOD

SCRUBBING FLOORS

MATERIALS REQUIRED:

Neutral Detergent

EQUIPMENT REQUIRED:

2 mop buckets
1 wringer
2 mops
1 dust mop
1 push broom
clean rags
1 floor machine with scrub brush or scrubbing pad
1 wet vacuum (optional)

INSTRUCTIONS:

1. Sweep or dust mop floor thoroughly.
2. Mix a solution of detergent in one bucket, two ounces per gallon of warm water (unless label specifies other dilution).
3. Apply detergent solution to floor. Do Not flood floor.
4. Allow detergent solution to work for you three to four minutes.
5. Scrub floor with floor machine and scrubbing brush or scrubbing pad, where necessary.
6. Pick up dirty solution with mop and second bucket with wringer, or with wet vacuum.
7. If floo is heavily soiled, rinse with clear water after scrubbing. Change water frequently.
8. Spot-clean baseboards with mop or with rags and detergent solution
9. Allow floor to dry thoroughly.
10. Clean and put away all equipment.

NOTES: If cleaning is being done in trafficked areas, put out "WET FLOOR CAUTION" signs to prevent accidents.

On hard (non-resilient) rest room and locker room floors, use wa stripper for better cleaning.

SERVICE ENGINEERING ASSOCIATES/INC.

For each significant item on the Project Calendar, a Project Method indicates manpower, materials and equipment required for the standard method. This material may be included in a training program.

**MATERIALS**

SPECIFICATION FOR WATER EMULSION METAL-LINK POLYMER FLOOR FINISH

SCOPE:
This specification covers a self polishing, slip resistant, all synthetic water emulsion floor finish intended for use on, and not detrimental to, sealed or finished wood surfaces, asphalt tile, linoleum, rubber, vinyl composition, painted concrete, terrazzo and cement surfaces. The product is to be used without dilution. Acid sensitive emulsions are excluded.

REQUIREMENTS:
1. The finish shall consist of all synthetic components. It shall contain no natural waxes.
2. The finish shall contain a minimum of 15% non-volatile material when determined at 105°C.
3. The finish shall consist of polyacrylic and polyethylene components. Polymers shall be metal-linked to provide excellent detergent resistance. It shall not contain polystyrenes.
4. The product shall be safe for use on all kinds of floors including asphalt, vinyl, linoleum, terrazzo, marble, sealed wood, cork, rubber, and other composition tile durfaces.
5. The product shall produce a colorless, rough, non-slip, water resistant coating having a very high gloss.
6. The finish shall be recoatable within 15 minutes after prior coat has dried. Second coat shall not whiten nor lift the first coat -- it shall produce enhanced gloss.
7. The product shall resist scuffing, powdering, and scratching under traffic to a satisfactory degree and shall respond to rebuffing.
8. The finish shall level well on application without streaking or puddling.
9. The finish shall resist wet cleaning with alkaline detergents without being removed from the floor and without destroying gloss. However, it shall be readily removable by a common detergent with ammonia added.
10. The product shall not be acid sensitive, and shall resist acidic cleaners and strippers, including acidic soft drinks and fruit juices.
11. The finish shall be non-yellowing on the floor and should be milk white in the original container, rather than tan.
12. The finish shall bear the U.L. seal of approval for slip resistance on label.
13. The finish shall be film forming at usual temperatures and shall produce no powdering and no crazing.
14. The finish shall be completely waterproof within 12 hours after application.
15. The finish shall be free from objectionable odor and must not develop an offensive odor upon storage in the original, unopened container.

SERVICE ENGINEERING ASSOCIATES/INC.

The Materials section of the program provides specifications for major chemical items, not only to eliminate the use of ineffective products, but also to reduce the cost of sanitation chemicals.

**MATERIALS**

FLOOR FINISH TESTING RECORD

Product Code ________
Area Identity ________
Date Applied ________

| Time of Evaluation After Original Application | Qualities Rated (1) | | | | | | | | | | | Effects of Maintenance | | | | Total Score at Each Evaluation |
|---|---|---|---|---|---|---|---|---|---|---|---|---|---|---|---|---|
| | | | | | | | | | | | | Buffing | | Spray-Buff | | |
| | Good Levelling | Dries 30-60 Minutes | No Cloudiness | No Discoloration | Gloss | No Slipperiness | No Powdering | Little Scratching | Little Scuffing | Little Black-Marking | Strips Readily (2) | Resultant Gloss | Removal of Scuffs & Marks | Resultant Gloss | Removal of Scuff & Black Marks | |
| As Soon as Dry | | | | | | | x | x | x | x | x | x | x | x | x | |
| 2nd Day | x | x | x | x | | | | | | | x | | | | | |
| 1 Week | x | x | x | x | | | | | | | x | | | | | |
| | x | x | x | x | | | | | | | | | | | | |
| | x | x | x | x | | | | | | | | | | | | |
| | x | x | x | x | | | | | | | | | | | | |
| | x | x | x | x | | | | | | | | | | | | |
| | x | x | x | x | | | | | | | | | | | | |
| | x | x | x | x | | | | | | | | | | | | |
| | x | x | x | x | | | | | | | | | | | | |
| | x | x | x | x | | | | | | | | | | | | |
| TOTAL RATING (3) | | | | | | | | | | | | | | | | |

(See separate Floor Finish Testing Instructions)

NOTES AND COMMENTS:

(1) Rating values: 2 = Good; 1 = Acceptable; 0 Poor
(2) Strippability: See Instruction 9
(3) Highest Total Rating usually will indicate best performance; but, product may be disqualified for unsatisfactory performance in one important category, such as "No Slipperiness" or "No Powdering"

F-2

SERVICE ENGINEERING ASSOCIATES/INC.

The Materials section may provide listings for manual supplies that may be used on a custodial cart, or, as shown above, a form for testing floor finishes.

**REFERENCE**

REMOVING STAINS FROM FLOORS

Use all chemical agents with extra care to prevent fire, skin irritation or injury, and damage to surfaces. Try to destain a small inconspicuous "test spot" before beginning a larger operation. Because of the many variables of surfaces, dilutions, application, and supervision, care must be exercised in the use of this chart.

A. BLOOD:
1. Asphalt tile, linoleum, vinyl, wood: rub with cloth dampened in clear cold water. If stain persists, dampen cloth with ammonia.
2. Concrete, marble, oxychloride cement, terrazzo: rub with cloth dampened in clear cold water. Bleach with peroxide if stain persists.

B. CHEWING GUM:
1. Concrete, linoleum, marble, oxychloride cement, terrazzo, wood: remove gum with putty knife. Apply alcohol, rub with clean cloth.
2. Asphalt tile: remove gum with putty knife. (The use of dry ice will make the gum easier to remove.) Do not use alcohol on asphalt tile.
3. Vinyl: remove as much as possible with putty knife. Rub with #0 steel wool dipped in all purpose synthetic detergent solution.

C. GREASE OR OIL:
1. Asphalt tile, linoleum: scrub with warm all-purpose synthetic detergent solution. Rinse with clear water.
2. Concrete: pour alcohol on spot. Rub on spot. Rub with clean cloth.
3. Marble, terrazzo, oxychloride cement: pour solvent on spot, cover with Fuller's earth and let stand for several hours. Repeat if necessary.
4. Vinyl, wood: pour kerosene on spot. Permit to soak for a short time. Wipe dry with a clean cloth. Wash with all-purpose synthetic detergent solution, rinse dry.

D. INK:
1. Asphalt tile, concrete, linoleum, marble, oxychloride cement, terrazzo: use warm all-purpose synthetic detergent solution. If stain persists, mix two tablespoonfuls sodium perborate in pint of hot water. Mix whiting to form paste. Apply to spot and let dry.
2. Vinyl: wash with all-purpose synthetic detergent, rinse, then dry. If stain persists, rub with cloth dampened with ammonia.
3. Wood: apply solution one quart oxalic acid crystals to nine parts warm water. Permit to stand until dry. Mop with clear water.

E. IODINE OR MERCUROCHROME:
1. Asphalt tile: warm neutral soap solution.
2. Concrete, linoleum, wood: apply alcohol and rub with clean cloth.
3. Marble, oxychloride cement, terrazzo: apply alcohol and cover with Fuller's earth.
4. Vinyl: wash with all-purpose synthetic detergent, rinse, then dry. If stain persists, scrub with scrubbing powder and warm water.

- continued next page -

SERVICE ENGINEERING ASSOCIATES/INC.

The Reference section contains material of use to supervisors and management in implementing the program, covering such subjects as stain removal (example above), care of equipment, custodial safety, etc.

**REFERENCE**

TYPICAL SMALL CENTRAL HOUSEKEEPING STORAGE AREA

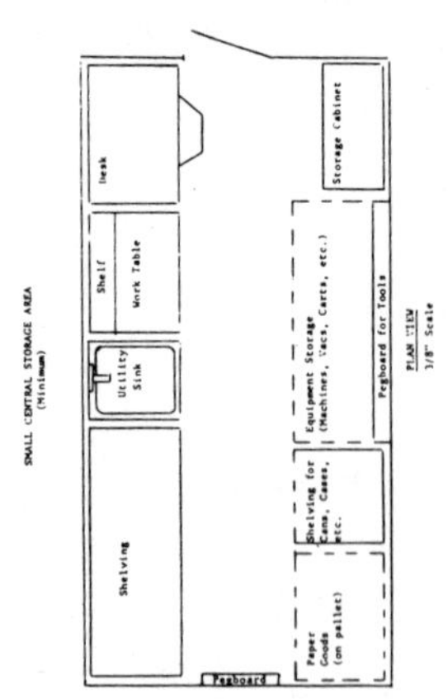


SERVICE ENGINEERING ASSOCIATES/INC.

Finally, reference may also be provided on physical factors such as design for maintenance during renovation or construction projects, and the layout of custodial storage, cabinets, and closet areas.

SERVICE ENGINEERING ASSOCIATES

# SPECIAL PROJECT RECORD

Building #26

Record the performance of each Special Project on the day it is completed. This record must be maintained accurately and up-to-date to assure regular performance of all Special Projects.

| SPECIAL PROJECT | AREA | FREQUENCY | JAN. | FEB. | MAR. | APR. | MAY | JUNE | JULY | AUG. | SEPT. | OCT. | NOV. | DEC. |
|---|---|---|---|---|---|---|---|---|---|---|---|---|---|---|
| Wash Mats & Runners | At Entrances | Monthly | | | | | | | | | | | | |
| Clean & Polish Hardware | At Entrances | Monthly | | | | | | | | | | | | |
| Clean Utility & Storage Rooms | All Areas | Monthly | | | | | | | | | | | | |
| Vacuum Radiators & Vents | All Areas | Monthly | | | | | | | | | | | | |
| Descale Fixtures (SPI 400) | Rest Rooms | Monthly | | | | | | | | | | | | |
| Machine Scrub Unwaxed Floors (SPI 203) | All Areas | Monthly | | | | | | | | | | | | |
| Clean & Rewax Composition Floors (SPI 200,350) | Rest Rooms | Monthly | | | | | | | | | | | | |
| Clean & Rewax Composition Floors (SPI 200,350) | Foremen's Offices | Every 6 Wks. | | | | | | | | | | | | |
| Clean & Polish Furniture (SPI 410) | Foremen's Offices | Every 2 Mos. | | | | | | | | | | | | |
| Wash Windows Inside (SPI 110) | All Areas | Every 2 Mos. | | | | | | | | | | | | |
| Wash Waste Receptacles | All Areas | Quarterly | | | | | | | | | | | | |
| Vacuum Light Fixtures | All Areas | Quarterly | | | | | | | | | | | | |
| Vacuum Walls & Woodwork | All Areas | Quarterly | | | | | | | | | | | | |
| Wash Windows & Screens (Contract) | All Areas | Semi-Annually | | | | | | | | | | | | |
| Strip & Rewax Composition Floors (SPI 250,350) | Foremen's Offices | Semi-Annually | | | | | | | | | | | | |
| Strip & Rewax Floors (SPI 250,350) | Rest Rooms | Semi-Annually | | | | | | | | | | | | |
| Seal Floors (SPI 320) | Concrete Floors | Annually | | | | | | | | | | | | |
| Damp Clean Light Fixtures (Engrs.) | All Areas | Annually | | | | | | | | | | | | |
| Damp Clean Radiators & Vents | All Areas | Annually | | | | | | | | | | | | |
| Vacuum Overhead | All Areas | Annually | | | | | | | | | | | | |
| Wash Walls & Woodwork (SPI 120) | All Areas | Annually | | | | | | | | | | | | |

SERVICE ENGINEERING ASSOCIATES

**SPECIAL PROJECT INSTRUCTIONS**

SEALING
CONCRETE FLOORS

SPI 320

| MATERIALS REQUIRED: | EQUIPMENT REQUIRED: |
|---|---|
| CONCRETE SEAL | 1 push broom<br>dust mop<br>seal pan<br>lambs wool applicator<br>clean rags |

INSTRUCTIONS:

PREPARATION

Floor seals reflect the condition of the floor beneath because of their transparent quality. For this reason, it is extremely important that the floor be put into good condition before sealing. Where the concrete is stained by oil or other materials, a series of treatments may be required to bring the stain to the surface and remove it. It is difficult to obtain a uniform finish unless the floor is carefully cleaned in advance and carefully dried. See Special Project Instructions for wet cleaning concrete floors.

SEALING

1. Be sure the floor is clean and completely dry. Since drying time for floor seals is critically influenced by humidity, it is suggested that seals not be applied on rainy days or days of high humidity, unless the work is to be done in a heated building. It is best to seal on the day following the floor cleaning.
2. Carefully sweep or vacuum floors to remove all mop strands and brush bristles.
3. Fill seal pan one-half full with seal and immerse applicator into seal. Press pad against side of pan so applicator is drip free when lifted from seal pan.
4. Apply seal in straight, even strokes 3 to 4 feet long. Do not try to make excessively long strokes. Carefully "feather" overlap strokes to avoid lap marks and to produce an even appearance. It is not necessary to apply pressure on the applicator; the weight of the applicator itself is adequate.
5. Shiny spots will indicate too much seal or spots where the concrete is more tightly bonded than others. Any excess seal should be spread out; all of the seal should penetrate into the floor.
6. After the floor is thoroughly dry, repaint any stripes or lines on the floor. Allow to dry thoroughly.
7. Apply second coat of seal if required.
8. Be certain that the floor is perfectly dry before re-using.
9. As needed from time to time, patch-seal particularly worn areas (such as aisles) using the same process as described above.
10. Clean and put away all equipment. Close containers tightly. Mineral spirits used promptly will normally remove seal from applicator pan and applicator so that they may be re-used.

NOTE: For drying time, refer to the concrete seal specifications sheet.



of every program, other items may be required or desirable. The program may include, for example:

1. Statistical reference material as obtained during the survey (fixture counts, square footages, personnel data, etc.).
2. Schematic layouts indicating floor types by color coding, areas of responsibility, traffic conditions, etc.
3. Comparison charts relating existing personnel and their cost to that which may be proposed.

4. An economic comparison of cleaning by contract as compared with cleaning by internal personnel.
5. Recommended product specifications, testing methods, purchase quantities, inventory levels, and estimates of materials cost.
6. Description of recommended equipment, specifications, and cost.

The following pages illustrate some of the types of information which might be contained in a given program.

In certain cases the program may include the results of questionnaires. A questionnaire may be distributed at the request of management to determine any given aspect of housekeeping, or it may be the result of an industry-wide inquiry about some specific problem.

Field Marshal von Hindenburg once said: "In war, only what is simple can succeed." Often, much the same is true in housekeeping. The program should be as comprehensive as necessary, but the parts that make up the whole should be conducive to rapid and faithful implementation through their simplicity. The most complex machines of our mechanical age are usually combinations of parts and components which are themselves rather simple. And the simpler the parts are, the better the machine because it functions more smoothly and requires less maintenance.

The preparation of instructions, another field which is becoming an art in itself, is a case in point. It is best to give directions to the worker in words and terms which are easily understandable. The more carefully his duties are defined and pinpointed, the greater will be the benefit to the organization and the lower the cost of achieving this benefit. Instructions should answer the questions of *what? when? where* and *how?* in advance. Choose words consistent with the level of the intended audience. Compare the following two columns of words. The left-hand column should be used for housekeeping instructions; the right-hand column should be saved for use elsewhere:

| | |
|---|---|
| get | obtain |
| put | position |
| do | perform |
| bad | defective |
| job | function |
| after | subsequent to |
| before | prior to |
| pour | decant |

Verbose sesquipedalianism, herein exemplified, detracts from the efficaciousness of the undertaking. Keep your words simple!

In modern life, most phases of activity seem to be accompanied by large quantities of paper work. It has been said that if a visitor from another planet looked into a typical office building, he would surely assume that all of the workers were engaged in the manufacture of small pieces of paper of various sizes, colors, and markings! Unquestionably we must avoid unnecessary paper work, and the war on paper should be continuous. Even so, the program should indicate the paper work necessary to its successful prosecution and, where needed, should actually propose the forms to be used. Not all of these forms will be needed in every case, even in the larger operations. Their use should depend on need and the benefit derived from them. The determination of forms is not complete without an indication of the number of copies necessary, their routing, and eventual disposition. Here are some possible examples:

Progress reports to management
Inspection reports
Grading reports
Materials-consumption records
Equipment records
Special project frequency records
Complaint records
Accident records
Work orders

## Personnel Determination

The heart of the program concerns personnel—the determination of their number, organization, scheduling, and other factors, all of which are interdependent.

The theoretical staff for performance of any function, housekeeping or otherwise, is calculated on the basis of dividing the total work load in hours by the number of productive hours per employee. The work load is a product of the number of units to which the work is applied, the frequency of application, and the time required for application.

Let us be a little more specific in relating this approach to housekeeping.

The determination of frequency of job performance and the desired level of housekeeping go hand in hand. That is, a determination of one dictates the other. In order to achieve a given level of housekeeping, it is necessary to relate frequencies of job performance to obtain that level—conversely, given frequencies will dictate a specific level. Thus we can see that our determination must begin with an expression from management of general objectives in housekeeping, which must then be related by the engineer to frequency. Often management will ask the consultant to propose an objective for housekeeping. In any case, the first step must be the determination of the objective, even if it is tentative.

The total number of work units, of course, is a product of the frequency and number of items or the measure of areas to be cleaned or otherwise maintained. The information required to develop these figures is taken from the physical survey of the premises. The application of time standards against the work units gives us the number of man-hours per year required for the total housekeeping operation. When this is divided by the number of annual productive hours per custodian, we obtain the theoretical number of custodians required.

To recapitulate, the calculation is based on the formula: Personnel-frequency × areas × standard times ÷ available productive hours per year. In using this formula it is extremely important to keep the units straight (hours, seconds, weeks; square feet, square yards, etc.) and to cancel them out properly in the calculation.

At this point we will have arrived at a *theoretical* calculation of the housekeeping staff. *It probably will not be correct.* The figure will require adjusting up or down on the basis of the many considerations and variables involved. It would be difficult to overstress the importance of these variables in determining the number of housekeeping personnel, and therefore in determining cost. It is also well to point out, at this point, that the general housekeeping program prepared for a typical plant will be of considerable value for a number of years. But when conditions, such as personnel, physical facility, management objectives, and production levels, change the program will require adjusting if its value is to remain high.

The forms utilized for the determination of work load are subject to considerable variation. They may consist of a separate sheet for each type of cleaning operation, for example, or possibly one sheet for each floor or each building with all the various jobs listed. The latter has the advantage of relating work load to a natural physical division, which would otherwise have to be done in a separate step. It has the disadvantage of mixing standards and units. Personal preference will dictate the specific form to be used, the column headings, the size of the form, and the method of extension. In any case, these factors must be recorded:

Area designation
Item or surface to be maintained
Measure or count
Housekeeping job
Annual frequency
Annual measure or count
Time standard
Total time

Let us look at some of the situations which affect the personnel determination and assignments.

The program must be based on a given efficiency and productivity for the housekeeping worker. Sometimes averages can be used, whereas other times the figures should be specifically for individuals. Although the program may recommend future work measurement and possible incentive techniques for certain jobs, figures must be determined at the time of the survey or during the survey analysis, for the preparation of the program.

## Special Projects and Relief

Special projects form the backbone of the preventive maintenance aspect of housekeeping. Some types of special projects are best performed by individuals, either with area or project assignments. Other types can most efficiently be done as a team effort. In certain cases, these teams can be augmented by a custodian on area assignment, where the project work is being done in that area and consumes only a short period of time. The special projects workers, incidentally, should always be as highly mechanized as possible. Special projects can arbitrarily be defined as those items which are performed less often than once a week.

Devise your program to provide the flexibility required to meet unusual situations. These situations may be caused by shortages in materials or equipment, accidents, emergency cleaning situations, or absence due to vacations, special holidays, family emergencies, sickness, or failure to report to work. A procedure should be set up in advance to handle each of these possible occurrences. The provision of a separate relief crew will solve the majority of this type of problem. Relief must be provided for each shift during which housekeeping work is performed, unless there are only a very few persons on a given shift performing policing operations, where the other workers can "cover." Relief personnel should always report directly to their supervisor at the beginning of the work shift, and, if not required for relief, they may be assigned to perform special projects work.

There is an inner-relationship between special projects and relief personnel, and it is these combined groups which Mr. Herman Winkler, Assistant Factory Manager of the Bendix Corporation, Scintilla Division, has so aptly termed the "expansion chamber." Thus, it is possible to adjust the size of these two combined groups to conform to a housekeeping staff that may be determined by the level of production. This permits permanent daily assignments and instructions to remain in force even during a changing situation. But, we must recognize that the housekeeping level will fluctuate with the size of this group, because it is extremely important that housekeeping be performed on a continuing level through the provision of relief, and, in addition, special projects are also important.

In order to eliminate interference between housekeeping personnel and others, as much of the work as possible should be performed after the busiest hours of the company. Thus, a safer operation will result, with more efficient working conditions both for housekeeping and other personnel. The advantages must be weighed against the drawbacks, because night cleaning work sometimes intensifies the problems of supervision, security, the ability to recruit workers because of transportation difficulties and the later hours, etc.

## Policing

When the basic cleaning work is done at night, it should be followed during the day by a policing operation. This is not primarily a cleaning operation, but one of waste collection and removal, spillage removal, refilling dispensers, etc. A large percentage of policing labor is consumed in the personnel areas such as rest rooms, locker rooms, etc. The policing operation is one of the more important, yet usually more disorganized, phases of the housekeeping operation, because the very nature of the work makes supervision difficult. This should be overcome by a very rigid policing schedule, with precise time indications for each work station.

Interestingly enough, the attitude of the public and factory workers alike concerning the appearance and level of housekeeping of a facility is often based on how well the policing function has been performed.

In the larger operations, the various housekeeping functions should be categorized and treated separately, even though certain jobs may be combined to be performed by a single worker or a team of workers. These divisions may include, for example:

Rest-room cleaning
Office cleaning
Dietary area cleaning
Machine cleaning
Waste removal
Chip collection
Power sweeping
Special projects
Relief
Nonhousekeeping responsibilities

In the smaller housekeeping departments we must work with indivisibles. For example, it may require 1.4 men to clean machinery, 2.7 men to clean the offices, etc. In view of this, in the smaller plants, for example, watchmen are sometimes given housekeeping responsibilities. These should be of such a nature that they can be interrupted without hazard or without wasting too much time, so that the watchman can perform his rounds. Some types of special projects are suitable for this purpose.

## Team Cleaning

There is a certain amount of controversy about whether cleaning operations should be performed by individuals on an area assignment basis or by groups on a project assignment basis. As is so with many other controversial issues, we find that we are involved here with "conjugate principles," where the advantage of either system can clearly be shown for certain cases. The issue of whether to perform cleaning on an individual or team basis, or some combination of the systems, must depend on a large number of other factors involved in the total picture. The proper choice is actually in the field of technique rather than technology.

An indication of the general advantages of each system is useful in considering the two possibilities. Area assignments have these advantages:

They give the custodian a proprietary attitude toward his work area, wherein he develops more pride in the results.

A very healthy competitive spirit can be developed in the cleaning of similar types of areas. Individual assignments lend themselves more to award or recognition systems, based on quality of performance.

The individual assignment pinpoints responsibility. Any deficiencies or failure to perform the work can be laid at the feet of the custodian involved.

Because of the fixing of responsibility, security control may be exercised more easily.

The custodian should be at his work station at all times, except for personal reasons, and therefore he can be easily and quickly located for general supervision, emergencies, etc.

Group cleaning may involve anywhere from two to six or more persons for each group. Some of the general advantages of team cleaning are:

The group can be highly mechanized and yet the investment in equipment for each custodian can be kept to a minimum. A floor cleaning group, for example, might consist of one custodian equipped with a mopping tank, one with a scrubbing machine, and another with a wet vacuum. If the work were to be performed on an area assignment basis, all of this equipment would have to be provided for the individual.

If a member of the team is a group leader or assistant foreman, supervision of the work can be intense and continuous. This may result in a work efficiency increase of 25 percent or more for the average worker.

If the team is performing special projects it assures the completion of this important work. Special projects performed by individuals often are postponed for too great a period of time because of the pressures of the daily repetitive work.

The team has advantages in cases of turnover because new personnel can be placed in the team and quickly trained by working along with the older more experienced employees, and under the direct guidance of the team leader.

Training is also simplified because the number of jobs which the team as a whole has to perform is limited, and the work of the individual within the team can be even further specialized.

This specialization permits a higher degree of efficiency because of the decreased make-ready and put-away time, and for other reasons as well. Specialization also often leads to pride in an employee's work.

The spirit of comradeship, and the mere fact of being a member of a team, improves morale, heightens interest, and reduces absenteeism.

Team cleaning assures completion of the project at hand in the shortest possible period of time. Thus it vacates the area for other types of work that may be performed and cuts down the hazard exposure time in the case of wet floors.

Team cleaning is less hazardous for the man doing the work. In case of accident, another worker can give aid or bring help. The specialization and intensive training further tends to reduce accidents.

## Other Factors

Many industries provide for the cleaning of production areas up to the aisles by production department personnel during approximately the last ten minutes of each work shift. This practice is consistent with the tendency of production workers to "unwind" near the end of the day. It has the advantage of promoting a greater housekeeping awareness for all such employees by involving them directly in housekeeping and thereby contributing to improved quality control and productivity. The worker will actually create less waste and soil when he realizes that it will be his responsibility to remove it. Whether this system should be used or not depends upon the type of industry, the method of compensation of the production worker, the difference in wages between production and housekeeping workers, and the type of waste and soil created. The amount of cleaning performed by nonhousekeeping workers directly affects the determination of the size of the housekeeping staff.

Job sequence has an important effect concerning cleaning time and quality of results. Instruction sheets should indicate the work and the proper order of performance to save as many steps as possible and to actually prevent the accumulation of more soil. For example, if the floors are dust mopped before the furniture is dusted, the custodian may complete his "cleaning" and leave a floor which has become dusty from furniture dusting. Or, if the wastebasket is emptied before the cleaning of the room is complete, extra trips must be taken to the waste cart.

There are situations where it may be more economical for some aspects of the work to be performed on a cleaning contract rather than by internal personnel. In a few cases, the entire work might be performed on a contract basis. The comparison of contract costs with internal costs must be very carefully made, and a decision must involve all of the many other factors already described. It is worth repeating that no contract cleaning will be successful unless close control is exercised over frequency, materials, methods, and quality of job performance.

Frequency deserves still another mention. In the usual housekeeping operation, it is found that certain frequencies can be extended with no apparent physical change in the overall result. Wax is often applied too frequently; vertical surfaces of furniture require dusting only weekly, rather than daily; acid descaling of rest-room fixtures should be done periodically, rather than daily. This type of consideration has an obvious effect on the number of personnel involved.

The cleaning of a multibuilding facility will require a different approach from a plant consisting of only a single major building. This may also necessitate an entirely different approach in terms of supervisory personnel or arrangement required. The arrangement of the personnel is open to many more possibilities, such as where team cleaning may be done all within a single building or may proceed from one building to the next.

The above items by no means represent a complete list of factors affecting the determination, organization, and scheduling of housekeeping personnel, but merely indicate some of the more usual situations.

Anyone can write some kind of housekeeping program. The basic difference between programs, and the results that the program will help obtain, will be a reflection of the qualifications of the individual or organization preparing it.

# 7

# Installing the Program

ALTHOUGH a great deal of effort and time will go into the survey and program document preparation phases of the program, these activities will not create results in themselves. In a sense, therefore, the housekeeping program only begins on completion of the written program document.

A page of the Bible or a page of *Mein Kampf* could be printed on a given piece of blank paper. Either would be meaningless to the individual unless it were read and then action taken in response to the content. Similarly, a housekeeping program is just a few ounces of used paper stock if its recommendations and schedules are not read and acted upon.

This chapter on installing your program, and the following chapter on implementing it, outlines some of the basic steps necessary to convert the stack of papers that is the program document to the objectives sought by management. The basic purpose of the installation phase is to "set the machinery in motion." It involves these steps: (1) reviewing the program document with management, (2) reviewing the program document with housekeeping supervision, (3) presentation to housekeeping workers, (4) interdepartmental liaison.

In attaining the housekeeping objectives, we will be moving from one level of efficiency and performance to another. If we compare this with a projected airplane flight from one city to another, we can compare the program document to the flight plan, the installation to taking off, the implementation to gaining altitude and the maintenance of flight to the destination.

Just as a take-off is a critical period during flight, so is the installation a critical period in the development of the housekeeping program. As in most endeavors, a good beginning augurs a good conclusion.

If the installation is being conducted by a consultant, he must be careful to establish the proper atmosphere. He must, in his own mind, and in the minds of those persons with whom he has contact, establish a personal identity with the organization. To the sensitive observer, every industrial organization has a unique "feel." This indefinable tone or general climate, we suspect, is a manifestation of such diverse causes as the level of business activity, the company's advertising campaign, the effectiveness of the industrial relations department, the status of the industry, the personalities of management personnel, the growth potential, the average age and service of the employees, specific grievances, and a number of other items as well. It is not an unusual experience to visit a plant one week where a smile and a friendly greeting is the rule and then the next week to work in a plant of the same general type where the same approach is met with silence or perhaps a bit of mumbling.

But to be of true value to the organization, the consultant must develop an understanding of these feelings and, if possible, the motives behind them. The empathetic projection of self into the situation must not prejudice the objectivity with which the consultant must approach his work. The whole approach must be a positive joining of forces and a sincere offering of assistance and understanding. This should be reflected in references to "our" programs and situations, the steps that "we" must take, and the benefits to "us." Always, where criticisms are unavoidable, they should be voiced as an expression of the positive benefits to be gained following the correction or reversal of a given situation.

During the installation, as well as in other phases of the program, the consultant must take advantage of the opportunity to learn. From each contact should come information and experience which will benefit the next contact. It is interesting to note that fundamental problems, no matter how great or apparently insoluble, do not look so serious if seen often enough. They lose their formidable aspect and yield to successively better corrective measures. The circus audience is afraid for the safety of the lion tamer; the lion tamer, however, is concerned about other matters, such as the health of his cats, or what he will have for dessert.

The first step in actually working with the program document is its presentation to, and careful review with, all involved parties.

This must be carefully done because we must avoid the resistance that such an imposing document can bring out at times as well as the confusion that may occur if the various phases of the program are not explained simply, carefully, and in a logical order.

## Presentation to Management

By the time the program is presented to management for consideration, a good part of the psychological aspect of the movement toward better housekeeping has already been assured. As a matter of fact, a management decision to conduct a housekeeping survey and to program the housekeeping operations is in itself the best possible sign of management's growing interest in this field.

In presenting your program document to management, you are strengthened by the knowledge that one of the most important campaigns in the battle for optimum housekeeping has already been won. Rather than a battle of men against men, this is a battle of men against conditions—and everyone wins.

## Recommendations

Before discussion of the recommendations and concept of the program itself, certain fundamental viewpoints should be expressed. It is true that management will assume most of these automatically, but they should be quickly reviewed for verification and emphasis:

1. All recommendations must be considered in the light of management's objectives and limitations. Not only must they have been proposed, but they also must be considered in terms of their practical application, rather than theoretical desirability. The practical application, however, can be a long-term or delayed project.

2. Recommendations may be given in many ways. The typical program carries recommendations in a specific listing of proposals, indicates frequency or scheduling in chart or table form, and infers recommendations on methods, materials, equipment, and other aspects of housekeeping in the work schedules. Because of the comprehensive nature of the program, many of these recommendations are interdependent and should be considered in the context of the whole program.

Acceptance, rejection, or modification of a given suggestion must involve consideration of other dependent or affected recommendations.

3. The program should be considered as a target to be worked toward over a reasonable period of time rather than a revolution or convulsion which would disrupt normal production operations. Thus, the program is a flexible, dynamic tool rather than a rigid, static instrument.

4. The final decision of the rate of implementation of the program must rest with management, based on its ability to effect the accepted recommendations. It is the responsibility of the engineer, based on experience, to suggest the speed of implementation. Typically, in the housekeeping field, improvements must be gradual and cumulative if they are to be lasting.

5. Management is neither legally nor morally obligated to accept any given program recommendation.

6. Management must set up the machinery for taking action. This is done best through a coordinator assigned to the program, who must naturally be given sufficient authority to perform this work effectively. It is well to remember that agreement on recommendations has no effect without executive authorization.

In his relationships with the client, the engineer follows a precise code of ethics. In addition, there are many unwritten responsibilities.

Comparatively little has been said about management's ethical responsibilities toward consultants. Fortunately, this rarely creates confusion or embarrassment, since most companies have had a number of experiences in working with consultants in various fields.

Management's attitude and approach toward the consideration of recommendations can have a very real effect on the progress of the program. For example:

1. Consideration should be given to suggestions with the attitude that a bad idea does not exist. Impractical, yes; impossible to implement, yes; premature, yes—but bad, no. Ideas should be considered as having a fragile and elusive life which, if approached negatively, will be snuffed out. The method of handling an unacceptable idea may have a very serious effect on the consideration of ideas to follow. The safest attitude with which to approach any suggestion, no matter how seemingly ridiculous or farfetched, is with the feeling "you may be right."

2. Management should accept recommendations as opportunities for improvement, rather than as implied criticisms.

3. Keeping score on the number of "good" versus "bad" recommendations in a given proposal can only impart a negative atmosphere, which invariably results in limiting effectiveness.

4. As the engineer must accept full responsibility for his recommendations, so must management accept responsibility for actions contrary to his recommendations.

5. Management should take an active interest in the attitude of subordinates

with respect to the implementation of the program. Where stumbling blocks show up, management should positively remove them or positively endorse them.

The consideration of the various recommendations, and the structure of the program as a whole, may require a good deal of management's time. Meetings, vacations, priority activities, and other factors have an effect on the amount of time required to consider each basic decision. In practice, a number of decisions are often made almost immediately, whereas others are delayed over a period of time, depending on the factors above, as well as the need for liaison, further investigation, etc. Implementation of approved items can begin where interdependence on other items is not a problem.

Revision of various pages or sections of the program may be required in the installation phases of the program, as well as later on perhaps, because of management's decisions on recommendations, reevaluation of objectives, changing production levels, the appearance on the market of improved equipment, etc. Thus, it is the revised draft which is followed in the actual operation of the program.

## Presentation to Housekeeping Supervision

Supervisory personnel within the housekeeping department should be made familiar as quickly as possible with the extent of their responsibilities toward the program and the steps necessary for its implementation. The best way to accomplish this is in a meeting which should be set up by the housekeeping department manager, following his review of the program document with the engineer. Chapter 9 deals with the housekeeping department manager and his vitally important role in the program.

A schedule should be set up indicating each action that will be necessary to implement accepted recommendations. Such a schedule should identify the item, name the individual responsible for its completion, and set a target date for its beginning and full implementation. Such a schedule can go far toward reducing the apparently complex program to a listing of simple and defined objectives.

Based on the program, management should express the objectives and limit of authority and responsibility of the housekeeping department in written form, circulating this to the departments which receive its service. It should name those general areas for which the housekeeping department is responsible as well as indicate specific areas of responsibility which may not fall to the housekeeping department, such as grounds keeping, changing lights, etc.

The installation phases of the program is the proper time to organize the paper-work and record-keeping functions of the housekeeping department. The forms put into use should be standard for the company as often as possible. Interestingly, there is a tendency for some housekeeping supervisors to seek paper work, often unconsciously, as a status symbol, thus avoiding the lower-status situation of dealing with custodians in plant areas. The supervisor must avoid every unnecessary record or piece of paper in order to be able to spend as much time as possible "on the firing line," where the most good can be done. Routine paper work should be assigned to a subordinate whenever possible.

A procedure should be set up so that all matters pertaining to housekeeping, including complaints, suggestions, commendations, and changes in personnel, will be referred to the housekeeping department manager. A permanent record should be kept of complaints and reviewed quarterly in an attempt to discover recurring situations or trends that require a more general rather than specific response. Such a record should indicate: (1) the date, (2) the source of the complaint, (3) the nature of the complaint, (4) housekeeping personnel involved, (5) corrective action taken, (6) date of corrective action.

Much of the written program material, especially for personnel assignments and responsibilities, can be reduced to chart form. If such a chart is made up in fairly large size, such as three feet high and four or five feet long, and posted in the housekeeping department, it provides instant reference information and helps to form an overall picture in the minds of the supervisors and workers. Such a chart, which might be titled "Housekeeping Operation Plan," can be divided into vertical columns, the headings of which could include:

Custodian name or number
Housekeeping area designation or number
Area usage or department name
Square footage (this column should be totaled)
Types of floors (approximate percentages can be indicated, an overall percentage)
Equipment assigned. This can include size designations, serial numbers, location, etc.
Work schedule. This can include an abbreviated description of the type of work, such as general cleaning, policing, projects, and the days and hours worked
Total weekly man-hours (this column should be totaled)
Custodial closet or facility to be used

The very preparation of this chart by housekeeping department supervisors will convey much of the structure of the program to them. A number of business equipment and supply firms manufacture scheduling charts which can be used for this purpose.

The machinery for the assignment and recording of special projects must be set up. Two basic types of forms will be required for this. The first is a register, each page of which should be designated for the building or major area involved and properly identified as to the period of time which it covers, such as for a given month of the year. Column headings might be:

Item number
Area designation
Type of project
Instruction sheet to be used
Individual responsible for completion
Target completion date
Actual completion date

For further control, a master record may be kept so that facility-wide completion of a general type of project can be controlled. In large installations, data-processing systems in use for other functions may be utilized for housekeeping special projects as well.

Standard work order forms may be used to assign the work, or special forms may be designed for use by the housekeeping department. Such forms would indicate the name of the person responsible for completing the work, the name of the person issuing the work order, and special instructions and equipment listings that may be required. These work order forms must be turned in periodically so that the register may be kept up to date through regular postings. Column headings may include: (1) type of project, (2) instruction sheet number, (3) name or number of area, (4) target completion date, (5) actual completion date.

It will be difficult for the housekeeping department to perform the operations assigned to it unless it has itself first "cleaned house." The time of installation of the program is a good opportunity for an actual cleaning of the housekeeping department office areas and storage areas. Also valuable at this time is the proper organization of the central storage area and custodial closets, as well as the establishment of inventory control procedures.

## Presentation to Housekeeping Workers

The natural human emotions involved, when a person learns that his work is being investigated, evaluated, and possibly rescheduled are fear, apprehension, and suspicion. These are perfectly normal reactions. As quickly as possible, convert them to a psychological climate suitable for effective program implementation. This is an important responsibility of the engineer. It is no place for subterfuge or evasiveness. Here, as in so many other places, the truth—clear and simple—is the most effective tool possible.

Hold an orientation meeting with all housekeeping department personnel, including their supervisors. Although many housekeeping training meetings will be given to less than the entire department because of working hours or duties, this meeting should, if at all possible, include all housekeeping personnel. A good time to hold such a meeting is at the time of a shift change, holding some workers over while asking others to come in early for this purpose. Representatives of management should be present at this meeting, especially the program coordinator, to indicate management's interest and sanction.

Following an introduction by the housekeeping department manager, the engineer should address the group, with the basic objective of seeking their cooperation. He must attempt this, bearing in mind these considerations:

1. To gain a positive immediate reaction, assurance should be given (if previously approved by management) that the program will definitely not result in any worker losing his position with the company. If reductions are to occur, they can be handled through normal attrition. Security is very important to the worker and this reassurance should be stated as early as possible.

2. It will relieve the workers if they are told that any feelings of apprehension or suspicion are perfectly normal and expected in most workers when they are first approached with a program of this type. But experience has shown that this attitude changes to one of interest and enthusiasm since the program invariably works to their benefit. The point can be reinforced by showing examples of literature from other organizations involving activities for personnel morale improvement, better equipment, employee cooperation, etc.

3. People do not like to be guinea pigs or feel they are in the hands of a novice. They must be reassured that the program, of which they are such an indispensable part, has been conducted successfully for other similar organizations, and these organizations should actually be named. The showing of literature, as above, is also helpful for this purpose.

4. Throughout, the engineer must exhibit the attitude that each worker is willing to provide a reasonable day's work. Any other position is untenable. Emphasis should be made of the fact that the program is designed to equalize the work load, so that each man is doing no more than his share of the work; and improved methods, equipment, and employee cooperation, as well as other aspects of the program, will be steps tending to make the work less burdensome.

5. It should be shown that the program represents a target to be approached over a reasonable period of time. Workers will not be expected to learn new

methods overnight, but will be given sufficient training and opportunity for practice in learning new techniques. Similarly, benefits will take time, and they should also exercise patience.

6. It must remain clear that management, and its appointed supervisors for the housekeeping department, retains all its prerogatives, and that the engineer is acting in an advisory capacity.

7. Because people are more at ease with a tangible object as opposed to an abstract conception, it is good to actually exhibit the program document. It should be pointed out that this document is nothing more than a collection of ideas and suggestions concerning housekeeping. It would be worthwhile to give a simple physical demonstration of a timesaving method to demonstrate the point. Chapter 10 further discusses the role of the custodian in the cleaning operation.

8. Finally, it is the worker himself who determines the success or failure of the program, as it is he who must perform all of the actual cleaning functions. Although here there is danger of being "corny," the worker must be encouraged to do his share as a member of the housekeeping team and of the company as a whole. It is well to encourage participation through suggestions and ideas.

## Interdepartmental Liaison

If the survey phase of the program did not permit an opportunity for a meeting with department heads in order to familiarize them with the objectives of the program and how they might play an important part in it, certainly the installation phase should accomplish this. As many affected department representatives as possible should attend such a meeting. Special requests for help from the housekeeping department is a subject suitable for such a meeting. In this case, for example, immediate housekeeping assistance should only be expected in actual emergencies. Requests for additional housekeeping personnel or special assistance should be made at least 24 hours in advance, in order to permit proper planning and efficient scheduling of personnel. Greater pressure than this to fill special requests can only result in wasted man-hours, serious postponements of work (which create their own pressures), and demoralization of workers, none of which is desirable.

Especially important during the installation phases are these contacts:

1. A general review of the program by the safety department to insure the inclusion of good safety practices throughout.

2. Contact with production foremen aimed at establishing a better understanding between production personnel and housekeeping personnel for waste handling and general housekeeping responsibilities in areas surrounding production equipment.

3. Close contact with internal methods engineering personnel for a possible synergism of ideas, if this is approved by management.

At this time, some of the first steps can be taken to pierce the invisible barrier that is often found separating direct and indirect labor groups. Such barriers tend to soften and dissolve through repeated contacts designed to impart an understanding of the interdependence of the two groups, and particularly through better production conditions which result from improved housekeeping.

The engineer is likely to improve his acceptance, and ability to enlist the cooperation of others, if he becomes familiar with the basic nature of the industry with which he is dealing as well as the processes involved in creating the finished product. He should be able to correctly use some of the specialized terms and jargon of the industry. For example, in the textile industry, he will hear and use such words as *beaming, carding, creel,* and *warp*.

The company may wish to issue news releases for local newspapers, trade periodicals, or to take advantage of the house organ to make an announcement of the program, its objectives, and potential benefits. Liaison will be required with the proper department, such as industrial relations, to accomplish this.

The engineer should document the major phases of the program, both as a report to management and as a memory aid to the housekeeping department and to himself. The completion of the installation phase is certainly a logical point for such a report. This installation report should include the following sections: (1) a review of activities, (2) observations or additional recommendations, (3) recommended agenda, (4) copies of prepared material such as inspection reports, implementation calendars, etc.

Throughout the program, we must bear in mind the *basic objective* of the organization. Just as in a hospital the care of the patient must be the overriding consideration, and in a school effective pedagogy should be most important, so should the earning of a profit be regarded in industry. The positive aspects of profitmaking in terms of more jobs, higher earnings, and better living standards should be stressed continually. There is no better place to begin working with this concept that in the installation phase of the program.

The installation phase of the program may require only a short time, or it may take an extended period. Generally, it will be difficult to point out a given time at which the installation phase is completed and the implementation phase is beginning, as certain activities of each phase will tend to overlap. In any case, the installation should be a good beginning.

# 8
# Program Implementation

THE effectiveness of the implementation of the housekeeping program will be limited by the quality of the written document. Conversely, the effectiveness of the entire program will be limited by the care and thoroughness of the implementation.

The fact that the program document represents a relatively long-range goal rather than an instantaneous achievement cannot be overstated. The transition that must take place to convert an existing situation to a situation which is proposed in order to accomplish management's objectives cannot, and should not, be accomplished overnight. There must be a "phasing in" so that decisions may be properly implemented, there will be sufficient time for the training of housekeeping personnel in their new duties and responsibilities with the least possible confusion, and housekeeping service may be provided for all departments without disruption. Sometimes such a transition must be taken in two or more steps, requiring the use of interim schedules and assignments until the final proposed material can be implemented.

How long a period of time will implementation require? It is important that the implementation of a program does not proceed so rapidly that the first gains are lost in the scrabble to make additional gains—or, in other words, cause "backsliding." Thus, the implementation must proceed at a rate so that gains can be made as quickly as possible while maintaining the improvements that are brought about. Unquestionably, the first year of a program is much more critical than the following years, since the first year will be the period of greatest gains, and probably the program as written will be 60 percent to 70 percent implemented within that year. But generally, three years are required to obtain the greatest possible benefit from a comprehensive housekeeping program, bearing in mind the costs involved, for these reasons:

1. To prevent backsliding, as mentioned.
2. To implement long-term projects, such as employee awareness campaigns, rating systems, etc.
3. To properly handle revisions and changes in management objectives because of information gained from early operation of the program.
4. In order to further refine the program in methods, materials, standards, etc.
5. To gain the advantages of a continuing training and morale-building program.

Another point: Situations exist where solving one problem results in the creation of one or more other problems, although the subsequent problems may be minor. Unless the implementation phase of the program is long enough to permit not only the full application of problem-solving recommendations, but also the resolution of any attendant minor problems that may be created, the program will not provide the full beneficial effect that is so desirable.

Implementation will provide cumulative improvements leading toward the objectives determined for the program by management. In a number of industrial plants where housekeeping has been sharply curtailed for fiscal reasons, and where there is an understaffing in the housekeeping department, recommendations are often made to management, and accepted by them, to establish the sole objective of raising housekeeping levels. Here the dollar savings which implementation brings should be returned to the housekeeping department. If the savings result from personnel performing one function of housekeeping, they may be transferred to another function. Where the objective has been the saving of dollars, and this is accomplished through a smaller housekeeping staff, any savings in personnel should be achieved through normal attrition.

## The Engineer

In the long run, the basic commodity of the industrial engineer is time. The mere possession of time assures nothing—yet, nothing can be done without time. After a number of years of training and experience in the conservation of time, it begins to get in a person's blood. It is the industrial engineer, no doubt, who may be shown in a cartoon fussing with his wife for being thirty seconds late for a party! As a general rule, the industrial engineer is a few minutes early for each appointment and begins to get nervous if a timed event begins to run five minutes over. In making time count, the reader may find the author's book HOW TO USE YOUR TIME TO GET THINGS DONE (Fell) useful.

In working with the program on the client's premises, the engineer should provide an example in working habits with respect to time. He should be punctual for all appointments, and his work should be well organized to save his time as well as the time of others. Meetings should be held to a planned time span and should not run over. In the last instance, where repeated difficulty is experienced in keeping meetings short, an interesting result will be seen if the meeting is held with all persons standing up!

Many persons become accustomed to working in an environment of disarray and generally poor housekeeping. Some people, over a number of years, become completely inured to such conditions, possibly as an expression of hopelessness. On the other hand, after years of specialization in housekeeping and sanitation, the engineer will find himself unable to sit down to a meal in a restaurant, or to enter an office, without mentally separating the conditions which are desirable from those which require further attention!

It is not necessary that a single engineer conduct all phases of the program. Specialization may have its advantages here, too. The various phases of the program may be handled by two or three different persons, each specializing in a certain aspect of the work. If the project is handled in this way, it is of extreme importance that each man becomes familiar with all phases of the program, because of their interacting effects.

## Maintaining Liaison

Lines of communication must be kept open. The housekeeping department becomes specifically involved with a number of personnel and departments during the implementation of the program. Short meetings should be held from time to time, as indicated in the previous chapter, to keep all interested persons up to date on imminent and completed changes and to give them an opportunity to ask questions and make suggestions. Every meeting should stress the importance of their personal participation and the resulting benefits to them.

Liaison will be required for special situations. Certainly the housekeeping department should receive the full support of the maintenance department because proper housekeeping results in lower building maintenance costs. For example, wall vacuuming and washing at proper frequencies extends the length of time between paintings. Proper floor care, and protection with suitable coatings, protects the surface from undue wear and avoids replacement of resilient tile or wood, the redensifying of terrazzo, or the patching of concrete. The maintenance trades should understand their responsibilities toward housekeeping in the performance of their daily work. With very little actual time or effort, maintenance personnel can create much better housekeeping conditions by washing their hands occasionally, spreading drop cloths to catch chips, drippings, and shavings, and by immediately removing any material which may cause damage to floors.

Accepted recommendations may include the formation of a housekeeping committee. Such a committee is very helpful in improving liaison through a more complete understanding of the program and the problems affecting it, and a quicker response to suggestions. Although a permanence of such a committee is desirable, the frequency of its meetings should diminish as the program becomes properly implemented.

## Continuing Improvements

A program document should certainly be considered a very big step in the direction of fulfilling the housekeeping department's objectives, but it should hardly be considered the last word. Improvements stimulate further improvements. New ideas are constantly being born; new equipment is reaching the market; new materials are being developed in laboratories. Improvements from outside sources should be properly worked into the program, and internal developments should supplement these. Program revisions will be required where such changes are fundamental or far reaching, or where a number of minor changes have accumulated.

Revisions may be also called for where, at any point during the implementation phase, management reformulates its housekeeping objectives. This can be compared with a change in destination in our analogy of the program with an airplane flight.

The application of work measurement, standards, and rating systems may be introduced at some stage during implementation, either through design in the program or by a separate proposal and introduction at a later date.

## Gaining Employee Awareness and Cooperation

In benefits provided versus cost in dollars, a well-conceived employee cooperation and awareness campaign is unquestionably a superb investment. If this activity is conducted in any way other than as part of a complete housekeeping program, thus having the effect of asking for help without offering it, such an arrangement may have a very limited effect. If, on the other hand, personnel are able to see improvements which the housekeeping department has made and have seen evidences of management's growing awareness of the importance of this subject, they are more likely to provide a positive response. Thus, consideration of such a campaign should be made early to permit proper planning and preparation, but its implementation should be carefully planned to coincide with improvements and other activities during the implementation phase of the program.

Even housekeeping programs now considered satisfactory can be improved through increased employee cooperation. Consider the effect on the custodial work load of a 10 percent improvement in attitude of a given plant which may have 1,000 employees. The effect approaches the equivalent of cleaning up after 100 less people! Such an improvement might permit, in our hypothetical plant, an annual saving of a few thousand dollars in wages, or perhaps the transfer of a custodian to special project work in order to raise the level of housekeeping.

Everyone wants a clean, healthy, and safe place in which to work. The important factor in obtaining employee cooperation is to bring this desire to the surface and keep it there. Pride, respect, and conscience in maintaining clean work areas will be the result. This natural employee attitude should be developed carefully, without forcing. For instance, a management directive which says "get this place cleaned up" will produce only very temporary results. Contrast this to a management policy which states "we recognize your desire for clean healthy working conditions and are doing everything possible to help you and our housekeeping department obtain them." Certainly everyone can respond positively to such an appeal.

The campaign must be devised and implemented with the cooperation and guidance of the industrial relations department. This type of campaign is one of the *raisons d'être* of the industrial relations department, and their attitude and response toward such a program will be enthusiastic and vigorous. They will normally "carry the ball." Literature and methods already put to good use by other organizations will benefit them in their work. Here again, the practice of assigning this project to an individual, providing him with the authority and means for coordinating the work, is most desirable.

It is a waste of money to attempt to awaken and sustain employee interest in cleanliness by a few half-hearted devices on a limited number of occasions. It is better to do nothing than to awaken interest momentarily and then let it die through failure to provide benefits and follow-through. Such an approach will only create new cynics and confirm old ones, with respect to housekeeping. Let the campaign continue over a long period of time. There is no reason why such a program should not be a permanent project of the industrial relations department. The intensity of the promotion will decrease, of course, as the months continue, but interest can be sustained through carefully timed applications of a number of different campaign techniques. These include:

1. Orientation booklets
2. Letters from management to department heads and supervisors
3. Letters from management to employees
4. Posters and banners
5. House organ and publicity
6. Leaflets and stuffers
7. "Calling cards"
8. Contests

### 1. ORIENTATION BOOKLETS

New employees of the company should receive an immediate impression and immediate expression of the organization's attitude toward housekeeping. Ideal for this purpose is a small booklet of ten or twelve pages to be given to each new employee. If desired, the general personnel policy booklet can contain a housekeeping section, rather than having a separate booklet. Periodic redistribution to older employees is valuable as well, particularly if the brochure is reprinted in different colors with a modified cover. The booklet would call attention to the steps that management has taken to provide well-maintained surroundings for its workers. It would point out how each employee can contribute to his environment, giving examples such as refraining from putting liquids into wastebaskets, snuffing out cigarettes underfoot, leaving desks littered at night, etc. The brochure might also call attention to a rating system, where each department or area competes for a prize or the possession of a good housekeeping cup. In terms as positive as possible, the company's housekeeping policies and regulations should be listed.

To lend authority and dignity to the booklet, it should open with a letter from the president or general manager. Interest can be caught and held by using drawings, photographs (including the "before-and-after" type), and by the judicious use of color and open areas. Typography should be limited, and the type face carefully selected. The booklet is a good place to tie in other programs interdependent with housekeeping, such as quality control, safety, and fire prevention.

### 2. LETTERS FROM MANAGEMENT TO DEPARTMENT HEADS AND SUPERVISORS

Obviously, the attitude of department heads, supervisors, and foremen toward the housekeeping program can materially influence its success in the areas for which they are responsible. An effective way of obtaining their interest and cooperation is through a personal letter. The letter should bear the signature of an official in top management and should be mailed directly to the individual. Periodically, supplementary letters should be mailed for follow-up.

### 3. LETTERS FROM MANAGEMENT TO EMPLOYEES

Employee response to personal letters from management enlisting their support is usually strong and positive. This is particularly true if the letter is "down to earth" and not officious. Such a letter rarely, if ever, brings a negative reaction. At worst, it falls, in the minority of cases, on deaf or passive ears, and even then response may be awakened through repeated efforts. Follow-up letters should be over the signature of various management personnel, to stimulate still further interest. It is a compliment to a person to ask his help.

### 4. POSTERS AND BANNERS

Posters and banners have the advantage of repeating their message each time they are seen. If such a message is left in place too long, however, it loses its effect and actually becomes annoying. Posters should never remain in place longer than a month. A good plan is to rotate posters every week. Although a given poster can be placed in a new location immediately, it should not be returned to its original place before one year.

If bulletin boards are provided, they should be well lighted and colorfully framed and located where most people will see them.

The National Safety Council has a standard series of colorful housekeeping posters available. Special posters and cartoons, which specifically relate to local problems, may be prepared by internal personnel.

### 5. COMPANY NEWSPAPER PUBLICITY

The company newspaper, magazine, or other periodical is a very good vehicle for housekeeping publicity. Editors usually welcome this type of material, as it provides for dramatization through photographs, case histories, etc., and because it ties in so well with other objectives. Here is a good place to build humor into the campaign in such a way that one employee will "kid" another for violating group-imposed rules for housekeeping. Part of the function of the house organ is to publicize other phases of the awareness and cooperation program.

### 6. LEAFLETS AND STUFFERS

Leaflets may be left on the desks of employess, or stuffers placed in their pay envelope, from time to time. Again, the humorous theme is best applied here through the use of cartoon situations. The watchwords to be used for this type of literature are brevity, color, and humor.

### 7. "CALLING CARDS"

Good results have been achieved in attempting to obtain employee cooperation on an individual basis by utilizing a personalized "calling card." These cards should be used by housekeeping supervisory personnel.

A 3-inch $\times$ 5-inch, typical card, of a bright color, is left on the desk or at the work place of a housekeeping offender, pointing out a housekeeping offense and requesting his assistance in creating better working conditions. When reporting to work the following morning, the offender finds this card which personally asks for his help. Many persons respond to this individual appeal for help. Where response is not forthcoming, the repeated appearance of a "calling card" on a given desk, for example, may attract the attention of the department head or foreman. If he has been properly motivated by other aspects of the awareness program, undoubtedly he will take a personal interest in the problem.

8. CONTESTS

The application of contests to this subject offers a wide number of possibilities, limited only by the extent of a person's imagination. Possible subjects for a contest are:

1. A competition for the best-maintained area
2. The determination of the best suggestion concerning housekeeping and sanitation
3. The phrasing of the most worthwhile slogan
4. The designing of the best housekeeping poster or banner
5. A statement of twenty-five words on the importance of good housekeeping

Awards might be rotating or permanent trophies, pennants, merchandise awards, or merely publication of the winning person's or department's name.

In order to improve employee cooperation, particularly as instituted on the department head level, it may be desirable to cost out housekeeping to the various departments. The first step in such a procedure is to set up an account numbering system and the forms needed to properly assign dollar costs to specific departments. A simplified method for performing this function is to assign a base, or fixed, amount for each accounting period for the repetitive work performed within the department and then to make specific charges for any project or special request performed in addition.

## Inspection

Inspection is a fundamental aspect of any complex and important activity, from producing electric components to building bridges or from dressing poultry to fitting watch jewels. Inspection is a basic function of the service program, providing a number of benefits:

1. It can be combined with a rating system in determining compensation or advancement.
2. It points out areas where cooperation from other departments has been weak.
3. It is an extension of the training and orientation activities.
4. Executives, supervisors, and employees, as well, either consciously or unconsciously, are making their own private inspections of the facility. Official inspections help to assure their satisfaction.
5. Inspections, by uncovering situations requiring correction, will avoid distracting and annoying complaints.

6. Inspection, as a form of supervision, will lead toward greater worker efficiency.
7. It will help measure the effect of various changes and improvements in the housekeeping program.
8. Since the program should be used for positive purposes rather than for criticisms or threats, morale can be raised by a judicious commentary and publication concerning exceptionally good work or method improvement by custodians.

We must be careful not to treat the inspection phase of the program too lightly. Its apparent simplicity is deceptive. The inspection form must be carefully prepared and faithfully used. A monthly general inspection is desirable, although more frequent inspections can be made for changing conditions or situations of a critical nature. For example, a superclean or "white" room should be inspected daily. A weekly inspection should be made of medical, food service, and personnel areas.

There are many possible varieties of the inspection form, but basically it is a check list and progress report. One form should be headed up for each area, building, or other physical entity. The various services and cleaning functions performed within this area are listed with a suitable indication either of acceptance or nonacceptance, or for grading by a method of descriptions or numbers. Where a condition requires attention, space should be provided to indicate needed action. The inspection form, or sometimes a separate letter, can cover the inspection of housekeeping materials, equipment, conformity to regulatory statutes, employee attitudes or comments, etc.

Copies of the inspection form should be retained by the program coordinator, the engineer, and the housekeeping department. It would be very desirable to provide management with a less frequent written inspection report covering the trend of the regular monthly inspections and supplemented with report information about general program implementation and standing. By all means, a comprehensive annual audit and report must be provided.

SERVICE ENGINEERING ASSOCIATES

# PROGRESS REPORT

PREMISES OF____________________

ADDRESS____________________

DATE____________________

Area A____________________

Area B____________________

Area C____________________

Area D____________________

Area E____________________

| Item No. | GENERAL | AREA A | | AREA B | | AREA C | | AREA D | | AREA E | |
|---|---|---|---|---|---|---|---|---|---|---|---|
| | | Approved | Req. Attn. | Approved | Req. Attn. | Approved | Req. Attn | Approved | Req. Attn. | Approved | Req. Attn. |
| 1 | Floors protected (waxed, sealed, etc.) | | | | | | | | | | |
| 2 | Floors and mats clean | | | | | | | | | | |
| 3 | Baseboards clean | | | | | | | | | | |
| 4 | Carpets clean | | | | | | | | | | |
| 5 | Walls clean | | | | | | | | | | |
| 6 | Windows, sills, drapes and blinds clean | | | | | | | | | | |
| 7 | Woodwork clean | | | | | | | | | | |
| 8 | Ceilings clean | | | | | | | | | | |
| 9 | Furniture and fixtures clean | | | | | | | | | | |
| 10 | Doors and door hardware clean | | | | | | | | | | |
| 11 | Glass in doors and partitions clean | | | | | | | | | | |
| 12 | Vents, radiators and pipes clean | | | | | | | | | | |
| 13 | Light fixtures clean and operating properly | | | | | | | | | | |
| 14 | Pictures and ornaments clean | | | | | | | | | | |
| 15 | Drinking fountains clean | | | | | | | | | | |
| 16 | Receptacles clean and properly arranged | | | | | | | | | | |
| 17 | Grounds, walks and parking area policed | | | | | | | | | | |
| 18 | Fire extinguishers, hoses and cases clean | | | | | | | | | | |
| 19 | Elevators, stairs and handrails clean | | | | | | | | | | |
| 20 | Rooms free from obnoxious odors | | | | | | | | | | |
| | REST ROOMS | | | | | | | | | | |
| 21 | Floors protected (waxed, sealed, etc.) | | | | | | | | | | |
| 22 | Floors clean and destained | | | | | | | | | | |
| 23 | Light fixtures clean and operating properly | | | | | | | | | | |
| 24 | Walls and ceilings clean | | | | | | | | | | |
| 25 | Doors and door hardware clean | | | | | | | | | | |
| 26 | Windows clean | | | | | | | | | | |
| 27 | Mirrors clean | | | | | | | | | | |
| 28 | Washbowls and washbowl hardware clean | | | | | | | | | | |
| 29 | All dispensers clean, filled and working | | | | | | | | | | |
| 30 | Sufficient number of dispensers | | | | | | | | | | |
| 31 | Receptacles clean | | | | | | | | | | |
| 32 | Stools and urinals clean | | | | | | | | | | |
| 33 | Rooms free from obnoxious odors | | | | | | | | | | |
| | CUSTODIAL & UTILITY ROOMS | | | | | | | | | | |
| 34 | Rooms clean | | | | | | | | | | |
| 35 | Lockers clean | | | | | | | | | | |
| 36 | Eqpt. and tools clean & stored properly | | | | | | | | | | |
| 37 | Lighting adequate and operating properly | | | | | | | | | | |
| 38 | Sinks and drains clean and operating properly | | | | | | | | | | |
| 39 | Flammable materials safely stored | | | | | | | | | | |



This type of inspection form uses a positive title and a non-numerical checklist.

# 9

# Housekeeping Standards

STANDARDS have given industry the means with which to produce goods in quantity.

Standardization in industry began with the concept of interchangeability of parts, developed by Eli Whitney in his firearms factory. Before that time, every gun was manufactured by a skilled craftsman, fitting one part to the other with his own measurements, tolerances, and technique. But with standardization came the opportunity to produce in mass quantities, thus gaining the efficiencies which can be applied to repetitive functions and permitting repairs and replacements from stock parts.

The growth of industry has been accompanied by a growth in the types and numbers of items being standardized. Most industries now have their own standards, societies, or associations in order to promote industry-wide efficiency. Other independent bodies, and government standards bureaus, establish and disseminate standards material on a wider scale.

Housekeeping is one of the last industrial functions to receive the application of standards. This has been primarily due to the consideration of housekeeping as relatively unimportant and to the unusually wide range of variables.

Because of these variables, standards must be developed for each installation, based on the type of production, the condition of the physical facility, the type of worker or applicant available, degree of congestion, turnover, management support and employee attitude, the type and amount of waste created, etc.

Standards developed for *general* use lead to misunderstandings and poor results when applied to a *specific* situation. Standards set up for one operation in a given industry should not be considered appropriate for another plant within that same industry.

But the development of specific standards does not necessarily mean that new time studies should be taken in each case; rather, it generally means correcting factors should be applied for the variables involved. The accuracy of the numerical value of the factors, of course, increases with the experience of the engineer.

Just as conditions will change, so must standards change to conform to them. The continued use of standards that have become obsolete through change will limit effectiveness or cause employee relations problems.

This chapter will discuss housekeeping standards for:

1. Personnel
2. Quality
3. Frequency
4. Methods
5. Materials
6. Equipment
7. Time

## Personnel Standards

In order to establish standards for personnel, they must first be identified. We are therefore immediately faced with the need to establish and standardize titles for personnel within the housekeeping department.

These titles, particularly on the supervisory level, must follow the pattern set for the company as a whole. Thus, in a given company, the head of a department may be called *manager, supervisor, superintendent,* or some other title.

For the sake of discussion, let us consider titles in a hypothetical organization:

| | |
|---|---|
| Housekeeping department manager | In charge of the housekeeping department. |

| | |
|---|---|
| Foremen | Salaried supervisors, one of whom may be designated as "assistant manager." |
| Group leaders | Working team leaders. |
| Custodians | The basic housekeeping worker. Custodians may be arranged in several categories or classes, depending upon the type of work performed. Hazardous duty or shift work. |

Standard qualifications for hiring custodians and other housekeeping department personnel will assure that personnel placed within the housekeeping department will be suited for the work. See "Selecting the Custodian," in Chapter 29.

Standard job descriptions may be arrived at in two stages. Descriptions for categories of workers may first be prepared and adjusted with use. Following this, individual job descriptions can be set up.

Both the qualifications and job descriptions will have a bearing on relative wages. The overall wage structure for housekeeping personnel, of course, is involved in management's consideration of relative wages and benefits for all departments.

## Quality Standards

As mentioned earlier, the determination of housekeeping cleanliness levels is difficult. Methods that permit direct measurement are involved and time-consuming, whereas more simple methods are subjective and judgmental in nature.

These are the basic methods for determining housekeeping quality:

1. Direct measurement. Through the measuring of dust count, light transmission, or surface wear, the quality can be expressed in numerical terms. These techniques, being expensive, are applicable only to critical situations such as super-clean rooms, precision assembly areas, etc.

2. Frequency relates directly to housekeeping level, all other factors being relatively constant. For example, a floor polished daily will be at a higher quality level than a floor polished weekly; a wall washed annually will remain cleaner than a wall washed every two years. An accurate recording of frequency of job performance, therefore, will in a sense be a measure of the level of housekeeping. This relationship holds more true, the more nearly all work conforms to methods and other standards.

3. Quality may be measured on the basis of mere acceptability or unacceptability of work against an arbitrary standard. Such a standard for window washing, for example, might require the absence of filming or streaking, visible soil on the

glass, and actual soil on the frame and casing as determined by the use of a clean dusting cloth.

4. A refinement of the use of arbitrary standards is to attempt to measure the approach to the ideal standard. Thus, if a perfect window cleaning job is classified as 100, ten points each may be deducted for streaking, soil on the glass, soil on the frame, unsafe conditions, failure to use a drop cloth, etc. Or, the work may be judged by an experienced observer as *excellent, good, fair,* etc.

The measurement of quality may develop as the program develops, beginning with the consideration of frequency as a guide and working toward the eventual use of numerical grading for general cleaning.

## Frequency Standards

Frequency standards should be as specific as possible. Although it is true that there are times when designations such as "after each use" are required for the maintenance of a given area or piece of equipment, the great majority of frequencies should actually be indicated as positive time periods. Designations like "as required" leave too much latitude.

It can be seen that indefinite frequencies lead to corrective maintenance rather than preventive maintenance. If it is the practice to maintain a surface only if it apparently needs maintenance, there will be several such surfaces needing maintenance at one time, and they will remain in this condition, progressively deteriorating, until the maintenance can be scheduled. Where a floor surface, through observation only, needs rewaxing, then the wax has already worn through and the floor has a bad appearance and is actually wearing away. We should no more wait until the wax on a floor has been worn away before rewaxing than we should wait until a machine has run out of oil before we lubricate it. Frequency determination is most useful where each cleaning job done on an intermittent basis is scheduled for each area where the work is required. Floor scrubbing, for example, might be required daily in one area, weekly in another, and semiannually in yet another. In arriving at frequencies for individual situations, it may be best to begin with a general frequency which is adjusted because of a given situation.

It is convenient to arrange frequencies on the basis of the type of area involved, such as office and other administrative areas, production areas, and storage or intermittent-use areas. If a standard frequency is established for each of these types of areas, they can be adjusted depending upon the traffic, soiling, congestion, and other factors found in the various locations within the general area.

Specific frequency and frequency standards should be adjusted as conditions change. A program should normally begin with a series of target frequencies, with the housekeeping department operating at those frequencies within a specified number of months. Once these target frequencies are achieved, some of them may be adjusted for performance more regularly. The means to accomplish this additional work will be provided by the increased efficiency of the housekeeping operation.

Frequencies should be set up so that occasionally it is necessary to perform a given job sooner than anticipated. A case in point might be the rewaxing of a lobby floor that had been set up on a monthly frequency. During a snow storm a good deal of sand and water might get in, quickly wearing away the protective coating. In this case, rewaxing might be required after only two or three weeks during this period. If frequencies are set up so that jobs are being done so often as to avoid any change of schedule, the work will be done too often on the average and will therefore be overly expensive.

Although typically we find housekeeping operations not performed frequently enough, there are cases where jobs are being done too often. In dusting furniture, for example, only the horizontal surfaces (which catch most of the dust) need be dusted daily; the vertical surfaces require dusting only weekly. Dust mopping of floors, as another example, need not be done daily in certain types of areas. The time that is saved by extending repetitive frequencies of this type may be used to accomplish intermittent project work which might otherwise be done too rarely or possibly not at all.

In setting frequencies and calculating time requirements for them, it should be remembered that some jobs eliminate the need for others. For example, it is not necessary to dust mop a floor which is going to be wet mopped unless the floor is exceedingly dirty, which would cause the mopping water to become ineffective too quickly.

Frequency is also affected by materials. Some polymeric floor coatings take on such a high gloss after initial application that buffing is not required for some time thereafter. Where buffing is desired on a regular basis, it will not improve the appearance of this type of material whatever for the first several days after application.

## Methods Standards

There is a best way to perform any given job. The best methods are not in use for most jobs.

The value of the method must take into account a number of factors. If a given procedure is extremely efficient, requiring a level of skill that cannot be achieved by most workers, it cannot be put into general use. If it requires a timing or coordination that is a mental or physical strain on the workers, it will be abandoned by them. The standard, therefore, must be one which can be met by the general worker who is to be assigned to that job. Workers who cannot meet the methods standard in question should be delegated to perform some job having a standard which they can perform.

Both materials and equipment have a direct effect on methods. Some methods depend directly on some specialized type of equipment or chemical. For example, autoscrubbing is a method of wet-floor cleaning that is primarily a function of the machine itself. In this case the method must be restricted to large uncluttered areas which are suitable for this type machine.

Some materials are designed to perform two or more operations with a single application. Just because a given material will perform a job faster than another is not sufficient justification for its use in a methods standard, however. A number of custodians use acid-type descalers in the daily cleaning of rest-room fixtures. Indeed, this does a very quick cleaning job, but has the disadvantages of presenting a safety hazard, damaging surfaces if improperly used, eroding floors if spilled or dripped, and being rather costly.

Methods also depend on organization. If all the workers in a given housekeeping department are set up on an area-cleaning basis, some of the efficient methods which can be achieved with team cleaning are lost.

The methods specification should primarily answer the question *how*. It should tell the worker how he should stand, how he should use the equipment, and how he should move. It should also indicate *what* equipment and materials are involved and the proper use-dilutions of all chemicals. It should tell what safety precautions should be taken. It is also desirable to point out *why* the job requires performance in a given way, or why it need be done at all, in some cases.

Both time standards and frequencies can be indicated on methods sheets for convenience in use by supervisors.

## Materials Standards

It is often easier to establish standards for housekeeping materials than it is to obtain standardized use of them.

Housekeeping personnel are prone to make unauthorized experiments with housekeeping materials. There is a natural tendency to utilize stronger materials than specified for a given job. Unless the materials-use situation receives supervisory attention, floor-stripping chemicals will be utilized for general cleaning purposes, for example. These chemicals are, of course, strong cleaners and good emulsifiers, and certainly do clean rapidly—but, in addition to cleaning, they remove wax or other protective coatings from floors, thus leaving them exposed to damage. Experimentation with products also can expose custodians to safety hazards. For these reasons, personnel should use only the materials specified for given jobs.

Even use-dilutions should be standardized. Overuse of a given material may not only cause more work than is necessary, but it also may damage surfaces or constitute a safety hazard.

The determination of material type will be based on such factors as water hardness, the degree and type of soil encountered, the type of floor covering, etc. It is desirable to operate the housekeeping department with as few products as possible, and dual-purpose products should be utilized for this purpose. Standardization of products brings these benefits:

1. Inventory levels may be cut because of the fewer number of products involved.
2. The larger quantity purchased of the fewer number of items leads to purchasing savings and improved storage conditions.
3. Training is simplified.
4. There is less danger or error in product selection and use.

If both frequencies and materials have been standardized, it is possible to determine consumption standards by multiplying the amount of material needed to clean a unit surface, for example, times the total surface, times the frequency per year. Under such *controlled* conditions, incidentally, increased consumption of materials will indicate increased frequency. Under uncontrolled conditions, it may indicate either increased frequency, wastage, pilfering, or misuse. Thus, a successful housekeeping program would be expected to cause an increase in materials consumption on the basis of improved frequencies and performing some cleaning functions that might never have been done previously.

## Equipment Standards

The most prevalent error in equipment selection for housekeeping purposes is the purchase of undersized machines. Any economic analysis will indicate that, in general, the largest and the most durable piece of housekeeping equipment that can be fitted to a given job is the best investment. The additional cost over the life of the equipment, even bearing in mind interest, taxes, and the like, is usually quite small.

It is also better to buy equipment which can be used for more than one purpose. The flexibility of a piece of equipment, owing to its style or availability of various attachments, often determines the amount of time in which it will be in use during the average day.

Of course, equipment must be selected that is safe and may be easily repaired or adjusted.

Once a suitable piece of equipment has been selected, it is best to reorder additional units of the same type rather than going from one supplier to another, even where there is some price advantage. In standardizing on a given type of equipment, these advantages are gained:

1. Purchasing savings where the equipment is bought in multiple units.
2. Not only is the training in the use of the equipment simplified, but also the training in its care and minor repair.
3. For major repairs, basic parts may be kept on hand where there are a number of pieces of identical equipment in use.
4. Suppliers are able to render more service because of the greater sale involved.
5. Interchangeability of parts and attachments results in a smaller overall investment.

## Time Standards

To many people, "standards" and "time standards" are synonymous. This is probably because so much attention is given to time standards in some operations and because time standards can lead to controversy.

It is of first importance to remember that time standards do not exist in a vacuum. The determination of time standards can only be successfully made after the establishment of all the other standards on which they depend.

Only when a given job is performed in a standard method by standard personnel using standard materials and equipment under standard conditions, can the standard times be accurate. In housekeeping functions, because of the wide variances, time standards are often indicated as ranges. Even these ranges do not comprehend all acceptable times under various combinations of conditions.

Inaccurate standards are one of the fundamental causes of misunderstanding between labor and management. The inaccuracy of standards and of standard times is almost assured when improper or inadequate standards have been established for personnel, methods, materials and equipment, and when conditions are not taken into account. Where conditions or the other standard factors differ from the normal, the time standard must be adjusted accordingly, if it is to be used at all.

Chapter 3 contains a partial listing of variables which affect the time required to maintain a given floor area.

Time standards, without a detailed definition of methods by which the standards were obtained, may mean little. Unless the method can be duplicated in practice, neither can the standard. Suitable allowances must be made for fatigue, make-ready, down-time, and personal time. Unless these are built into the time standard, compensating factors must be applied.

There are several methods now in use for the determination of housekeeping time standards:

1. Actual time study by qualified personnel undoubtedly will provide the most reliable information. Since a large variety of housekeeping jobs would require study, the actual making of a time study may be restricted to only the more important jobs.

2. Internal records are useful if they have been kept carefully and the work was performed under proper supervision, with the use of standard methods, etc. The mere use of historical time rates, where these other standards have not been applied, will tend to perpetuate and officially authorize inefficient work.

3. Various time standards information has been published from a number of sources. The use of a synthetic set of housekeeping standards by adopting one of these listings, or by taking the average of several listings, is unsound. Very few serious independent investigations have been made of housekeeping time standards. Much of the information published is a restatement or a duplication of other previously published information which, in itself, may have been estimated in large parts. Standard times developed for other than your own operation should be used only for comparative purposes, or as a check list, and never applied to the work at hand.

4. Mere time estimates can be useful in trying to apportion labor to an infrequent or unusual job. Such estimates cannot be used for grading the work.

Fixed-time standards for small work units are difficult to put into use in housekeeping operations. Their value is in the establishment of larger work units. For example, the amount of time necessary to dust a cigarette urn of a given size is of little practical value, except when it is utilized in the establishment of a standard for the complete dusting of an office area of unit size.

An experienced analyst will find it unnecessary to completely develop time standards for every housekeeping job as applied to every surface to be maintained under various conditions. Rather, he is able to prepare a standard for a series of given situations and then apply weighing factors against these standards to compensate for the variables which may act on them. Some weighing factors will be positive, others negative, depending on whether the variable would indicate a greater or lesser time standard. The weighing factors must be applied judiciously since they were developed as units, which, however, interact when more than one factor is applied. Experience (and mathematics) provide control figures, maximum and minimum, beyond which the standard cannot be adjusted by weighing factors, since the result would go beyond practical achievability.

# 10

# Planning for Better Maintenance

THE design of a new building utilizes the combined talents of a number of specialists in order to achieve the best possible results. We do not wait until construction begins—or has been completed—to consider such things as electrical distribution, drainage, sprinkler protection, floor loading, fenestration, heating, or equipment layout.

Is there any reason, therefore, to complete a new structure, or to renovate an old one, without considering many of the aspects of housekeeping and sanitation? Yet, this is usually the case. Surely we do not want a motto such as "Plan Ahead, Except for Housekeeping!"

How many housekeeping department managers have bitterly considered the serious housekeeping problems which would have been avoided if they had been given an opportunity to advise on a simple point! How many supervisors realize that a number of cleaning duties are necessary only because of the lack of housekeeping planning!

Many companies have constructed new plant facilities in order to eliminate the inefficiencies which were unavoidable in an older structure. But, unfortunately, the housekeeping problems are carried directly from the old building into the new through lack of planning in this field.

The reason for the lack of action in this direction may be a failure to understand the economic importance of housekeeping. Or, it may not be well known that housekeeping can be planned.

Unquestionably, the greatest possible return on the housekeeping dollar will be realized where the housekeeping program is developed during the planning stages of plant construction. The service engineer and the housekeeping department manager can work with the architect and plant engineer in selecting materials and surfaces. Together, they can locate and determine the size of custodial storage and operating facilities and provide other conditions that lead to good housekeeping.

Once again, we must turn to management for direction. The benefits of housekeeping planning can never be realized without management's interest in this direction. The architect, plant engineer, housekeeping department manager, and service engineer all have an obligation to attempt to awaken and sustain this interest.

Some architectural and engineering firms are awakening to the importance of maintaining the facilities they construct. As an example, Carl W. and Robert T. Clark, A.I.A., architects-engineers of Syracuse, New York, have provided an extension of their usual architectural services through the provision of maintenance manuals. They point out to the building owner that "there is a sort of magic in the beauty of a new building: the clean freshness of new paint and wood finishes, the immaculate floors, the smooth precision of the mechanical equipment. Whether you look at the building through the eyes of a citizen, an owner, or as an architect, your thought is the same: 'if only it would stay this way!' "

Yes, architects are beginning to give housekeeping new thought. They will provide greater services through the use of consultants to improve the cleanability of the structures they design. The provision of a complete housekeeping program may be part of a "package," or it may be an optional service. Through such services, the progressive deterioration that normally follows quickly after building completion can be retarded. Many "new" buildings become "old" in a surprisingly short period of time—there is no reason that the newness cannot be prolonged for many years through proper housekeeping planning and action.

Also beneficial would be seminars for architects and engineers, either undergraduate or postgraduate, so that they might become familiar with the possibilities of housekeeping planning and programming.

The opportunity to plan for better housekeeping for a new facility comes only once. And once that opportunity is lost, it is lost forever.

Fellow science fiction enthusiasts will remember a clever story by Clifford Simak called "Dusty Zebra." In this story, a small machine collects dust and transports it into another space-time continuum. More likely to be developed eventually, in other stories of tomorrow, are homes which keep themselves clean with supersonic, electromagnetic, and other soil-loosening and attracting devices. For these homes, the interesting point is that the structure is designed to work *with* the device.

No matter what our industrial future holds, it certainly can be a better future through planning.

A listing follows of *some* of the things that can be done to provide better housekeeping. Their actual use must be decided on the basis of the other factors involved in a specific project as well. They are arranged as:

Custodial facilities
Floors
Furnishings
Openings
Rest rooms
Walls and ceilings
General

Of course, the most complete planning will involve the preparation of a comprehensive housekeeping program. If it is prepared during the planning and construction phases of the facility, it provides a remarkable opportunity for "building in" and conducting an efficient housekeeping operation.

## Custodial Facilities

1. Provide an adequate centralized area for the exclusive storage of cleaning materials and equipment.

2. Be sure custodial closets and lockers are adequate in number and size and are properly located.

3. When utility sinks are used, they should be low and fitted with a metal lip to prevent the damaging of the enamel.

4. Provide adequate sources of both hot and cold water throughout the plant for custodial use.

5. Locate sufficient electrical outlets for custodial use in corridors and large rooms. Outlets located near doors are helpful in case the outlets inside the room are inaccessible.

6. Where possible, electrical receptacles mounted in floors should be avoided. They are easily damaged by liquids (such as detergents and waxes), can cause accidents, and are difficult to clean around. They are also rather unsightly.

7. Provide for the laundering and treating of dust mops and cloths in the

central storage area. Provision for the cleaning of wet mops and venetian blinds is also desirable.

8. Provide adequate shower and locker facilities for the custodial staff. Make sure that such facilities are conveniently located.

9. Built-in vacuum systems are desirable in certain special cases.

10. Adequate storage facilities should be provided for all types of waste materials, such as paper, scrap metals, food waste, etc.

11. Bulletin boards should be provided for the posting of housekeeping information and instructions.

## Floors

1. No floor should be selected which requires hand maintenance or care by other than the average custodian.

2. Try to avoid the meeting of floors at different levels—they should be flush.

3. Where the difference in levels between floors is slight, a ramp having a nonskid surface will be much safer than stairs. It will be easier to clean and will permit easier equipment handling.

4. Special "floating" floors should be designed for data-processing areas so that wiring, ducts, etc., may be run underneath the floor. Removable portions of the floor make the area underneath easily accessible.

5. Special floors must be specified for superclean rooms or "white rooms."

6. Subfloors should be as smooth as possible in order to avoid depressions and raised areas. These irregularities cause resilient finished floors to wear away quickly and make them extremely difficult to maintain properly. Concrete subfloors should be machine troweled wherever possible.

7. Concrete subfloors should be properly cured, and any irregularities corrected, before the application of composition tiling.

8. Floors which are continuously wet, such as in the poultry-processing industry, should be provided with trench-type drains rather than individual spot drains. The floors should be sloped gently toward these trench drains.

9. For resilient flooring, vinyl asbestos provides a serviceable floor at reasonable cost.

10. Prepare for adequate lead time on floor installations so that newly laid resilient tile will not be scrubbed until thoroughly set, which requires from three to six weeks.

11. Floor patterns should be selected carefully, as experiments have shown that floor designs and color combinations can range in effect from therapeutic to annoying.

12. The use of floor sealers should be specified in construction contracts where floors are to be of wood, concrete, or terrazzo. This permits full protection of the floors before occupancy, as well as a much more efficient application of seal because of the lack of congestion.

13. Corridors should have no recesses in the wall or projections into the corridors. Drinking fountains in corridors should be avoided, if possible.

14. Utilize rubber or composition cove bases, rather than wooden baseboards, to provide a scuff-free surface and rounded joints which are easily cleaned.

## Furnishings

1. Attempt to minimize the amount of furniture to be moved when cleaning by the use of built-in or wall-mounted furniture.

2. Furniture made of metal, plastics, or composition materials are more easily maintained than wooden furniture.

3. Fabrics are dust catchers and should be eliminated when possible. This includes curtains, drapes, upholstered furniture, etc.

4. Ornamental decorations catch dust and are normally difficult to clean. On the other hand, decorations such as murals, photographs, or paintings are normally easy to clean if simply framed.

5. Signs and calendars should be removed from walls, glass partitions, and rest rooms where possible as they are unsightly.

6. Bulletin boards should be provided for all employees for purposes of promoting housekeeping, safety, fire prevention, as well as for other purposes.

7. Place waste receptacles immediately adjacent to areas where waste is created.

8. Waste receptacles should be of adequate size and provided with covers when used for odor-producing materials, perishable food, or dusty waste.

9. There is greater loitering and abuse of facilities in poorly lighted areas than in well-lighted areas of similar nature and traffic. Stairwells, corridors, locker rooms, rest rooms, and custodial closets should be well lighted.

10. The proper and frequent cleaning or replacing of air-condition filters by designated personnel decreases the cleaning requirements for all surfaces which require washing and dusting. Soil is trapped in the filter rather than being distributed throughout the facility.

11. Vending machines, where possible, should not be placed immediately against a wall, which prevents cleaning behind them.

## Openings

1. Walkways leading to doors should be of hard material and properly drained to prevent mud and water from being tracked into the building.

2. Use matting at doors and between areas where it is necessary to catch soil.

3. Use flush doors rather than paneled doors, as they do not collect dust.

4. Curbs or high thresholds should be avoided at doors, as they prevent the moving of carts and other equipment from one floor area to another. Where curbs must be used because of water on the floor, provide ramps to permit equipment access.

5. Translucent or tinted glass is easier to maintain than transparent glass.

6. Where caulking is used, it should be smooth and continuous.

7. Proper window-washing equipment should be designed into the building. This is particularly important in multistory buildings where a railing may be

installed around the roof from which a permanent retractable scaffold is suspended.

8. Blinds, shades, and other window coverings should be avoided whenever possible.

9. Marble window sills, though rather expensive, require very little maintenance and are quite decorative. Their use should be considered for administrative areas.

10. Use aluminum framing rather than wood or painted steel.

## Rest Rooms

1. Rest rooms should be organized to permit a circular flow of traffic. Wash basins should be located near doors, and mirrors on walls opposite the wash basins.

2. Rest rooms should be located away from busy areas so as to equalize the traffic load on floors.

3. Each rest room must have a floor drain for emergency spillages.

4. Brightly lit rest rooms stay cleaner, with less abuse of facilities and a reduction in loitering.

5. Use porcelain fixtures.

6. Rest-room fixtures and stalls should be hung from the wall or ceiling to leave the entire floor area open for ease of sweeping and mopping.

7. Be sure to build in "grab bars" in shower areas or in other places where slipping is possible.

8. Ash trays should be located between urinals so as to prevent the throwing of cigarette or cigar ends into the fixtures.

9. Lockers should be provided with slanting tops for easy cleaning. They should be placed on a concrete or ceramic base so that soil will not be trapped underneath, and so that the locker floor area may be mopped or scrubbed freely.

10. Paint walls with light-colored, light-reflective materials. Enamels are suggested.

11. Paper towel and tissue dispensers require less attention when they are of a large size. The tissue dispensers holding one roll in reserve, in addition to the roll being used, are desirable.

12. Dispensers and cabinets should be large enough to require a maximum of daily service, if possible.

13. Utilize a type of soap dispenser that may be refilled without spilling.

14. Use dispensers having metal or plastic containers.

15. Do not use bar soap, as it is a medium for cross-infection. It also has a bad appearance. Use liquid, powdered, or paste soap instead.

## Walls and Ceilings

1. Materials should be selected which can be properly dusted as well as wet cleaned.

2. Walls should be utilized for the hanging of as many items as possible, permitting rapid cleaning and avoiding damage to the fixtures. These items include rest-room fixtures, partitions, waste receptacles, water fountains, cigarette urns, etc.

3. Glazed tile should be used on the walls of custodial closets, some hallways, stairways, rest rooms, kitchens, etc.

4. Use paints that are durable and washable, or use other washable materials such as plastic coverings.

## General

1. The cleaning of a new facility should be worked out in advance of construction. Most contractors perform a final clean-up, but this usually needs to be augmented by internal cleaning before occupancy.

2. Arrangement of facilities can do much to determine the custodial work load. A number of cases are on record where office equipment, for example, was rearranged to permit a considerably more efficient performance of the housekeeping work without disturbing the efficiency within the office.

3. Whenever possible, inactive areas should be sealed off. Departments with greatly reduced personnel should be moved to smaller areas.

4. Arrange for smooth traffic flow throughout the facility so as to shorten the number of steps that have to be walked. Try to avoid a traffic pattern that will produce overwear in certain areas.

5. Air-conditioned areas are easier to maintain because of the dust control that is provided.

6. Attempt to eliminate internal window sills, ledges, and all dust-catching surfaces.

7. Provide round corners where possible, so that they will not catch dust.

8. Be sure that plenty of conduits are installed in the building for running all types of wiring, so that wiring need not be exposed.

For a fuller treatment of this important subject, the reader is invited to read the book "Building Design for Maintainability" (McGraw-Hill Book Company), by this author.

# 11 Contract Cleaning

AM I getting my money's worth in housekeeping? More and more managers are asking themselves that question. There are growing pressures to conduct an efficient and productive function from within as well as from without. And contract cleaning is turned to as a possible alternative.

The concern is not only about the dollars and cents of sanitation, but its *effects* as well. Ineffective housekeeping can lead to deterioration of surfaces, damage to equipment and supplies, an excessive accident rate, fire hazards, an environment not conducive to high productivity, and an increased incidence of cross-infection. So, pressures to improve housekeeping can rise from every department head in the organization.

Let's not deny there are plenty of capable men and women operating internal housekeeping departments in an effective way. Even though any complex job can be improved to a certain extent, in general they are giving their managers a good return for the money they entrust to them in the conduct of their department. But the fact is, there are also plenty of housekeeping departments that are operated inefficiently and which need considerable improvement.

In order to really compare housekeeping as performed by one's own department to that which may be provided by a contractor, we need to take a closer look at just what housekeeping *is* today. Let's review:

An effective housekeeping program must begin with management sponsorship and support. To begin with, administration must build a *positive image* for the housekeeping worker in his job. The concept of the custodian and maid of yesteryear, as slovenly workers carrying pails of dirty water and torn mops, must be replaced with that of qualified workers performing a valuable function. Then, management must encourge interdepartmental participation in the housekeeping program, so that other departments are *alleviating* the housekeeping problem rather than *contributing* to it. Finally, facility-wide cooperation must be stimulated—and *will* be stimulated—by administration's interest in this activity.

To develop a housekeeping program, the facility must first be analyzed to determine the basic housekeeping objectives and limitations; and to provide statistics concerning the number of square feet of each type of area to be maintained, the frequency of job performance, fiscal and personnel requirements, and the like. This statistical information, along with a complete set of floor plans, is studied from the standpoint of the optimum organization and system to provide adequate protection of surfaces, proper appearance, and a healthful and productive environment.

An organization is developed, and staffing set up to accomplish all the jobs at their indicated frequencies. Personnel are then trained for this work in those methods which will provide the greatest productivity; and this instruction includes orientation, classroom, and on-the-job training. Certificates should even be given for completion of the training course which in itself is a good morale building device. The performance of special projects, those infrequent jobs that are really the backbone of the housekeeping work, are scheduled on a *preventive maintenance* program. The workers are provided the proper custodial facilities, such as closets, cabinets, and mobile carts for the work. They are furnished the right type of housekeeping chemicals.

The work is mechanized with some of the numerous types of mechanical equipment now available for the housekeeping department—things like automatic scrubbing machines, drum head adapter vacuums, pack-vacuums, light fixture louver washers, eraser cleaners, garbage can washers, venetian blind cleaners, fountain-type window washers, and the like.

The modern housekeeping program further includes planning for better housekeeping, for renovation or construction, so that these facilities may be cleaned for many thousands of dollars less per year.

Now that we have taken a second look at a modern housekeeping program, the question is: Who is to set up and perform such a program—an internal housekeeping department, or a contract cleaner (or some combination of the two)? Naturally, the contract cleaner will tell you he can solve all of your problems. But can he? Let's take a look at the contract cleaning industry and how it developed to its present state.

## Development of the Industry

The first type of contract service to be offered on a widespread basis was window cleaning. This took hold in the larger cities where multistory buildings presented problems of safety and efficiency. The contract window cleaner was able to obtain specialized cleaning and safety equipment, and carefully train and inspect personnel for this work, thus relieving internal housekeeping personnel of this difficult and hazardous duty. Many general cleaning contractors demonstrate this origin by the use of the words "window cleaning" in the title of their organization, although it may now only constitute a relatively small part of their service and income.

From window cleaning came a branching out into other services through satisfactory contact with the client. A number of pest-exterminating organizations have gone into contract cleaning because of their major orientation to contract labor services and, again, because of healthy client contact based on their original service. Sanitary supply firms enter this field through the desire to establish a captive market for their products and equipment, and in order to diversify from their extremely competitive field. Housekeeping department managers and executive housekeepers have seen the money which can be saved in improving the efficiency of their own operation, and enter this field on the basis of local need or by actual invitation from prospective clients.

Finally, some people enter the contract cleaning field with absolutely no experience of any kind in housekeeping or sanitation, in answer to business advertisements that offer large profits.

Not only are small organizations springing up in many cities, but some of the giants in the industry have set up branches in cities throughout the country. Some of the larger firms are offering franchise arrangements and permitting entry into the contract cleaning business to people with very little capital and no experience.

## Types of Services

The typical contract cleaning organization offers a wide range of services. Most establishments provide two or more types of service. This not only gives the investigator of these services the opportunity of deciding whether or not to use contract cleaning, but he can also consider the questions of "what kind" and "how much." Some of the services include:

1. Single daily projects, such as waste collection and removal.
2. Single intermittent or periodic projects, such as washing outside window surfaces.
3. A contract for a single project, such as the cleaning of overhead areas.
4. Supplementary services to assist the regular housekeeping organization in completing an unusually difficult project, or in meeting some deadline.
5. Selected types of services, such as night cleaning only, or the cleaning of administrative areas only, with the balance of the work being performed by internal personnel.
6. The complete cleaning of a given portion of the facility, such as outlying buildings.
7. Complete cleaning services with all jobs and responsibilities being undertaken by the contractor.
8. Variations and combinations of the above possibilities.

### ADVANTAGES

That *some* clients find benefits and advantages in contract cleaning is undeniable. The rapid growth of the industry could not have been sustained without an increase in the number of organizations using contractors and retaining their services from year to year. These organizations profit from one or more of the following:

1. Cost is by far the most important factor. Although contract cleaning can be successful where the wages paid by the contractor to his personnel are the same or even greater than those paid by employers

who operate their own housekeeping departments, it flourishes where the contractor is able to pay considerably less.

2. Avoidance of hazards to the organization's own employees. This is especially the case in such jobs as outside window cleaning in multi-story buildings, overhead cleaning, cleaning with hazardous chemicals such as the acid-cleaning of new masonry, the descaling and scouring of rest room fixture traps, or equipment cleaning.
3. Permits the performance of extra work without an increase in the basic staff. Contract cleaning personnel can be hired for seasonal cleaning.
4. Avoids the investment in specialized or even standard housekeeping equipment, thus reducing the total capital investment.
5. Contractors can be called on for emergency cleaning at short notice.
6. The use of contract cleaners may remove management from problems with unions.
7. The payment of *higher* rates by the contractor may mean the securing of a more qualified worker, and thus a more effective program.

### DISADVANTAGES

But, some of the organizations continuing to use contract cleaning are dissatisfied, to a greater or lesser extent, for one or more of the following reasons:

1. By far, the most important is the regular deterioration of quality. Both the client and the contractor contribute to this situation. The client often short-sightedly selects the lowest price bid without an evaluation of what this means in terms of cleaning quality obtainable at the price. The contractor, on the other hand, may continue to decrease quality to "as much as the traffic will bear" in order to increase his profits.
2. There is a disparity in basic objectives which may be hard for some to reconcile. The contract cleaner, being in a legitimate business, must make a profit. A reputable contractor will make a reasonable profit on every contract, but a disreputable contractor will give as little cleaning as possible in order to make the largest profit. The objective of the client is just the reverse; namely, to obtain as much cleaning as possible for his dollar.

3. Where work is done in an organization by personnel not belonging to that organization, there may be some loss of management control.
4. An organization is made up of people, and the progress of the organization depends on the attitude of its people. This proprietary attitude is lacking in contract cleaning.
5. Problems in security control may become troublesome where this is a prime issue.
6. There is a loss of flexibility in meeting changing conditions, with some types of contracts.
7. Inter-departmental liaison and cooperation may become more difficult.

One of the greatest problems in this field is the purchase of contract cleaning services on the basis of price. It is similar to the problem involved in furnishing cleaning materials on the basis of price. Specifications in both cases are rather difficult to word and evaluate. Quality deteriorates steadily.

## The Decision

How does one decide whether or not to use contract cleaning services, and how does one avoid the pitfalls?

The decision to use contract cleaning—or not to use it for that matter—should be based on a sound analysis of all the variables involved. Consultants are often able to give an objective viewpoint to such an investigation.

The study must include an economic analysis of the competitive benefits and limitations, as well as a discussion of relatively unmeasureable factors such as morale and safety.

One should carefully investigate a potential contractor and should require information of the following types:

1. Reference accounts. In addition to being permitted to list the more successful reference accounts, the contractor should also serially list all accounts if a small concern, or the last twenty accounts if a large concern.
2. The contractor should indicate the extent of his experience.
3. Financial references.
4. Insurance coverage, both for cleaning personnel and from a public liability standpoint.
5. Qualifications used for hiring, screening and training personnel.

6. Methods and extent of supervision to be used.
7. Method of payment, along with the period of termination notice required.
8. Professional qualifications and background of the organization, and its principals.

One of the circumstances that mixes reputable with disreputable contractors with disturbing regularity is caused by the fact that it is so easy to get into the contract cleaning business, and because there are no regularatory agencies governing the conduct of this type of service. (The same problem, of course, applies both to the sanitary supply industry and the consulting field, although in the latter case there is stringent state regulation if one is to use the title "engineering"). There are even opportunities for entering the field on a franchise basis with a capital investment of less than $5,000, and even this can be paid out of profits over a long period of time.

## Types of Contracts

Many, if not most, of the contracts now being written practically guarantee that no reputable contractor will bid on the job!

This may be because the contract is for too short a time; or the specifications are written in an ambiguous or loose way, permitting all types of interpretation; or because there are requirements that can obviously not be met.

It is not unusual for contractors to bid on a one-year job on the basis of "what will it take to get this job?"; and if the contract is awarded to the bidder, he then subtracts his profit and overhead, and gives his customer the amount of cleaning possible with the funds remaining. The blame lies as much with management as with the contractor in cases of this type, for failure to properly specify the work in all of its aspects, as well as for the naive entering into a contract that, to the skilled investigator, is perfectly impossible of performance.

There are several types of contracts possible for cleaning:

1. A fixed periodic cost (either as a grand total or per unit area), based on a quality definition and/or frequency specifications. The periodic rate type of contract is the one most frequently used; it is available in two forms, the most common of which being with *uncontrolled*

input. This means that the contractor has agreed to provide a service of a given nature, but is not required to provide any given number of man-hours, and is usually not required to have on hand any given numbers of types of equipment.

This type of contract (the type most frequently used!) is by far the most troublesome and the one least likely to bring a meeting of the minds, and therefore satisfaction to both customer and contractor; but if this form is to be used, these aspects of the specifications become very important:

a. A definite frequency of job performance, with complete elimination of phrases such as "as needed" or "as required."
b. The method of performing the work should be described carefully. If the floor is to be mopped daily, one should specify whether this should be with a one-bucket, two-bucket, or three-bucket system, and whether or not the floor will be rinsed.
c. The nature of the chemicals and materials should be specific.
d. The nature and amount of supervision should be controlled.
e. A description of the type and extent of reports and inspections should be made.
f. The contractor should provide legal and fiscal information. One of the problems inherent in this type of contract is the provision of a contract document so voluminous that it frightens off many of the reputable but smaller cleaning contractors (repute, of course, may have nothing to do with size), and also makes policing of the contract a very difficult matter.

2. A variant of the periodic rate contract has a *controlled* input. This means that the contractor is required to provide a minimum number of man-hours, in addition to all the other requirements named above. For many jobs, this would be tantamount to having everyone bid on that *minimum* number as the *actual* number of man-hours to be provided. Where the periodic rate contract is to be used, we strongly recommend a controlled input.
3. A specialized type of contract is the management fee. In this case, the customer provides his own personnel, equipment, and chemicals, while the contractor provides the management and, perhaps also, the supervision for the work. This is a hybrid type operation which may be based on a union contract requiring the personnel to remain

on the organization's payroll, or perhaps because of management's interest in seeing that all personnel in the organization receive the same benefits and opportunities for transfer to other departments.

4. In an effort to avoid the pitfalls of the periodic rate, a number of organizations have turned to a contract on a cost plus percent profit basis. This has the advantages of providing a perfectly variable, flexible program, as the customer is free to determine his needs on the basis of changing quality requirements, physical plant, production or occupancy, weather, and the like. On the other hand, this system provides an incentive to the contractor to increase cost in order to earn a larger profit.
5. Finally, a contract may be established on a *cost plus fixed fee* basis wherein the contractor is reimbursed for all costs but paid an unchanging periodic fee for the management of the program. Here we have the advantages of a completely flexible program along with the elimination of any incentive to spend more money. As a matter of fact, since the contractor is guaranteed a reasonable profit as long as he continues to perform a meaningful service, this becomes one of his most valued jobs and he is thus anxious to retain it as long as possible. Because the cost plus percentage profit or cost plus fixed fee contract preferably requires management to be able to develop its own program in detail, the use of a consultant can be of special value here.

We were once asked by a client to prepare two reports for his facility, one showing how the facility should be cleaned on contract, and the other showing how it might be cleaned with his own staff, so that he could compare systems and costs. We pointed out to him that the reports would be identical! Since there is a one best way to clean any given facility, in terms of supervision, staffing, equipment, chemicals, methods, and the like, then that one best sanitation program should be followed whether the work is done by a contractor or by internal staff. Whatever type of contract is used, it is important that these factors be given careful attention:

1. Contracts should be written for longer terms, to interest the more reputable contractors. Annual re-letting of bids should be abandoned, in favor of a minimum of a two-year contract with a one-year option to renew. Contracts up to five-years in duration should prove beneficial.

2. Regardless of the length of the contract, a cancellation clause should always be provided, either of 30 or 60 days length, depending on the situation.
3. Very close inspection should be given the work; in larger facilities, this will amount to one or more full time salaried employees on the customer's payroll who will check method of performance, time cards, quality achievements, utilization of chemicals, safety, and the like.
4. Again, in any type of contract, penalties should be assessed for failure to perform the work or to provide the proper number of man-hours. Penalties should be especially rigid for failure to provide the proper supervision, since this is of such importance to the conduct of the job.

**Controlled Input–Variable Bid**

Especially for organizations which come under public scrutiny and are controlled by a body composed primarily of lay people, such as a public school or college operating under a board, we have had a great deal of success with a variation of the controlled input type of contract.

There are some problems with controlled input: it tends to "promote" the less qualified firm to the same level as a more qualified firm, since they are proposing on precisely the same thing. It may decrease the incentive to reduce costs by a more reputable firm, since that firm has nothing to gain by such an activity.

On the other hand, if one does not develop a standard control, any bidder can come in with a low bid–it may not be responsive to the needs of the organization at all, even though it cannot be refuted. One is then subject to the question "why did you not select the lowest bid?"

Our consultants developed a form of contract where there may be a variable bid around a standard controlled input. The property owner (or his consultant) makes a measurement, develops time standards, prepares man-hour requirements, based on standard procedures and frequencies, and from this develops a pro-forma budget just as in the standard controlled input system. But the purpose in this case is to be able to evaluate the variable bids, described below.

In this system, the contract bidder is allowed a variation or "spread" on the man-hour bid–this variation might be 10% or 15% higher or lower than the control figure. This allowed variation on the initial bid is to permit the contractor flexibility in terms of his own system of incentive payments, cost savings ideas, motivation programs, supervisory coverage, new techniques, etc. Once his figure is named, it becomes the new control figure and he must thereafter stick with it. In order to develop these figures, the bid sheet would show the minimum standard and maximum allowable manpower, and finally he would enter his proposed figure.

The contractor's proposed staffing would not only be the total figure, but, in a public school system for example, he would show the figures for each school (or in the case of a college, for each building).

Further, the contractor must furnish in his bid individualized work assignments for each of the positions shown that he has developed for his customer. Each assignment must be described in writing along with a graphic designation of its extent (of course, the property owner provides the drawings). This descriptive material provides the owner a system of monitoring assignments.

Further, the bidder should provide an organization chart, a system of supervisory coverage, training program plans and also his qualifications for performing the work.

For the owner to be better able to determine the bidder's qualifications, the contractor may be required to supply information on length of employment for personnel of various categories, turnover rates, absenteeism patterns, training history, supervisory development system, etc.

A final determination of whether or not to use contract cleaning should be made on sound business principles considering the relative cost and quality benefits on either side. Emotional viewpoints should be avoided. Contract cleaning is, simply, a legitimate alternative consideration open to management.

Questions asked of the bidder are important in determining his capacity to handle an important contract (continued next page).

A. Class of Employee

| | MANAGEMENT | SUPERVISORY | NON-SUPERVISORY | TOTAL |
|---|---|---|---|---|
| Number of Employees | | | | |

B. Distribution of Years of Employment by Class of Employee

| Length of Employment | MANAGEMENT | SUPERVISORY | NON-SUPERVISORY | TOTAL |
|---|---|---|---|---|
| 0-1 year | | | | |
| 1-2 years | | | | |
| 2-3 years | | | | |
| 3-4 years | | | | |
| 5-9 years | | | | |
| 10-19 years | | | | |
| Over 19 years | | | | |

C. Personnel Turnover for (Non-Supervisory Employees Only)

| Years Employed | No. of Hrly. Employees | |
|---|---|---|
| One or less | __________ | Added to Payroll |
| | __________ | Dropped from Payroll |
| 2-4 | __________ | Dropped from Payroll |
| 5-9 | __________ | Dropped from Payroll |
| 10-19 | __________ | Dropped from Payroll |
| Over 19 | __________ | Dropped from Payroll |

D. Absentee Pattern for the Year (Record a check (✓) in each appropriate box)

No. of Scheduled Work Days Absent 1976

| No. of Non-Supervisory Employees | 0 | 1 | 2-4 | 5-9 | 10-19 | 20-49 | 50-99 | 100-199 | Over 199 |
|---|---|---|---|---|---|---|---|---|---|
| 1 | | | | | | | | | |
| 2-4 | | | | | | | | | |
| 5-9 | | | | | | | | | |
| 10-19 | | | | | | | | | |
| 20-49 | | | | | | | | | |
| 100-199 | | | | | | | | | |
| 200-499 | | | | | | | | | |
| 500-999 | | | | | | | | | |
| 1000-1999 | | | | | | | | | |

E. Employee Training    Average Hours of Training Provided/Year/Employee

| Length of Employment | MANAGEMENT | SUPERVISORY | NON-SUPERVISORY |
|---|---|---|---|
| New Employee | | | |
| 1 - 2 years | | | |
| 2 - 4 years | | | |
| 5 - 9 years | | | |
| 10 -19 years | | | |
| Over 19 years | | | |

# 12

# Consultants -- Who Needs Them?

WHO needs consultants? In order to answer that question, we must first determine what *is* a consultant . . .

The Thorndike-Barnhart dictionary puts it very simply: It tells us that a consultant is "a person who gives professional or technical advice."

So, who needs consultants? The first individual that comes to mind is the President of the United States. He has a great number of things to do, and a limited time in which to do them—so he uses a number of consultants. It is interesting that no one seems to associate his extensive use of consultants with any weakness or failing on his part.

There is another person I can think of who needs consultants, and needs them badly.

That's me.

In order to keep up with our challenging, dynamic field of housekeeping management, and to save time in other areas to do so, I need help, and lots of it.

My consultants include an accountant, a lawyer, a doctor, a psychologist. They include the sanitarian, the labor relations specialist, the time standards specialist, the hospital administrator in or associated with our own organization. My consultants are also a few hundred authors of books and articles.

And, since it is inevitable that we learn from each client, I consider them as my consultants too.

Who needs consultants? Let's see if we can narrow it down by the process of elimination. Who DOESN'T, then, need "professional or technical advice?"

1. The person to whom everything is known, and who applies that information in a perfect way. No human being, of course, can fit this category.
2. Contented hermits who are illiterate.
3. Prisoners in death row who have already talked with their ministers.
4. Hospital patients in terminal coma.

That seems to leave at least everyone who can read this chapter!

Why should we all need help from others? It is because we are living in the age of the information explosion. More information will become available in the next ten years than in all the centuries of human life before, *put together*. Libraries being built for *twice* their present space needs will be overflowing with books in just 120 short months. A former century of thought and invention is now being compressed into a thousand days.

This great mass of facts and viewpoints has created an age of change. Of all the scientists, engineers, inventors, researchers—people who make change—who ever lived, 90 per cent are at work *right now*, changing things for you and me—a fascinating thought.

And it is this flood of information and change that has brought—demanded—an intense specialization. Where once the engineer was a glorified mechanic who built bridges and designed tools, now we have Ceramics Engineers, Textile, Cyrogenics, Vibration, Industrial Engineers. Even they are specializing: the Industrial Engineer might work in Time Study, Methods Analysis, Hospital Systems, Rate Setting, Operations Analysis; and a few are even doing work in Housekeeping Management!

So, it is not a question of *if* to use consultants, since your mere reading of articles demonstrates that you already do, but rather, *when* and *whom* to use.

The difference between a consultant who makes a charge for his time and the one who doesn't, is the same sort of difference you find between a profession and a hobby. Take the case of the registered engineer who consults. He has said that he legally puts his license to practice on the line with every recommendation, and with every contact with his client or his client's people, just like a lawyer or a C.P.A.

When is it time for you to consider paying for your consultation? Here are some possibilities:

1. When your organization, and you personally, are strong enough to implement recommendations, and to sustain improvements; or

2. When you are growing fast, and need the equivalent of an experienced staff you can't afford to develop and keep full time; or

3. When you have made all the internal improvements you can, and feel the need of objective analysis; or

4. When you can't get from management and other departments the interest and support you must have to function effectively; or

5. When you want to improve morale with a demonstration that management is willing to invest in a program of motivation and training.

And whom do you use? When you need professional help (such as a doctor or lawyer), you look neither for the cheapest (whose advice you would probably never follow anyhow) nor the most fashionably expensive. You look for the one who can best help you at a reasonable cost.

You should not be interested in "sharing the savings"; if your doctor made your two sick children well, you wouldn't give him one of them!

Nor would you be interested in an "audit" in most cases: you wouldn't ask your doctor for a five-minute examination not to exceed $10, and expect to be happy with the results, if you had anything other than a head cold. The "audit" type of service often provides too shallow an investigation to make a serious determination of the possibilities for cost and quality improvement.

The mixing of consulting with product sales and contract cleaning activities limits the objectivity with which your condition can be evaluated.

The consultant you choose must live and act a business and professional philosophy that is consonant with your own. Here is the way in which the client-consultant relationship should evolve:

After interest is shown by a potential client, a proposal should be developed to assist the client in meeting his objectives. In order to make a proposal and provide a later service that is objective, contract cleaning services should not be provided by the consultant, nor equipment or chemicals sold, nor such sales or services by others promoted. Thus, the consultant may objectively serve:

1. Clients who do their own cleaning.
2. Clients cleaned on contract.
3. Contract cleaners.
4. Government.
5. Organizations and institutions.

The fee must be calculated as a direct function of the engineering time involved, plus expenses, rather than "what the traffic will bear." Options might be offered concerning implementation services, and later "on call" services.

Again, sharing the savings is undesirable: quality measurements are often obscure; some things do not show up for a while, and the consultant cannot be objective.

We feel the consultant's services should comprehend these factors:

1. The program should be operable with existing supervision and labor wherever possible.

2. The consultant should work himself out of a job within a reasonable time; usually after one year of implementation.

3. All assignments should consist of a reasonable day's work.

4. The service should benefit, wherever possible, all three levels of personnel involved: management, supervision and custodial help.

5. Use should be made of the "work design" concept of creative analysis, rather than specific criticism.

Who needs consultants?

You don't—if you are able to cope with the information explosion unassisted;

AND—if your lines of communication to management are without obstruction;

AND—if management is confident in the objectivity of your recommendations;

AND—if management understands the balance between housekeeping quality and cost, and has given you sensible objectives;

AND—if you have a static organization, and plenty of time to make changes;

AND—if you are able to train and motivate your supervisors and workers adequately and regularly;

AND—if you are able to stimulate management to act on your recommendations;

AND—if you have been able to develop the proper status and compensation for your department and its members;

AND—if you have had broad experience in *all* phases of the work, in many different organizations and locations.

A manager, by definition, is a person who gets things done through others.

*You* turn to *your* consultants for help, not because you are in any way inadequate, but simply because you have enough understanding and self-confidence to be a good manager.

Who needs consultants? The good manager.

**Comparing Housekeeping Consultants**

The consultant you use for your custodial department should be able to answer "yes" to all of the following questions:

1. Does housekeeping consulting represent the fundamental service provided by the consultant, rather than a sideline?
2. Is the work for each client directed by registered professional engineers?
3. Is the consultant entirely independent of a supplier and contract cleaning organization, without interest in such a firm?
4. Is the consulting firm owned entirely by its own consultants, without outside interest?
5. Are fees based only on time involvement by professional consultants?
6. Are services and fees specific, eliminating the need for additional services and fees?
7. Has the consulting firm worked at least ten years in this specialized field?
8. Do the personnel of the consulting firm spend full time in professional consulting activities?
9. Does the consulting firm broaden its fund of experience through seminars and other educational activities?
10. Does the consultant provide a comprehensive reference list of work done within that firm?
11. Has the consultant contributed to the literature of housekeeping management?
12. Does the consultant completely avoid the use of stop watches on the client's premises?
13. Does the consultant avoid the use of short-interval scheduling, relying rather on the development of reasonable work loads?

# 13

# Selecting a Training Consultant

If you became involved in a situation dealing with maritime law, there is no regulation that says that you must use an attorney, or get help from anyone for that matter. You could handle it yourself, but you would probably find that the amount of time it took to research the field and to "re-invent the wheel" would be far too expensive in terms of your own time and money, as well as yielding results that may not be as good as you desired.

Or, if you wanted to patent an invention, there is no law that says that you must use an attorney to help you – you could handle it yourself. If you did choose a lawyer to help you, it would not be a divorce or tax lawyer, it would be a patent specialist.

The analogies refer almost directly to the question of using a consultant to assist you in developing a Custodial Training Program. You simply want help because the work will be done more quickly, and considering the value of the time of your own staff (which one must consider!), probably less expensively—and the input will involve a lot of experience that would make things go more smoothly.

In selecting a training consultant, you would look for some of the same things that you would in selecting a physician:

1. Does his field of specialty relate to your situation? You want an orthopedist to handle a bone problem, and you want a custodial consultant to handle your housekeeping training situation.

2. Does he have the depth of experience on which you can rely? Experience gives one a practical feeling for what can be done, how quickly, and how well it is received by the workers.

3. What is his reputation? Is he known by a number of people who can provide references relating to the quality of the work?

4. Does his personality fit yours? That is, is there agreement in terms of objectives, limitations, and approach?

5. Is the fee charged for the service reasonable?

6. Does the specialist remain independent of influence by other organizations?

The above questions, while useful in considering the selection of a consultant, are not the only factors. For example, if the firm to do the work is only a year or so old, it may not last out the duration of your project. Preferably, you should look for an organization that would be at least five years old, and therefore stable enough and with enough backup to assure you of continuity.

Considerations in the selection of a consultant for training would be the same as if you were making a complete analysis and recommendation for the building sanitation operation (you may wish to consider this, since training has an inter-relationship with other functions, such as the method of job assignment, procedures for cleaning, the equipment and materials involved in procedures, etc.). Actually, most consultant work in housekeeping is performed by general management and accountant consulting firms, whose personnel know litle about this subject. Many are the complaints that the housekeeping supervisory personnel had to "educate" their consultant about what it was that he was studying! It is not enough for a training consultant to know how to take pictures – he should know the cleaning business, so that he can take meaningful pictures, and those which workers do not recognize as "fakes." A training consultant should be able to answer test questions that you might pose to him, such as: "What benefits do the new high-speed floor machines offer in spray buffing with metal-interlock polymers?" Or, "What are the drawbacks of vertical scheduling in a building using area assignments augmented by project teams?"

### Objectives

The objectives of a custodial training program would include:

1. Safety to custodians and others alike.
2. Protection of health.
3. Conservation of capital investment through surface protection.
4. Improved morale of all building occupants.

5. Efficiency of housekeeping performance.
6. Development of the relationship between supervisors and workers.

The best means of obtaining all of these objectives is with a personalized program developed for the individual organization – utilizing pictures of workers in that facility using the equipment and supplies made available (of course, the consultant should advise on additional or different types of equipment and material which would improve efficiency). A "canned" training program would typically show other types of people doing work in other buildings in different ways using the wrong equipment, and one spends a great deal of time "apologizing" for the information which is being given to them that does not relate to their situation.

These are the steps to follow in the development of a training program:

1. Determine standard procedures to be used for the various cleaning activities (this would invariably involve a determination of equipment and chemicals as well).
2. Determine the form of visual material. Experience indicates that slides provide great flexibility in making changes. They are available at a reasonable cost, and lend themselves very well to supervisory development, as well as to retention of information by the worker.
3. A script is prepared for shooting the pictures, based on the standard procedure, and include such items as make-ready, safety precautions, clean-up, etc.
4. Volunteer workers are selected to act as models (remember to obtain a release form).
5. At least twice as many pictures should be taken as are needed, so that only the best shots will be retained (in organizations with their own training department, this can be done in concert with them).
6. The slides are keyed to the scripts.
7. The consultant shows management and supervisory personnel how to properly utilize the slides, in addition to putting on some demonstration courses himself.

### Pre-conditioning

Actually, most custodial workers receive audio-visual training on cleaning activities daily. Millions of dollars are spent annually by some of the nation's largest companies to provide this training to workers, and it is given in a colorful, energetic, and enthusiastic way by some of our nation's best salesmen. I am referring to television commercials! But typically what we see on television, in terms of how to mop a kitchen floor or how to clean a bathroom sink, is quite different, if not the opposite, from the system of cleaning, as well as the chemicals and equipment involved, that we desire our personnel to use. (Television will show, for example, the push-pull system of mopping, which is very fatiguing and of limited productivity – and the copious use of scouring powder for restroom facilities and fixtures, which is abrasive and creates problems of its own.) A good example of the forcefulness of this training is the fact that a number of workers will bring products from their home to use on the job at their own expense!

Our training must be good enough and interesting enough not only to portray the proper system to the worker, and to show him its benefit, but to overcome the exposure he has had to television and other forms of commercial advertising.

### Cautions

Sometimes it is difficult to sell management on the desirability of using a consultant for the development of a custodial training program. Even in educational institutions, where everyone is receiving instruction, the custodial department may be the only group of people who are not! It helps to have good literature from the consultant to be able to show to management along with a proposal which has a fixed total fee (no "extras"). Training is an exercise in selling, and the consultant should be skilled in salesmanship, so he can be used to assist you in selling this idea to your management.

Do not expect results too quickly with a training program. Many workers have a "wait and see" attitude and may not respond right away until they are truly convinced that the purpose of the program is to help them find easier, safer ways to do the work (which it most certainly should!). Repetition and continuity are of utmost importance, as is professionalism. It is often impressive to sanitation workers that management has obtained a consultant to work with them. It helps to convince them that they are doing important work. So does the printing of an attractive certificate that accompanies completion of the first phase of the training program. Custodial training should be a continual activity, and should not come to an end after a few months or even a year – this would simply convince some of the workers of what they might have suspected all along – that management was not serious about this.

The eventual evaluation of the results of the program would be made in terms of these factors:

1. Improvement in the absenteeism rate.
2. Improvement in the turnover rate.
3. Improvement in the quality of worker performance.
4. Improvement in the worker-supervisory relationship.
5. Improvement in supply costs (there may be a decrease in chemicals con-

sumption, if it has been wasted in the past, or a legitimate increase, if work is being done at an improved frequency).

6. More attractive custodial closets and cleaner equipment.
7. More presentable-looking employees.
8. A smaller number of complaints.

Yes, a training program combining the experience and expertise of a specialist consultant with your own staff will definitely be a good investment!

# 14

# Energy Conservation and Housekeeping

ENERGY conservation in general is a subject that has become very important and will become ever more important in the years ahead. It has ramifications in all aspects of building operation (and life in general, for that matter), with specific applications to custodial operations.

## Night Versus Day Cleaning

An apparent obvious opportunity, and also a controversial question, concerns daylight versus night-time cleaning. It is an enticing thought to change cleaning from an evening or night shift to a day shift operation, to save both lighting and other energy costs, such as heating and air-conditioning. As with many other subjects, there is no "pat" answer–it must relate to the facility in question. Here are some factors to consider:

1. If the major consideration is saving on lighting, then one is only talking about roughly 15% of the total energy cost for a typical building. There are other ways to save lighting than day cleaning, and these are discussed below.

2. For air-conditioning and heating, during custodial activities, the same intensity is not required as for clerical and other administrative functions. Also, consideration should be given to the overall operating hours of the building, and the cost of shut down and start up as compared to continuous operation, with perhaps a reduction of heating or air-conditioning during the night hours. The use of plastic window covering material, blinds and curtains, in-

sulation, and other energy saving systems might do far more than changing cleaning hours.

3. Public opinion is an important factor in the night-time lighting of a building. With all the publicity and pressures on the citizen to save energy, he becomes irritated, if not incensed, at the sight of a building ablaze with light at night-time. Window coverings (mentioned above) help to solve this problem, and the future will undoubtedly see many more windowless buildings, which would solve both energy, the cost of window cleaning and re-glazing problems as well. But the basic solution, if night-time cleaning is desired, is in "vertical scheduling", and/or part-time work, both discussed below.

4. Unquestionably the efficiency of building cleaning is considerably reduced when daytime custodial work is performed. Most estimates (including the experience of our firm consulting for many facilities) indicate a loss of efficiency in the range of 20% to 30% for work done during the day. The reasons include:

- Workers often have to do partial jobs because of the traffic of other personnel, such as mopping half of a corridor at a time, which must be chained off, and even then people will track through it, causing additional work.
- Very often a cleaning worker will have to visit an office or area several times before he can obtain entry—he might have been barred from his work by a conference going on, the user of the office making a long distance call, the exposure of confidential documents and the like.
- The custodian is often called away from his work to perform other duties, and in this sense may be confused and demoralized by having "too many bosses".
- Where the work is done on an evening or night shift, a greater security against vandalism and theft is provided, and this is lost if all cleaning work is done during days.
- Even those organizations which attempt to do all cleaning during daylight find that it cannot be done, and certain work is still scheduled for nights, such as shampooing carpets, stripping and refinishing waxed floors, and the like.
- A significant proportion of cleaning work is performed by "moonlighting" personnel and some of our best motivated labor will be lost if we avoid evening shift work, since their second job would be naturally a daytime job in most cases.

5. Housekeeping work during the day is a more hazardous activity to the building user, because of wet floors, projecting mop sticks, the use of floor machines, exposure to chemicals and similar items.

### Conservation of Lighting

Where cleaning is done on the day shift, it would be well to have a part of the time available where there is no conflict of activity, such as having the cleaning crew start two hours before the other work force, or end two hours after, so that wet cleaning could be done in corridors, some project work completed and the like.

Conservation of lighting energy can be accomplished in numerous ways, in addition to changing the cleaning work shift. Some examples:

1. When light fixtures are kept clean and free from dust, they may give out as much as 25% or more illumination, since particles of dust and other soils tend to absorb light. It is also important to keep the reflectors not only clean but painted an appropriate reflective color. Light bulbs and fluorescent tubes themselves need to be cleaned if they are to last a significant period of time. (Caution: Do not clean incandescent bulbs while they are burning, as they may break and injure the worker.) Some fixture cleaning can be done from the floor using a vacuum (such as pack vacuum) with extension wand and bristle brush attachment, while other work has to be done through washing in place or in a fixture cleaning solution tank. (The last case should be reserved only for exceptionally difficult cleaning, such as in an industrial situation with sticky and greasy soils.)

2. The intensity of illumination required for clerical work is not necessary for custodial activities, and a switching arrangement that permits every other row of lights to remain burning for cleaning work is perfectly satisfactory, and immediately saves 50% of the lighting energy.

3. Further, the extent of illumination may be reduced in general through the use of lower-wattage bulbs and tubes. Rather than 40-watt four foot flourescent tubes, consideration should be given to the use of 35-watt tubes (with necessary fixture adjustments).

4. The use of reflective surfaces in buildings provide better general lighting since light is absorbed by dark colors. Examples of this are the use of semi-gloss enamel wall paints, light-colored plastic wall coverings, light-colored vinyl asbestos floor tiles and terrazzo floors, and the lighter colors of ceramic floor and wall surfaces. These selections should not completely negate the advantages of design for maintainability, such as the need for other than a white grout between ceramic tiles.

### Part-Time Cleaning

Many organizations (including probably the majority of contract cleaning firms) do cleaning with part-time personnel, often from 6 to 10 P.M. or similar hours. These hours permit a worker, having a daytime job ending at 5:00, to arrive at work after having a bite to eat, and still get home and obtain a reasonable nights sleep.

From the energy conservation standpoint, it means that the lights of the building are only on during those hours. In many locations and also at certain times of the year, there is still enough daylight that little interior lighting would be required until the latter part of the work shift.

Where an organization is cleaned with full time personnel, it is possible to make a transition to part-time personnel through transfers to other departments, and switching over floor by floor, based on normal worker attrition due to retirements and other terminations.

### Vertical Scheduling

It is generally accepted that custodial work is more efficient where the custodian is assigned to a given geographical area, and is responsible for all repetitive work in that area. (Project work, such as stripping floors and washing walls, is performed by a team trained and equipped for that purpose, to augment the area-assigned workers activity.) In a hypothetical case, where a four-story building is maintained by four workers, it is natural to assign one cleaner to each floor. In this case, where the cleaning is done at night, all the lights are always on. This is called horizontal scheduling.

An alternate to this is vertical scheduling, where the assignments are made from top to bottom of the building rather than from side to side. In the case above, the four workers would each be assigned one fourth of the top floor for his initial activity—they would not be working as a team or gang, but would complete the repetitive work in, let's say, the north wing of the building for one worker. Perhaps at break time, the lights for that floor would be turned out, and after break the lights for the next floor down would be turned on. This tends to help pace the work without getting involved in the distastefulness of short-interval scheduling.

Thus, with vertical scheduling, in the four-story building in question, only one floor is lit at a time. The same principle could apply to a building of any size, with a greater or lesser number of people working in area assignments on a given floor, depending on the size of that floor. It is possible that a twelve-story building might only have two floors lit at a time.

As with everything else, there are pros and cons. A spin-off advantage of vertical scheduling is better personal protection (such as from an accident) since fellow workers are near by—but this very proximity can lead to certain problems of over-socializing.

Other values of using vertical scheduling with several personnel on the same floor of a given building at the same time, include:

1. The personnel feel much safer from a possible personal attack.
2. It is possible to use a working group leader very effectively in this situation, since he would be responsible for all personnel on the floor, under the general direction of a foreman or supervisor. The group leader, in order to have time available for this, would be assigned approximately a 15% lesser time responsibility than the other workers.
3. Similarly, it is easier to supervise people in this fashion, since they are all on the same lighted floor.

The benefits of vertical scheduling far outweigh the shortcoming, and these can be handled by proper supervision, which in itself is simplified by this technique.

### Combined Activities

Where the advantages of several approaches to energy saving can be obtained simultaneously, such as with vertical scheduling, part-time cleaning, reduced lighting intensity and the like, it is clear that there is no mandate for performing cleaning work only during the day, and suffering consequently a considerable reduction in performance. The additional pay for shift differentials for evening work have been shown to be an excellent investment in terms of productivity.

# 15

# Job Descriptions

JOB Descriptions are considered by some people to be unnecessary red tape, and also a waste of time and good paper. Yet, properly prepared and properly used, they can be valuable working tools.

Job Descriptions for "cleaning and maintenance" workers, like all Job Descriptions, serve one or more of these purposes by defining the nature of the work to be done:

1. Information and instruction for the worker.
2. A reference for supervision.
3. An aspect of a union or Civil Service Agreement.
4. A basis for compensation.
5. Provision for continuity in case of turnover.
6. A requirement by some governing or accrediting body.

But beyond the above benefits, there are other advantages of having job descriptions:

1. They assist in developing compliance with government regulations, such as concerning equal pay and equal opportunity.

2. There may be a benefit in fire prevention and safety, as the Job Description can provide information that indicates possible hazards.

3. Job Descriptions indicate to the worker management's interest in their position, and a display of management interest is always of great importance to service workers who sometimes consider their status low level.

4. Job Descriptions are useful in planning training programs, as they indicate the type of information that needs to be relayed to the worker.

5. For the industrial engineer or methods improvement specialist, the Job Description may trigger thoughts on improved efficiency.

6. Vocational counselors rely on job descriptions to advise workers, such as those who are handicapped, disadvantaged, or inexperienced.

7. Finally, and this could be most important, in some cases Job Descriptions may improve employee relations by providing a basis of understanding concerning job responsibilities, and thus may limit grievances relating to them.

Job Descriptions may be confused with other activities, such as job assignments. There may be some overlap, but the assignment describes in general the amount of work to be done and specifically where it is to be done, while the Job Description tells of the nature of the work.

It should go without saying that the Job Description cannot be used to circumvent the law. One good example is the equal pay law—some organizations have tried to defend a discrepancy in pay between men and women doing essentially the same work by referring to Job Descriptions which separate these—but the equal pay law not only supersedes Job Descriptions, it supersedes stronger instruments as well, such as union contracts, Civil Service agreements and state laws.

There are many forms of Job Descriptions, but in one order or another they typically include the functions of the job, the type of work to be performed, equipment and materials to be used, may also contain a description of job factors and therefore perform functions of a job evaluation as well as description.

The factors most often considered are these:

1. Educational background required.
2. Training requirements.
3. Mental effort.
4. Visual effort.
5. Physical effort (and dexterity).
6. Responsibility for the material or product.
7. Responsibility for tools and equipment.
8. Responsibility for the safety of others.
9. Work surroundings or conditions.
10. Unavoidable hazards.
11. Effect on the activities of others.
12. Effect on the conservation of physical facilities.
13. Initiative requirements.

In preparing Job Descriptions, it is important to analyze the "job" rather than the "worker". It is a common mistake to keep in mind an especially capable or successful worker in preparing Job Descriptions, who does not represent the typical worker for the position.

To their sorrow (usually felt after conditions have changed) some organizations attempt to make a Job Description complete and fully detailed. This leads to the often heard triumphant cry made by some workers on being instructed to perform a duty, "but that's not in my Job Description!". The workable Job Description is somewhat flexible, and should typically conclude with a statement such as "the worker will perform other duties as assigned by his supervisor from time to time." Some people prefer to have the wording "other related duties."

The Job Description should also instruct the worker to comply with any assignment, if it is in the nature of an emergency as expressed by a supervisor or member of management. The emergency might relate to personal safety (such as water on a finished floor), damage to surfaces (such as water on carpet), etc. The rule is that the job should be done first, then any questions asked or appeals made later.

Job Descriptions should not be taken lightly—they are an important working tool, and can be harmful or helpful. Each organization should develop its own (perhaps with help from others) and should not just "lift" a set used somewhere else. Nevertheless, the following examples may serve as a starting point.

## CUSTODIAL SUPERVISOR

I. Characteristic Routine Work

Definition:

This job entails the supervision of the housekeeping function for buildings and facilities. Employees in this job classification will report to a Shift Supervisor or Services Superintendent. This person will direct the daily activities of all housekeeping personnel assigned to a particular "Zone".

Examples of Routine Work:

- Designate shift and area work assignments for custodians and Group Leaders
- Frequent evaluation of the performance of all custodial personnel assigned under their jurisdiction
- Conduct frequent inspection of all assigned areas to insure that all work is being carried out in a proper and orderly manner
- Develop and maintain a current training program for all zone custodial personnel
- Conduct classroom-type and on-the-job training for all zone custodial personnel

- Establish and maintain inter-departmental relations with personnel, maintenance, and purchasing
- Evaluate and assist in the selection of new cleaning products, materials, and equipment
- Assign relief custodians to fill in for absentee workers
- Maintain all pertinent records, reports, and other paperwork as needed
- Develop and maintain a projects list for the Project and Relief crew
- Provide a consultation service for management when new construction is planned
- Perform all necessary duties relating to student registration
- Assist in scheduling and maintaining space for conference groups and other special functions
- Perform other related work as required and instructed

II. Qualifications

Required Knowledges, Skills, and Abilities

- Knowledge of all cleaning methods, materials, and equipment
- Ability to instruct housekeeping employees in the performance of manual tasks requiring moderate strength
- Must possess a thorough knowledge of the operation of all mechanical cleaning equipment
- Must be able to schedule repairs for all above equipment
- Ability to prepare written reports, and other correspondence relating to the job as necessary
- Ability to establish and maintain good working relationships with subordinates, as well as with other supervisors and management personnel

Minimum Training and Experience:

High school diploma or equivalent. Any training and experience gained through attendance at special schools or seminars desirable. Two years experience in custodial or general maintenance work–of which at least one year spent exhibiting a proven ability to supervise and lead relatively large groups of people.

## GROUP LEADER

### I. Characteristic Routine Work

Definition:

This is routine manual labor associated with the daily cleaning of buildings. Employees in this job classification are under immediate supervision and perform various duties requiring moderate physical strength. Typical routine duties include, but are not restricted to: dust mopping, sweeping, wet mopping, vacuuming, dusting, waxing, cleaning restrooms, removing trash, polishing furniture, and simple operation of basic mechanical cleaning equipment.

In addition to the above listed tasks, this person will be responsible for: conducting on-the-job training for custodians; ordering and distributing supplies and materials; relaying instructions from the Custodial Supervisor; assisting the Custodial Supervisor in employee evaluations; determining the need for special project work to be performed; leading a group of custodial workers and participating in non-routine work listed below, scheduling set-ups for special functions.

Examples of Routine Work:

- Police public areas
- Dust mops, sweeps, wet mops, spray-buffs floors
- Vacuum clean rugs and carpets
- Scrub and clean restroom fixtures and keep supplied with towels, soap and other items
- Dust and clean walls, doors, windows, woodwork, and other above floor surfaces
- Dust, clean, and polish furniture
- Clean elevators and stairwells
- Keep simple records and make simple reports
- Police around outside of buildings
- Notify Supervisor of equipment and facilities needing repair or replacement
- Replace accessible light bulbs
- Perform related work as required and instructed

### II. Characteristic Non-Routine Work

Definition:

This is manual labor associated with the recurring, but non-routine cleaning and maintenance of buildings. Employees will be under immediate supervision and will perform various tasks requiring moderate physical strength. This work is normally of the project-type nature.

Examples of Non-Routine Work:

- Stripping and refinishing floors
- Scrubbing floors with machines
- Vacuuming furniture and drapes
- Window washing
- Wall vacuuming and washing
- Wall, vent, and radiator washing
- Light fixture vacuuming and washing
- Waste receptacle washing
- Various duties related to student registration
- Snow shoveling and removal
- Carpet shampooing
- Related work as required and instructed
- Set up rooms for conferences, lectures, and special events
- Move furniture

III. Qualifications

Required Knowledges, Skills, and Abilities:

- Knowledge of cleaning methods, materials, and equipment
- Ability to perform manual tasks requiring moderate physical strength
- Ability to operate mechanical cleaning equipment
- Ability to follow simple oral and written instructions in the conduct of the job
- Ability to establish and maintain working relations with Supervisors, fellow workers, and students
- Ability to conduct simple inventories of cleaning supplies and products
- Must be capable of filling out requisitions and related forms
- Must possess the necessary communicative skills needed to relay instructions from the Supervisor

Minimum Training and Experience:

At least two years experience in performing cleaning work and other related manual labor—training and experience as may be gained through attendance of schools and seminars—must have demonstrated, through actual experience, the willingness and ability to give exceptional job performance.

**CUSTODIAN**

I. Characteristic Routine Work

Definition:

This is routine manual labor associated with the daily cleaning of buildings. Employees in this job classification are under immediate supervision and perform various duties requiring moderate physical strength. Typical routine duties include, but are not restricted to: dust mopping, sweeping, wet mopping, vacuuming, dusting, waxing, cleaning restrooms, removing trash, polishing furniture, and simple operation of basic mechanical cleaning equipment.

Examples of Routine Work:

- Police public areas
- Dust mops, sweeps, wet mops, spray-buffs floors
- Vacuum clean rugs and carpets
- Scrub and clean restroom fixtures and keep restroom supplied with towels, soap and other items.
- Dust and clean walls, doors, windows, woodwork and other above floor surfaces
- Dust, clean, and polish furniture
- Clean elevators and stairwells
- Police around outside of buildings
- Keep simple records and make simple reports
- Notify Supervisor of equipment and facilities needing repair or replacement
- Replace accessible light bulbs
- Perform related work as required and instructed

II. Characteristic Non-Routine Work

Definition:

This is manual labor associated with the recurring but non-routine cleaning and maintenance of buildings. Employees will be under immediate supervision and will perform various tasks requiring moderate physical strength. This work is normally of the project-type nature.

Examples of Non-Routine Work:

- Stripping and refinishing floors
- Scrubbing floors with machines
- Vacuuming furniture and drapes
- Window washing
- Wall washing and vacuuming
- Wall, vent, and radiator washing
- Light fixture vacuuming and washing
- Waste receptacle washing
- Various duties related to student registration
- Snow shoveling and removal
- Carpet shampooing
- Related work as required and as instructed
- Set up rooms for conferences, lectures, and special events
- Move furniture

III. Qualifications

Required Knowledges, Skills, and Abilities:

- Knowledge of cleaning methods, materials, and equipment
- Ability to perform manual tasks requiring moderate physical strength
- Ability to operate mechanical cleaning equipment
- Ability to follow simple oral and written instructions in the conduct of the job
- Ability to establish and maintain working relations with Supervisors, fellow workers, and students

Minimum Training and Experience:

Such training and experience as may be gained through schools, some experience in performing cleaning work and other manual labor, or any equivalent combination of training and experience.

# 16
# The Quality-cost Balance

STANDARDS of quality and cost of custodial maintenance are inevitably interwoven, yet each encompasses a number of different factors which can be considered separately.

## Who Establishes Standards?

In any environment, there are four groups of people who are competing, albeit perhaps unwittingly, to establish housekeeping standards: (1) management, (2) occupants, (3) the custodial employes responsible for performance of housekeeping work, and (4) those providing funds for all purposes including maintenance. The last group seldom involves itself directly in questions of maintenance standards.

In actuality, the occupants do little to establish the standards of housekeeping to be maintained, but they certainly help to establish the difficulty or ease with which standards are achieved. The degree to which this group cooperates can be controlled to a significant extent by the attitudes of both administration and custodial personnel. Personnel hired to perform custodial work should not be left to judge the standards of maintenance and performance.

Basically, then, the establishment of maintenance standards must fall to management; but this means that the superintendent of buildings and grounds, plant engineer, physical plant director, or the parallel individual of whatever title, has the principal responsibility for recommending and justifying to his superiors the appropriate standards of housekeeping maintenance. Let us examine the factors with which he is concerned.

## How Are Quality Standards Set?

Various individuals have invoked such instruments of measurement as radioactive tracer elements, fluorescent components, and light reflectance; but no one nor any combination of these technics has served to provide over-all quality criteria. Thus, it becomes necessary for standards to be set through words, demonstrations and other non-objective approaches. The most basic level of housekeeping maintenance which could be tolerated would be one providing at least adequate hygienic protection for the inhabitants of the environment. This could scarcely be accepted as sufficient, however, for added to it must be considerations of protecting property—a tremendous and hard-won capital investment—and protecting the desired public image. Finally, the level of housekeeping maintained by any organization cannot fail to affect its basic mission whatever it may be.

Assuming that a given housekeeping job is performed adequately each time, the quality of cleanliness maintained will obviously depend upon the frequency with which the task is repeated.

Thus, for example, the maintenance of acceptably clean desks and chairs in a classroom will depend primarily upon the frequency with which that furniture is dusted. Of course, a corollary rule is that the frequency necessary will depend upon the rate of soiling.

In any given environment, it would of course be possible through experimentation to determine the frequency with which the elemental tasks must be repeated in order to produce a given quality of result. It is much simpler, however, if the experience of some individual or group can be drawn upon to establish a frequency likely to give satisfactory results, which frequency may then be varied as observation indicates.

## How Is Quality Translated to Cost?

We are constantly reminded in our simplest daily negotiations that in this life we pay for what we get. It is true, therefore, that if we seek reasonable quality standards for housekeeping maintenance, we must expect to pay a reasonable price for the goal. In their efforts to determine whether they have received their money's worth, some make the mistake of attempting to compare directly their own housekeeping costs with those experienced by someone else.

*It is futile to attempt a direct comparison of cost per square foot between two different environments when it is impossible to standardize all other factors.*

In the first place, the quality levels sought or actually achieved by two different facilities are never identical. In the second place, there are seldom enough factors taken into consideration to make possible a valid comparison.

At a meeting of the Environmental Management Association, a member of the cost study committee of the New York Chapter presented the results of a study comparing the costs in three different office buildings. The conclusion of the committee was that it was wholly impracticable to make a meaningful comparison of the cleaning costs as such. There were too many variables involved such as labor rates, fringe benefits, supervisory overhead, and so forth.

The committee did feel that a comparison could be made of the annual labor hours required to clean one square foot of a specific type of area; but even here an analysis of the report indicated material variations in the performance factors that would directly affect the quality of the work. The acceptance of so-called average per-square-foot costs as a standard makes no more sense than the acceptance of the "average shoe size" as being the proper one to buy.

If we accept the formula that quality equals essential tasks times the frequency with which these tasks are performed, we can readily see that the addition of one more factor will give us work cost. In other words, cost equals task performed, times frequency, times cost per performance. Of course, cost per performance can further be broken down into time consumed, times wages per unit time.

We have thus established a beginning point for the formulation of cost standards. In other words, we are ready for job measurements or the engineering approach to cost control in housekeeping. The trouble is that we have now reached the point where the unwary frequently are lured into the pitfalls of attempting shortcuts and over-simplication.

## When Is Equipment Too Old?

In the provision of equipment for custodial maintenance, management is frequently reluctant to abandon equipment already on hand which, ostensibly, has considerably more useful life. Where these considerations arise, the following rules can be helpful in the achievement of a sound decision:

1. Consider only present and future requirements, not past requirements.
2. Do not consider book value, but rather use real worth of the items in the present program.
3. Consider present worth for capital recovery; in other words, consider trade-in value, or resale value.
4. Consider likelihood of better equipment becoming available in the future. In some situations, this may lead to the decision to delay purchasing.
5. Cost out labor requirements for each alternative. For example, consider the cost of doing the job manually versus the capital cost of mechanized equipment, plus the cost of labor which will still be required to operate that equipment.
6. Be sure to consider the productive capacity of each method under consideration.
7. Include the interest cost in considerations of capital expenditure.
8. Consider whether there is any factor, such as tax savings, which might dictate present expenditure, in addition to the saving which would be experienced through using the equipment.
9. Convert all costs into common terms, such as annual cost, total present cost, cost per item.

## Other Factors

There are important factors other than square footage that will affect the actual maintenance job.

These include the amount of congestion, the amount of traffic likely to bring about soiling or to interfere with the cleaning operation, the cleanability of surfaces, and the number of fixtures or the amount of furniture present. Some of these factors are measureable only within varying degrees of accuracy. The over-all job still will be affected by the peculiarities of the situation. By this we mean the distance of custodial closets and sink facilities from the area of work performance, the convenience of transporting tools and equipment, the existence of soil factors such as boiler and incinerator chimneys, or limitations such as the availability of only male or only female employes.

## How Are Time Standards Set?

How do we actually arrive at a reasonable time standard for the basic steps or tasks involved in a given job? Obviously, this can be done through the use of the stop watch, but a single-performance timing is valueless if we are seeking a reasonable average time and not the time required by the best or the worst worker with whom we are likely to deal.

We must seek performance through random sampling. This would mean a large number of timings of a given operation performed by many different persons. Our sampling must be broad enough to ensure its validity. Taking into consideration a given essential job and the desire that our observations be dependable, the number of samplings will have to be increased in inverse ratio to the length of time required for performance of the task. In other words, the measurement of a small unit is much more critical than the measurement of a very large unit.

Each task should be reviewed from the standpoint of determining whether it is being efficiently done.

Any cleaning task is likely to involve the transportation of equipment, the make-ready of solutions, time lost through waiting for areas to be cleared of traffic, and the removal of materials and equipment to another area upon completion of the task itself. Thus, the analysis of any over-all job may be a very complex procedure, and might even involve the use of the flow process chart technic. If an effort is being made to change the manner of performance of a task, an approach involving time measurement units for each element of the task may be needed.

## Are We Doing the Right Jobs?

When the time analysis has been completed, it is desirable to look critically at the performance system to be employed in actual practice. This involves a "work design" concept, as described by Dr. Gerald Nadler. Let the work time involved in performance of the task be represented by the horizontal distance between the upright sides of a triangle, and the base line of the triangle represent the existing work requirement.

Theoretically, we might desire to eliminate labor costs completely, through automation. In our graphic work design concept, we would have reached the apex of the triangle, where work time is zero. Obviously, we can scarcely hope to reach this ideal system. We may not even reach an ideal workable system, but we'll settle for one much better than the present or the usual approach as being the most practicable system. Once we have analyzed our work load and determined that it can be theoretically carried by a given number of persons, we are ready to establish our manning needs. In most instances, unless we are dealing with a completely new organization with no existing manpower, we are faced with the problem of adapting somewhat less-than-ideal personnel to the job at hand. For this reason, we will probably start out by setting down the names of the people available, determining what they are actually doing (so that we can later make a comparison between that and what we wish them to do), and recognizing any limitations inherent in the potential performance of those individuals.

## Organizing to do the Work

It is a truism that no organization is likely to function better than the supervision afforded it. Thus, it is desirable to look at the job at hand and to determine how we can best use the supervisory personnel available and what others will be needed.

We are now ready to establish our working organization, remembering that it should extend down all the way from the top point where policy affecting sanitation maintenance is originated. We should define clearly the lines of supervision, and, finally, we should indicate exactly the assignments of all positions in the organization. The individual duties should be spelled out to include the area of responsibility, a full statement of the tasks to be performed, and description of the tools available for use on the job.

An essential part of the job description is an indication of the frequency with which each task must be repeated.

While the work load for the average custodian will be made up of tasks to be performed daily or at frequent intervals, other cleaning operations must be repeated at less frequent intervals to assure maintenance of the selected level of cleanliness. These are what we normally term "special projects." Except in the smallest of organizations, there usually will be a special project team established, and it will cover this list of tasks on a rotating basis. The scheduling, of course, must be provided by the housekeeping supervisor. In the development of a program of this type, management or administration undoubtedly will be concerned about a cost comparison between existing or past organization and the new program organization.

## Area Cleaning vs. Team Cleaning

In the performance of work, the question frequently arises as to whether individuals should be assigned to do area cleaning jobs, or whether they should be assigned to teams that will perform certain functions throughout the entire establishment. In actual practice, both approaches have advantages, according to the particular task under consideration.

Area cleaning requires less travel time, pinpoints responsibility, tends to instill a proprietary attitude, provides for the handling of special problems, establishes measurable performance, tends to minimize socializing, limits personality conflicts, provides smaller staff in a given area, sometimes gives greater flexibility, does not place an upper limit on the rate of performance, sometimes reduces the need for supervision, and is compatible with the balancing of work loads.

On the other hand, team cleaning requires less equipment, provides for the use of specialization of skills, requires shorter training period, requires less make-up ready time, simplifies communication, standardizes methods, sometimes requires less supervision, reduces interference time because of ease of scheduling jobs, may reduce utilities cost since the entire team will be working in a single area, and may help achieve controlled rates of performance.

For some jobs area cleaning is to be preferred, while for others team cleaning is more advantageous.

We are all familiar with the advantages of the use of a small team for heavy scrubbing, or for cleaning and waxing floors. On the other hand, the crowding of a number of individuals performing a variety of tasks into a given area at one time does not constitute proper team effort.

Regardless of how the task of standardization is approached, it is not only a desirable undertaking, but it is truly essential to the attainment of administrative goals and control of maintenance costs.

# 17

# The Optimum Cleaning Department

DESPITE what many advertisements claim, there are no magic solutions, no shortcuts, to developing the ideal sanitation program. You can't "solve all your housekeeping problems in five minutes," eliminate all floor care difficulties with "a new magic wax" or use any other one or two pragmatic steps and expect to obtain the type of improvement that is possible in so many cleaning operations.

Rather, a comprehensive program must be developed encompassing all variables relating to sanitation. Thus, we find the optimum housekeeping operation has seen serious, continuing attention given to a large number of items, many of which involve management.

## Management's Responsibility

Management must first be acquainted—we can even say management must be "educated"—to the relationship that exists between *quality and cost* in housekeeping. Many executive housekeepers or supervisors have had the experience of having someone in management come to them and say "we must get this place cleaned up," with a resulting crash program of hiring additional personnel, letting supplementary cleaning contracts, performing overtime work, and the like, until things seem to look the way management wants them. Yet, some months later, that same manager may return to the housekeeping department with a demand that "you've got to save a good deal of money; we're spending too much for cleaning!" The fact is, in general, management has no method of determining just how much cleaning is obtainable for a dollar spent in this direction. It is up to the housekeeping head to prepare the figures to demonstrate, for example, that the cost of cleaning is *not* directly proportional to the production level or number of employees in an industrial plant, nor to the number of patients in a hospital, nor to the number of students in a school, nor to the number of guests in a hotel. True, there are relationships, but not *direct* proportions. Further, he must be able to demonstrate mathematically the quality effect of either an increase or decrease in dollars spent for cleaning, through the required changes in frequency resulting from fluctuations in man-hour provisions. Such information must concern itself with both daily cleaning, as well as the less frequent periodic cleaning. Where a fluctuating staff is unavoidable, it becomes the department head's responsibility to arrange the work so there is the least disturbance to the long-term quality result.

Then, management must define interdepartmental responsibilities. An example of this is re-lamping, where in a given organization the housekeeping department might be required to re-lamp a given fixture on one occasion, yet the re-lamping may be done by electricians on another occasion. In such a case, the responsibility is not fixed and frictions and complaints will occur. "Gray areas" of this type should be eliminated by written directives from management.

Of course, it is management's responsibility to serve itself as well as its housekeeping department by better planning for construction and renovation, so that new surfaces, fixtures, or equipment are not installed that are overly difficult to maintain, without making provision for this excessive maintenance cost. Unfortunately, architects cannot be relied on, for the most part, to show an interest in the maintainability of the items they specify. A review of the architectural guide books shows an almost complete absence of any information on this subject. Thus, management is required to provide the architects and interior dseigners with a check list of preferred surfaces, colorations, specifications, and the like. The potential savings here are impressive.

The housekeeping department badly needs official management recognition and support for the service that it is providing. It is where the cleaning operation is considered a "necessary evil," and where no management attention whatever is given to the subject, that we find not only a low quality level, due to demoralization of both the workers and the department head, but also excessive costs, such as might stem from damage to surfaces through neglect or indifference. A very positive way in which management can become involved is through the development of an interdepartmental cooperation and employee awareness program. Although such programs require a little thought, they involve only nominal amounts of money; yet the results can be significant. We have observed cases of cooperation programs securing benefits to a housekeeping department equivalent to the addition of several personnel to that department—and at almost no cost. Such a program would involve posters, circulars, letters to employees, articles in house newspapers. These are most effective when officially recognized by the top manager in the organization.

Management, and only management, can provide, further, a condition which can be called *organizational integrity.* Readers with navy experience know that *watertight integrity* means to avoid leaks, which if not controlled can lead to the destruction of the ship. We also want to avoid "leaks" to the cleaning department; we must not pull people out of this department to perform every other conceivable job that no other department wants to do or has no time to do. For example, a cleaning operator running a power sweeper may be interrupted from his work during a given day to assist in unloading a truck; then to deliver a package to another plant; then to run an errand for the plumber. These interruptions tend to prove to the custodian that, if he can be interrupted to handle anything else that comes up, then his cleaning job must not be important! If utility jobs are to be done, then utility perseonnel should be utilized for doing them; where cleaning work is to be done, then custodians should be used, and used for nothing else except in the very smallest of operations.

Management has the responsibility to consider contract cleaning as a legitimate alternative consideration, but not as a panacea or an "easy way out." Management owes its own department, then, the support it needs to develop and achieve its objectives, while making legitimate comparisons with the contract cleaning service.

The cleaning department should have its activities coordinated with other departments, such as grounds, and maintenance. The quality objectives in the one should balance those in the other. It is difficult to motivate a maid to perform a satisfactory policing job in an office area when she has but to look out the window to see grounds that are badly littered; the custodian cannot understand the need to remove spots from a wall that obviously requires plastering or painting.

Finally, management will never achieve its optimum cleaning objectives without a realization that housekeeping represents a legitimate *investment* toward the achieving of the organization's fundamental objectives, rather than the proverbial "necessary evil." Sometimes this requires "selling" management on the benefits of a sound housekeeping program:

1. Protects health
2. Promotes safety
3. Eliminates fire hazards
4. Improves morale
5. Avoids damage to materials
6. Extends surface life
7. Protects equipment
8. Improves productivity
9. Improves quality control
10. Provides good public relations
11. Brings optimum cost

## Organization and Scheduling

The basic requirement for any organization is fixed lines of authority and responsibility, wherein each worker knows to whom he reports, and each supervisor knows the limit of his authority. Where we have more than one supervisor able to give orders to an individual cleaner, we have a condition worse than where the cleaner would have no supervisor at all; he is constantly demoralized through conflicting instructions, and is soon looking for another job. A good test of this is to attempt to draw an organization chart, not representing the *desired* organization, but the actual *existing* conditions. If this chart is confused and complex, this can be a good clue to the need for reorganization.

The structure of the department, as well as all other aspects of its operation, should be developed through a well-ordered analysis of all variables, rather than through the simple attempt to improve a few facets of the operation. By far the best results will be achieved utilizing the synthetic approach, wherein a hypothetical development is made to solve observed problems and conditions, without reference as to how the work might have been done in the past. Then an attempt is made to install this ideal operation, making whatever changes are necessary because of personality problems and other conditions.

No organization can function properly without open lines of communication. The custodian and maid must be able to approach their supervisor —and go on to successively higher levels of management when they are not able to obtain satisfaction—with problems, observations, and suggestions. Here, we must refer back to management once again for the provision of a policy that will provide effective communications in both directions.

The department should perform its cleaning functions through a *preventive* maintenance program rather than a *corrective* one. If we wait until floors apparently need stripping before taking any action, for example, and this need is apparent to the public, we will have waited too long. Floors need stripping, then, immediately *before* they appear to need it, as judged by trained personnel. This approach requires a scheduled frequency of special projects, utilizing a calendar or data-handling system. Since special assignments and crash programs are often inefficient, the preventive program, while providing a considerably higher quality level, generally costs no more than a corrective system. Surely we want to avoid the operation of a department whereby one is sitting by a "hot line" telephone, waiting for complaints and urgent calls to come in, then sending out cleaning personnel to "put out fires"!

Our ideal housekeeping department will have an optimum number of cleaning workers; this means that if one additional worker were added to the payroll, his cost would exceed the benefit—direct and indirect—to be gained from his services, while, conversely, the reduction of the staff by one person would cost the organization more than his total wages, benefits, and support cost in terms of damage to surfaces, decrease of production, quality control, and the like. No serious consideration can be given to optimum number until all aspects of the productive work—method, equipment, chemicals, frequency, custodial facilities, and so on—are optimized. Most definitely we want to limit the work done by each person to a reasonable day's work over a reasonable period of the day.

Where cleaning workers will come to their supervisor and complain of an excessive work load, an analysis of the program often indicates that the complaint really relates to an *imbalanced* work load. When Henry, the custodian, complains "Boss, you've given me too much work to do!" what he probably means, possibly without quite realizing it himself, is "Boss, you've given me more work to do than you gave Jack!" Balanced work loads do not mean that each person should clean three rest room fixtures, twenty-three desks, and two and a half corridors; rather, it means that the effort required to perform each job assignment should be essentially equivalent, within the physical limitations of the worker.

In the scheduling of the work, consideration should be given to the use of both female and male housekeeping workers. The men provide advantages in being able to use larger equipment, performing high-up work, and providing a greater flexibility; while women can do a better job in office cleaning, and present fewer problems in terms of turnover and recruiting. The department should be structured, and the cleaning program designed, to utilize the proper balance of men and women dictated by the local labor market or pre-existing organizational limitations.

In staffing the department, attention should be given to the recruiting and selection procedures. It is as bad to hire over-qualified workers as it is to hire those who are under-qualified; in the first case, the workers will not remain on the payroll very long, and a considerable turnover will be experienced, while in the latter case the workers will not be able to perform properly. It becomes the responsibility of the department head to place limits on the type of people who will be hired into the department, within organization rulings, of course. It is far better to concentrate on improving the morale and productivity of existing workers, when the department is understaffed, than to hire people where experience indicates that they will not stay around very long. In staffing, consideration should be given to the use of part-time workers, special shifts for housewives who might work only four or five hours a day, mentally retarded personnel, and other relatively untapped sources of labor.

In scheduling work, it is first necessary to determine frequency, the basic determinant of cleaning quality. A carpet vacuumed daily will of course be at a higher quality level than one vacuumed weekly; although it must be remembered that the time requirement for each performance of a cleaning job often increases as the frequency is lessened.

Finally, the work should be performed at the proper time. Typically, this is during those hours when the facility is at its lowest occupancy level. In an office building, this should be a late evening shift; in an industry, the third shift; in a school, evening shift if there are no evening classes, but the third shift if there are; in a hospital, an evening or night shift for administrative areas, but the day shift for patient areas. The times must also be selected with a consideration for transportation problems.

## Supervision

The housekeeping supervisor is, inescapably, the backbone of the cleaning operation. Without him, we have little to work with. Although he is expected not to perform any of the cleaning work himself, naturally, it is his responsibility to motivate everyone else in the department to do the work in the way that management has developed it.

In the larger organizations, a pyramid of supervision is required. With a cleaning force of one hundred or more, for example, we may have a department head to whom several supervisors report; to each of them might report two or three group leaders, as well as all the actual cleaners. In a still larger organization we might have assistant department heads on each shift, or assistant supervisors reporting to each area supervisor. In smaller operations, a full-time department head, who is also the supervisor, is certainly justified for any department of ten or more personnel. For less than this number, the supervisor might also have additional responsibilities, such as security, grounds, maintenance, and the like. But even in a small operation, it is desirable to have some position of advancement for cleaners, such as to "group leader" or even to a position which might be called "senior custodian," which while not providing any authority, does give an additional compensation reward and title for exceptionally good work and service, and might also be used as a training aide for new workers.

The first requirement of the supervisor is a high motivational ability; this is by far more important than technical proficiency, experience, age, or any other such factor, since all of these things can be overcome in time, but without motivational ability the supervisor becomes a figurehead.

Experience, especially, must not be overrated. Much previous experience in housekeeping, in a typical case, involves methods that are now obsolete or will soon be obsolete, and a lengthy experience may merely indicate a repetitive history of use of these old techniques. On the other hand, a history of experience exchange is desirable, where the supervisor has taken the opportunity to visit other facilities, takes membership in such desirable organizations as the Environmental Management Association or the National Executive Housekeepers Association, has read materials and attended seminars on cleaning, and the like. Surely after such a person has been hired, this experience exchange should be continued and even intensified. So much is being changed in the cleaning field that it is relatively impossible to keep up to date within the confines of a single physical facility.

In time, the supervisor must become quite knowledgeable about the techniques of cleaning, not only in order to be able to schedule the work and estimate the time requirement for it, but also to be able to command the fullest respect of his workers. And, in this age of change, the supervisor could hardly be expected to move forward without a receptivity to change—not a toleration, but an *embracing* of new ideas.

## Methodology

It is clear that the worker does not instinctively find the best way to do a job; this is as true for the cleaning worker as it is true for any other type of operative employee. In the establishment of method, management has a simple choice: either to prescribe the method itself and see that it is utilized on a standardized basis, or to abdicate this responsibility and let the workers determine for themselves what is the proper method. It is the very same principle that applies to time standards: either management is going to determine the rate at which jobs will be performed, or, in the absence of this direction, the workers will determine it for themselves. This leads us directly to Parkinson's Law, where "work expands so as to fill the time allowed for its completion."

The utilization of effective methods can best be guaranteed when a theme of "let's work smarter, not harder" (assuming a reasonable effort) pervades all training and methods improvement. Naturally, the worker is principally interested in the effect of change on him as an individual, rather than such unimportant (to him) things as his institution saving a few thousand dollars a year, or his company making another few thousand. Fortunately, there are a large number of new cleaning techniques that provide higher quality with the same or a lesser effort.

There are times when team cleaning provides advantages. In the stripping and rewaxing of floors, for example, a team of two or three men (depending on the size of the area to be done) is more efficient than the same number of men working individually. But we cannot assume that team cleaning is best for everything, and try to use it on a complete basis for all cleaning functions, just because it works well in some areas. We must give similar attention to the advantages of job specialization (such as in rest room cleaning) and area assignments, where the workers can develop a proprietary attitude.

In making these assignments, it is extremely important to provide a number of personnel who will not be assigned on a repetitive basis to fixed jobs, but who are available for relief of absent workers so that each job assignment will be performed daily; and when not required for this relief, these special personnel will be able to handle project and utility work.

Finally, correct methodology must involve due consideration to custodial facilities, cleaning equipment, and sanitation chemicals, described below.

## Custodial Facilities

The first thing a construction contractor does when building any type of structure is to prepare a temporary housing for his office, personnel, and storage requirements; in other words, he provides his work facilities. In a housekeeping department, however, we may give little thought to the provision of custodial facilities, which are often after-thoughts when all other requirements have been taken care of. Thus, we find the central storage area in an inaccessible part of the basement, or the custodial closet under a back stairwell. Where adequate area is provided in original construction, it is often appropriated by other "more important" departments. With serious morale and motivation problems to begin with, it is doubly poor practice to require the custodian to do his work without giving him the means for the job.

The first requirement is a central storage area, where the bulk chemicals, large equipment and spare parts can be stored, and where such duties as dust mop treatment; lubrication, repair, and cleaning of equipment; venetian blind cleaning; and other items can be performed. Having adequate space and storage racks for purchase of chemicals in drums, as opposed to having to buy them in one gallon containers, can save twenty to thirty cents per gallon in purchase price.

The basic work station is the custodial closet—or "janitor closet," shortened to "J.C.," as seen on blueprints—wherein is stored the supplies which the custodian is to use in his assigned work area. Janitor closets are still being provided in brand new buildings with a typical floor area of ten square feet; this is supposed to take care of the utility sink, all chemicals, supplies, parts, mopping outfits, and motorized equipment! *No more than one percent of the architects in this country have any conception of how cleaning is being done in the buildings they design, and what provision for these supplies and equipment should be made in custodial closets, and where these closets should be located.*

Where inadequate provision for custodial closets has been made, it is possible to purchase prefabricated cabinets, which are better than nothing. Some of these are well designed, and even have water and drain provisions.

A specialized aspect of these custodial facilities is the cleaning cart, which becomes a mobile work station to carry chemicals, manual equipment, a waste receptacle, and the like, thus avoiding numerous trips from the work area to the custodial closet. These carts are normally quite inexpensive and last a number of years, especially with repainting on occasion. Of course, the custodial closet should be designed to contain these carts when not in use.

The provision of custodial facilities should include a distribution system for expendable materials and supplies. We should avoid the need—as well as the excuse—for maids and custodians to return to the housekeeping department office at various times during the day for additional supplies. These should be placed in their custodial closets or cabinets, perhaps on a weekly basis, to fulfill all their expected needs for a reasonable period of time.

## Cleaning Equipment

No one would hire a painter to redo his house, and provide him a toothbrush with which to do the job; yet this is done regularly in housekeeping operations. We require a custodian to clean a large floor area, yet give him a small bucket, an inadequate mop, and the wrong type of wringer with which to do this job; or if we do provide him mechanical equipment, it is often of the wrong size or type. The provision of adequate custodial equipment can provide one of the most dramatic quality improvements over a relatively short period of time that is available to us.

We must first concern ourselves with the *number* of items of equipment. If we have scheduled assignments for each worker, this schedule will comprehend certain equippage; and if this is not provided, then perforce the assignment must then be changed, which in turn will change the method and frequency, and thus the quality. The companionship of equipment and assignment, then, is inescapable.

Concerning *size,* the provision of equipment of inadequate size is the mistake most often made in its selection. A 17″ single-disk floor machine is *not* essentially the same size equipment as a 19″ machine; calculations of the area of the brush in this case will indicate that the machine which is two inches larger in diameter will be about 25% more efficient, other things being equal. Purchasing agents, naturally, are anxious to save their organizations as much money as possible, but must be acquainted with what does and does not constitute an "or equal" situation. And, in purchasing, we must bear in mind the advantages of standardization. If we have a dozen floor machines, and they come from six different manufacturers in seven different sizes, we have created a considerable problem in stocking various brush attachment plates, pad sizes, replacement switches, and the like. In pricing competitive bids for such equipment, it would not be amiss to give preference to a quotation that represents a move toward standardization.

There is also the question of *type* of equipment. An interesting example of this is with carpet vacuums, where the type ranges from a light-duty surface vacuum, through a very aggressive pile-lifting vacuum, with various grades of surface and pile lifting vacuums in between. In some cases, more than one type of vacuum will be needed, perhaps a surface vacuum for daily use and a pile lifting vacuum for intermittent use, especially after the dry-cleaning process, for example.

Coming once again to the preventive versus the corrective approach to cleaning, a specialized type of equipment is entrance matting. The provision of suitable mats or runners at entrance areas can trap a good deal of soil and water, where it can be quickly and easily removed, rather than have this soil and water tracked throughout the building. Both rental type matting systems and those which are owned and cleaned internally can be useful. An extension of this program involves carpeting elevators and stair landings, to trap soil in these areas. Naturally, the policing program must comprehend attention to these devices.

## Sanitation Chemicals

Cleaning chemicals are just another aspect of the "tools for the job." Returning to the analogy of the painter doing your house, you recognize that if a good deal of labor is going to be spent in this work, you will want a good quality of paint, so you don't have to do it over again very soon; would this not apply to a floor finish as well? It is naturally demoralizing to cleaning workers to be required to put down a finish which they recognize is going to require a good deal of buffing, and have to be stripped much more frequently than otherwise would be required; that it might not be scuff resistant, and might be too slippery, thus presenting an endless source of complaints.

The chemical must be safe, both from the standpoint of avoiding hazards to the public (such as a slippery wax), as well as hazards to the user (such as a harsh cleaner that is damaging to the skin). It should be easy to use, and not require excessive stirring or working. Convenient packaging is helpful, so that excessive time is not used for decanting and measuring.

Although chemicals amount to just a few percent of the total cleaning costs, their consumption should be controlled, not only to save money, but also to provide proper use dilutions. After all, over-use of a detergent, by way of example, can cause so much foaming that an extra rinsing operation is required, or a detergent film is left on the surface which looks dingy and attracts soil. Measuring cups and dispensing pumps, as well as portion packaging, are all useful consumption-control devices.

## Morale Building

While the morale problem is a serious one in the housekeeping department, it is hardly insurmountable. There are many approaches which can be taken to alleviating this problem, many of which involve little or no expenditure. A number of these have already been discussed in terms of management's responsibilities: the development of inter-departmental cooperation; publicity programs; employee cooperation drives; recognition of the value of the work being done by the cleaning prsonnel.

Beyond these benefits, many more steps can be taken. Classroom training sessions are excellent morale builders, and while half the value of such programs is in terms of imparting information on better ways of doing things, probably the other half is in terms of improving morale; that's why training classes should be regular and continual, rather than haphazard and then discontinued after a time.

The provision of uniforms is an excellent morale builder. Although it is difficult to provide an economic justification for uniforms, nevertheless those organizations which have provided them to custodial personnel almost always feel that it has been a good investment because of the changed attitude of the workers concerning themselves as individuals and the department as a whole.

Along with the uniform can come an individual identification. Plastic badges may be worn over the shirt pocket, carrying the department name and the name of the worker; or there may be an embroidered patch sewn on in this position. Shoulder patches might also be used to identify the department, and patch or sleeve markings—bars or stars, for example—can be added to indicate length of service or the fulfilling of certain training requirements, the position of group leader or senior custodian, or the like.

The recognition can be taken a step further through the awarding of certificates for attendance at training sessions, or for the winning of a contest as "custodian of the month."

The very tools with which to do the job mentioned above—custodial facilities, equipment, and chemicals—represent in themselves considerable morale-building investments. Along these lines, when purchasing equipment, the additional cost of providing better-looking items should be considered seriously. This might involve buying a floor machine with chrome-plated or baked enamel parts, rather than a machine that is merely painted.

Since an instinctive need of the worker is to know where he stands, we find that inspections, when handled on a positive basis, are actually morale builders. These inspections should be performed as much as an opportunity for commenting on and rewarding good performance, as for uncovering opportunities for improvement.

## Summary

The housekeeping manager, whether he be running the housekeeping department in an office building, industry, university, hospital, public school, or other facility, or whether he is a resident manager in any of these organizations for a contract cleaner, is, in a sense, in business for himself. He has a given amount of money to spend and an objective to attain in terms of quality. Like a business, he is beset by competition on all sides: from individuals who want his job, both inside and outside his organization, to external organizations who want to disband his whole department. In any case, he is able to develop his own security and serve his organization best, by professionalizing his work, avoiding the building of empires (which are so vulnerable) and by developing an optimum department through a consideration of all the variables involved, as described above, rather than through piecemeal and sporadic improvement.

Things are not going to stay as they are for the housekeeping manager, any more than they are for anyone else in this age of lightning-like change. We have the simple choice of living in the future or living in the past, of moving ahead or dropping back.

Have you made your choice?

# 18

# A Housekeeping Self-audit

THIS housekeeping self-audit can serve both as a qualitative indicator of the effectiveness of your cleaning program, and as a check-list which may help you identify items which need more attention.

Follow these steps:

1. Fill in the score with a "0" if your answer is a definite "no"; put the maximum number of points indicated if your answer is a definite "yes"; or enter some figure in between depending on your standing relative to the ideal.
2. Force yourself to answer each question before going on to the next one.
3. If the question does not relate to your situation, assign it the maximum score.
4. When finished, total your score, and divide by 3.2 to determine the percentage score.

In order to help you consider each point objectively, the questions are arranged in a random order.

No scoring of this type can tell you exactly how efficient or effective your housekeeping operation is, since so much depends on the objectivity of the answers, as well as on the weighting of the maximum points, which, while established for a typical situation, may require considerable adjustment for your operation. Further, the cost must be related to the size of the staff, and that can only be evaluated through a specific analysis.

Roughly speaking, a score of 90% is good; over 80% is acceptable; while 50% to 70% is typical. The difference between a low score and 90% gives some indication of the improvement possible.

## The Audit

| | MAX. POINTS | YOUR SCORE |
|---|---|---|
| 1. Do you operate your housekeeping department under an annual budget? | 3 | — |

A budget provides a predictable expenditure, while at the same time giving a base against which to measure economic performance. Establishment of the Housekeeping Department budget should always involve participation of the department manager. The budget should be flexible enough so that, as conditions on which the budget was based are changed, the budget may be changed correspondingly.

| | MAX. POINTS | YOUR SCORE |
|---|---|---|
| 2. Are housekeeping services costed out to other departments, both for standard repetitive work as well as for special requests? | 2 | — |

By assigning actual housekeeping costs for both regular and special request work to individual departments, your own operating efficiency and value to the overall organization may be

| | MAX. POINTS | YOUR SCORE |
|---|---|---|
| better interpreted. It can also help to pinpoint unreasonable demands on the time of Housekeeping Department personnel, excessive or insufficient cleaning frequencies, etc. | | |
| 3. In calculating the budget, do you use a creative approach, rather than extending past performance? | 4 | — |
| Budgets should be designed as a function of a positive, comprehensive program. Extensions of past performance merely tend to perpetuate past practices. The budget should be a target figure, capable of achievement through effort and thought, but not through "coasting." | | |
| 4. Have you a way of comparing your housekeeping costs with those of similar organizations? | 2 | — |
| While no two organizations have the same requirements in cleaning and sanitation, the same personnel, or the same physical facility, nevertheless cost comparisons can help to turn up areas of potential improvement. Such a comparison might suggest turning away from, or turning toward, contract cleaning for a specific job, for example, or it may suggest the desirability of jobs currently being handled by other departments being pulled into the Housekeeping Department, or vice-versa. | | |
| 5. Do you periodically evaluate housekeeping jobs currently performed by outside contractors to see if they could be handled at lower cost by your own Housekeeping Department? | 5 | — |
| Because a particular area or job has been handled by a cleaning contractor for a number of years does not mean that this will always continue to be the most economical and efficient way of han- | | |

| | MAX. POINTS | YOUR SCORE |
|---|---|---|
| dling this work. Changes in personnel, organization, mechanization and efficiencies within the internal Housekeeping Department can shift the balance toward performance by your own Housekeeping Department. | | |
| 6. Do you periodically evaluate housekeeping jobs currently performed by your own Housekeeping Department to see if they could be handled at a lower cost by outside contractors? | 5 | — |
| The contract cleaning industry is dynamic and aggressive; it is moving into broader service fields and geographical areas. As this industry gains knowledge, experience and reputation, it will better be able to handle certain types of work under proper controls and specifications. (Any change from internal cleaning to cleaning by contractors must be made with extreme caution because of the difficulty in evaluating the effectiveness of the work done by others, difficulty of reinstating an internal department, and effect on the remaining work force.) | | |
| 7. Have you studied current housekeeping work procedures to see if savings could be effected? | 12 | — |
| A surprising number of outmoded methods are to be found in the typical housekeeping operation. Where management does not provide efficient work procedures, then the worker either develops his own procedures or perpetuates a procedure learned from an older worker. The housekeeping field is now moving very fast, and improved methods are appearing regularly. | | |
| 8. Has full advantage been taken of the economics to be obtained by mechanizing your housekeeping? | 10 | — |

| | MAX. POINTS | YOUR SCORE |
|---|---|---|
| Housekeeping is one of the last indirect functions where mechanization has received attention. Even in industries which are highly mechanized or, in some cases, automated, a number of the housekeeping functions are still being performed manually where they could be mechanized. Equipment includes such items as auto-scrubbers, power-sweepers, pack vacuums, baseboard scrubbers, wall washers, hand washers, venetian blind cleaners, window cleaners, etc. | | |
| 9. Do you have systems or devices for controlling the consumption of cleaning materials? | 3 | — |
| Although housekeeping materials account for only 5% to 10% of the total housekeeping budget, nevertheless wasting of these materials can consume a considerable amount of money which might have otherwise been usefully employed. Equally important, overuse of product can actually increase the work load on housekeeping employees by increasing the time required for rinsing, or leaving a filmed-over surface which soils easily. Control devices include drum pumps, finger-tip pumps for small containers, controlled packaging, aspirating proportioners. | | |
| 10. Have you standardized on materials so that you are using a minimum number of cleaning products? | 2 | — |
| Only a half-dozen chemical products are required for normal daily repetitive cleaning, and a few others for intermittent special projects. A larger number of products than this results in intensified training requirements and can cause confusion and even costly error in mis-application or even replacement of damaged sur- | | |

| | MAX. POINTS | YOUR SCORE |
|---|---|---|
| faces. Finally, standardization permits purchase in larger quantities at lower cost. | | |
| 11. Do you have suitable time standards for housekeeping jobs? | 12 | — |

Time standards permit the scheduling of both repetitive and special project work so that it may be performed efficiently, balancing the individual work loads so that each housekeeping worker is given a reasonable day's work. This contributes not only to overall operating efficiency but to improved personnel morale as well. Preparation of meaningful, weighted time standards requires a comprehensive approach to the entire housekeeping operation, involving standards for methods, materials and equipment; as well as a consideration of the organizational structure, the labor climate, the type of personnel involved, the physical facility, etc. The implementation of a housekeeping program on the basis of time standards alone perpetuates outmoded practices in other phases of the operation.

| | MAX. POINTS | YOUR SCORE |
|---|---|---|
| 12. Do you have a basis for comparing your frequencies of housekeeping job performance with those found at other similar organizations? | 2 | — |

There is a strong relationship between the frequency of housekeeping job performance and the level of housekeeping obtained. Thus, a floor which is mopped and buffed daily could be expected to be cleaner and look better than one which is mopped and buffed only twice weekly. Inadequate frequencies cause poor appearance or even damage to surfaces; excessive frequencies may waste effort and money. Frequencies should be balanced by taking time from one function and

| | MAX. POINTS | YOUR SCORE |
|---|---|---|
| applying it to another where it may be more needed. | | |
| 13. Do you know whether the man-hours devoted to housekeeping are in line with your actual man-hour requirements? | 10 | — |

As in Parkinson's Law, has the work expanded so as to fill the time allowed for its completion? Many people spend a full day at a job which should require hours less. Work sampling techniques may be used to determine the portion of a worker's time spent in productive effort, although the effort may be more or less efficient. Time, methods, and other standards weighted for use in your own facility are necessary to establish the proper staffing and scheduling.

| | | |
|---|---|---|
| 14. Is cleaning performed on a scheduled basis? | 8 | — |

Where housekeeping services are provided on a corrective basis, or "putting out fires," it can cause considerable confusion and deterioration of morale. Such an inefficient method of corrective maintenance is rarely efficient from a long-term standpoint, because the work is performed only after something unpleasant has occurred.. Housekeeping should be performed on a preventive maintenance basis with major projects being performed, insofar as possible, immediately before the need has become apparent. This provides the greatest efficiency and protection of the investment in the physical facility, while also providing the best operating conditions for the entire work force.

| | | |
|---|---|---|
| 15. Have specific performance levels been established for housekeeping workers? | 6 | — |

| | MAX. POINTS | YOUR SCORE |
|---|---|---|
| The custodial worker does his best when he knows exactly what is expected of him each day in terms of performance, both from a quantity and quality basis. The worker should understand approximately how much time should be involved in the completion of the major segments of his work. The assigning of specific time intervals for the beginning and completion of each job has certain advantages, but these may be counterbalanced by the morale problem involved in having each job timed. The assignment of a reasonable work load, and a description of its method and order, suffices quite well for most cases, where there is a continuing training and inspection program under qualified leadership. | | |
| 16. Do cleaning schedules allow for seasonal variations? | 4 | — |
| A housekeeping organization should be flexible enough to permit special attention being given to problem areas during inclement weather. Examples are snow removal, cleaning of mats and runners, mopping up water. Scheduling should permit the performance of these duties without creating a crisis in regular daily housekeeping service. | | |
| 17. Do you have a system for rating levels of cleanliness? | 3 | — |
| Describing a level of sanitation and rating a given physical facility against this description, requires a good deal of thought and experience. The work must be described both in terms of quantity and quality, and the analysis of results requires a system for measuring or observing both. | | |
| 18. Do you have written instructions for housekeeping work? | 7 | — |

| | MAX. POINTS | YOUR SCORE |
|---|---|---|
| Written instructions have many advantages: supervisors are properly able to interpret work assignments and methods; training of new employees is simplified; morale is improved because the employee knows precisely what is expected of him; flexibility is achieved in the simplifying or transferal of people from one area to another. | | |
| 19. Have definite lines of authority, and reporting functions, been established for the housekeeping department? | 8 | — |
| Workers should know precisely to whom to report or ask questions in cases of uncertainty; from whom instructions come; and who will provide training and inspection. Of course, equally as bad as lack of supervision are unclear lines of authority where a worker may be receiving instructions from more than one supervisor. | | |
| 20. Are custodians not performing tasks other than those involved in housekeeping and cleaning? | 4 | — |
| When housekeeping workers share their time between cleaning and other duties, the result is often a lack of interest or proprietary attitude concerning any of the jobs. Efficiency is impaired because the "make-ready" and "put-away" times, which are non-productive, are increased considerably when workers turn from performing one type of job to another. | | |
| 21. Is adequate supervision, and relief for supervision, available? | 12 | — |
| The proper proportion of supervisors to maids and custodians will be based on such factors as turnover, travel time requirements, type of cleaning being done, scheduling (area or group | | |

| | MAX. POINTS | YOUR SCORE |
|---|---|---|
| assignment), quality of the worker, etc. Particularly in the field of housekeeping is the quality and intensity of supervision related to the results obtained. | | |
| 22. Has a statistical approach been used with respect to the assignment of housekeeping tasks? | 3 | — |
| Determination of square footages, fixture counts, surface types and condition, etc., contribute to "defining the problem." | | |
| 23. Are persons outside the permanent housekeeping staff permitted to perform cleaning duties only with proper training or supervision? | 3 | — |
| In many cases, cleaning of one type or another will be done by persons not in the Housekeeping Department. Nevertheless, this personnel should attend housekeeping training sessions, and their supervisor should depend on the housekeeping department for recommendations concerning methods, equipment, and materials. In some cases it is also desirable for the housekeeping department to make a periodic inspection of these areas maintained by other personnel. | | |
| 24. Do you have a definite plan for reduction or expansion of the cleaning force to compensate for increases or decreases in activity? | 5 | — |
| In a number of organizations, the number of housekeeping personnel authorized is related directly to the production or occupancy level: this may be an arbitrary management decision over which housekeeping has no control whatever. The housekeeping program should be formulated in such a way that a fluctuating staff will not unduly disrupt housekeeping operations. | | |

| | MAX. POINTS | YOUR SCORE |
|---|---|---|
| 25. If a new building or wing is constructed, do you have an accurate way of pre-determining personnel requirements and housekeeping costs? | 4 | — |

Many new facilities are staffed on a trial-and-error basis, through the use of a small staff which is increased until the work is apparently performed satisfactorily. Or, staffing may be accomplished through the application of some average square footage determinations, based on prior history. Pre-determination based on adequate standards, however, assures an efficient work force without danger of damage to the facility, endangering the health of other employees, or affecting the quality level or productive output.

| | | |
|---|---|---|
| 26. Is an effort being made to improve worker skill and efficiency through a regular training program? | 12 | — |

No one would think of asking a carpenter, for example, to do worthwhile work, using new materials, equipment, and methods, without suitable orientation and training. Yet, cleaning personnel are asked to do this every day. A continuing training program for housekeeping is necessary not only to improve worker skill, avoid accidents, and conserve materials, but also because it provides a vehicle for morale building in a job where morale is often a serious problem.

| | | |
|---|---|---|
| 27. Do your workers wear uniforms? | 3 | — |

Uniforms improve morale because of their identification with a specific department, and the impression created that the worker is a specialist. Uniforms are generally a good investment if their

| | MAX. POINTS | YOUR SCORE |
|---|---|---|
| use does not create a problem with other departments. A name badge is important too. | | |
| 28. Does each custodian fully understand his responsibilities? | 5 | — |
| Demoralization results when a worker is given a nebulous job to perform. This may occur through using him as relief for a worker performing a different type of assignment; when there is considerable turnover and orientation time is not adequate; where supervision is insufficient; or where a well defined program does not exist. | | |
| 29. Have you ever tested custodians to determine their understanding of their jobs and their duties? | 3 | — |
| It is difficult to evaluate the effect of a training program without periodic testing to provide a history of improvement, and to evaluate the type and extent of additional training required. A complete testing program will include both written tests (or oral if the subject is illiterate) and practical performance tests using normally assigned materials and equipment in a typical work area. Testing can become, if desired, an integral part of an employee rating and qualification system. | | |
| 30. Has an attempt been made to determine the time custodians actually spend in productive work? | 4 | — |
| Industrial engineering techniques may indicate an excessive portion of the total time being devoted to transportation, personal time, absence from the work station, etc., and may indicate the need for action. Carefully done, a work sampling can be | | |

| | MAX. POINTS | YOUR SCORE |
|---|---|---|
| taken without disturbing or arousing hostility in employees. | | |
| 31. Is recognition or reward given to custodians who do exceptionally good work? | 3 | — |
| Where good work goes unnoticed or unmentioned, the natural result is for all workers to gravitate toward a barely acceptable level. Recognition, being one of the fundamental human drives, can bring positive and lasting results when well handled. | | |
| 32. Are written qualifications used for hiring custodians? | 2 | — |
| Where housekeeping is mistakenly considered a job which anyone can perform adequately, employees may be hired who are wholly unfit for the work and who may cause considerable financial losses through supervisory time, retraining, errors and damage to surfaces. In times of stress, the tendency to hire almost anyone who applies for a position in the housekeeping department should be resisted as much as possible. The housekeeping department manager's suggestions concerning qualifications should be given serious consideration. One of the dangers, however, is setting qualifications too high with the risk of hiring employees who may become bored with the work, thus increasing the turnover rate. The housekeeping department will be judged by the performance of its poorest workers. | | |
| 33. Are rewards given for good suggestions? | 2 | — |
| Where an organization-wide suggestion system exists, housekeeping department personnel should be encouraged to participate. Where | | |

| | MAX. POINTS | YOUR SCORE |
|---|---|---|
| there is no such overall program, it may be desirable, at least for a period of one year, to set up a departmental suggestion system, particularly if a comprehensive program involving a number of reassignments is in progress. | | |
| 34. Are physical examinations given to prospective custodians? | 3 | — |
| A thorough physical examination should be given prospective employees in order to eliminate the hiring of disease carriers. Housekeeping personnel work in areas such as rest rooms, locker rooms, food-processing areas, around drinking fountains, etc., where cross-infection can become a serious problem. | | |
| 35. Do you conduct regular housekeeping inspections? | 5 | — |
| A fixed inspection program has the advantage of providing a history of improvement, while providing a tool for measuring quantity and quality of work performance. It also has the supervisory value of placing the supervisor in the work area periodically, providing contact with the workers where they may ask questions or receive instructions or commendations. Inspection is a necessary aspect of training, while testing, mentioned earlier, also represents a type of inspection. | | |
| 36. Are custodians hired and trained before new building occupancy? | 3 | — |
| Pre-training custodians for the maintenance of a new facility assures efficient work, adequate service, and protection of health and surfaces immediately rather than after a frustrating period of trials and errors. | | |

| | MAX. POINTS | YOUR SCORE |
|---|---|---|
| 37. Do you have an effective inventory-control system? | 3 | — |

Good control of housekeeping materials assures early investigation of over- or under-consumption; discourages pilferage, or unauthorized use by other departments; assures proper chargeout to other departments for authorized use; and guarantees a reasonable supply of materials on hand with which the housekeeping personnel will perform their work.

| | | |
|---|---|---|
| 38. Do your custodians feel that management is interested in their work and their problems? | 6 | — |

An active interest in housekeeping by management does much to improve the morale of all housekeeping personnel, while at the same time assuring desirable inter-departmental cooperation and facility-wide personnel awareness.

| | | |
|---|---|---|
| 39. Have you recently reviewed your housekeeping practices from the standpoint of safety? | 4 | — |

A sound safety program without good housekeeping is impossible. Each is a part of the other, and each should be evaluated in view of its inter-relationship with the other. Many claims adjusters estimate that over half of all building accidents involve housekeeping in one way or another.

| | | |
|---|---|---|
| 40. Are custodial facilities adequate? | 5 | — |

Custodial closets and cabinets, when properly located and equipped, can increase efficiency significantly by cutting the travel time from work station to work area. Note that the custodial cart falls in this category, as it is a mobile work station.

MAX. POINTS | YOUR SCORE

41. Is the housekeeping department consulted when plans are made for new construction? 5 —

The efficiency of cleaning a new facility can be materially affected during the planning stage, generally without adversely affecting the beauty of the structure or its cost. Involved are such things as selection of floor type and color, type of paint, using wall-mounted fixtures, location and size of custodial facilities.

42. Do you have a good materials-distribution system? 2 —

A good system of supplying housekeeping workers with their materials and equipment, on a periodic basis, avoids the need for workers to make repeated trips to the supply room, which also avoids the excuse to be absent from their work area.

43. Are adequate records maintained to insure proper protection of surfaces at proper intervals (such as washing walls)? 3 —

Records of the completion date for special projects, as well as setting up target dates for future projects, are indispensable to a sound preventive maintenance program.

44. Does your organization have a continuing awareness and inter-departmental cooperation program related to housekeeping? 3 —

An improved attitude on the part of all employees concerning personal housekeeping responsibilities can materially decrease the work load on custodians, particularly in policing functions, permitting more time to be used in basic cleaning.

| | MAX. POINTS | YOUR SCORE |
|---|---|---|
| This requires a continuing, rather than a one-time, campaign. | | |
| 45. Do you require your suppliers to provide demonstrations and other services? | 2 | — |
| Most suppliers have services and literature, and provide demonstrations on request to their larger customers or potential accounts. | | |
| 46. Are records maintained as to the nature and source of complaints concerning housekeeping? | 2 | — |
| The periodic review of such a complaint record can turn up repetitive problems that may indicate corrective action with personnel, materials, management liaison. | | |
| 47. Is housekeeping equipment cleaned after each day of use? | 2 | — |
| Proper care of equipment, leading to an extension of its life and effectiveness, is a matter of habit. The habit must be formed through guidance by supervision. | | |
| 48. Do you regularly read current literature concerning housekeeping? | 3 | — |
| Books and magazine articles are regularly appearing in this field. It has been estimated that available information is doubling every ten years —it is easy to fall behind! | | |
| 49. Do you have a suitable central supply area for the housekeeping department? | 3 | — |
| A good central supply area, conveniently located, provides reserve supplies of materials and expendable equipment, has facilities for minor repairs, lubrication, storage of equipment, and | | |

MAX. POINTS | YOUR SCORE

provides for such functions as diluting concentrated chemicals, treating dust mops, etc.

50. Are regular reports made to management on housekeeping progress? 3 —

Management should be kept informed on budget, personnel, complaints and their rectification, suggestions, overtime, turnover.

51. Are you in contact with other responsible persons in the housekeeping field? 5 —

Correspondence with persons in similar positions, and attendance at seminars and meetings, help to keep one's technical and managerial abilities sharpened and objective.

52. Have you developed specifications for housekeeping materials and equipment? 5 —

For the larger operation, specifications can be a means to improving quality while significantly cutting costs, particularly where there is not a requirement that the lowest price always be accepted. Actual performance under normal use conditions must be the governing factor.

53. Have you posted an organization chart for the housekeeping department? 3 —

The organization chart can help to clarify lines of authority and responsibility, while helping to build a departmental spirit. Photographs help.

54. Has noise control been considered when selecting housekeeping equipment? 2 —

Often, with very little additional cost, equipment can be selected which reduces much of the noise that is common to many housekeeping

| | MAX. POINTS | YOUR SCORE |
|---|---|---|
| duties. Noise-controlled equipment includes such items as mopping outfits, floor machines, vacuums, waste containers, carts. | | |
| 55. Do you have a specific program for improving or sustaining morale in the housekeeping department? | 8 | — |
| Custodial morale and status is one of the fundamental problems in the housekeeping department. Positive steps should be taken to help relieve this problem, on a continuing basis. | | |
| 56. Is good liaison maintained between the housekeeping department and the executive charged with fire prevention? | 4 | — |
| Many a building has been brought to the ground in ashes because of poor housekeeping. The housekeeping department must play an active role in the organization's fire prevention practices. | | |
| 57. Do you have a housekeeping committee? | 2 | — |
| Such a committee, which may meet quarterly to monthly, provides improved inter-departmental liaison (in a hospital, the cross-infection committee can serve this purpose if it includes the executive housekeeper). | | |
| 58. If a labor contract provides "bumping" into the housekeeping department because of seniority, is the worker trained in housekeeping before beginning work? | 5 | — |
| Many organizations mistakenly assume that anyone can perform housekeeping functions, despite the fact that rather sophisticated materials, equipment, and techniques are now being used | | |

| | MAX. POINTS | YOUR SCORE |
|---|---|---|
| in cleaning. New persons coming into the housekeeping department must be adequately trained. | | |
| 59. Are the operations of the housekeeping department considered a part of your organization's overall public relations activity? | 4 | — |
| Housekeeping personnel not only personally come in contact with visitors, suppliers, and customers, but their activities have an effect on others' opinion of the organization. | | |
| 60. Is the housekeeping department manager able to communicate with higher levels of management easily? | 6 | — |
| Open lines of communication assure early correction of problems or potential problems, improves morale, and often replaces corrective action with preventive action. | | |
| 61. Is your housekeeping primarily performed by a centralized housekeeping department? | 6 | — |
| Where cleaning is done by scattered groups of persons reporting to various department heads, much money and effectiveness is lost through the inability to standardize on equipment and materials, methods, etc. Supervision is often ineffective because of conflict with other responsibilities, and the workers themselves are often indifferent toward housekeeping as it represents a secondary duty. | | |
| 62. Are inactive areas sealed off when not in use? | 3 | — |
| Sealing off inactive areas, and giving them attention only from a policing and fire prevention standpoint, relieves the pressure of the cus- | | |

| | MAX. POINTS | YOUR SCORE |
|---|---|---|
| todial work load, so that more time may be spent elsewhere. | | |
| 63. Have you considered the use of a housekeeping consultant? | 5 | — |
| A qualified consultant is able to bring objectivity and wide, specialized experience to the housekeeping field, which is now changing quite rapidly. A consultant loosens lines of communication and focuses attention on one of the last remaining indirect functions to receive industrial engineering analysis. He should be able to assist in training and implementation. | | |
| 64. Have working hours been carefully selected? | 5 | — |
| Housekeeping work can be much more efficiently performed when done during other than normal business hours (except for policing functions); this also gives the benefits of reducing potential accident hazards. | | |
| 65. Are you confident that the housekeeping department contains the correct number of workers? | 12 | — |
| The optimum staff should contain that number of persons where an increase of one would not be economically justified by the amount of savings from all sources, or a decrease of one would cause a greater loss than the cost of maintaining him on the staff. | | |
| 66. Are sensible titles in use? | 2 | — |
| Titles should indicate status, responsibility, and should have a positive connotation. Example: "Custodian" rather than "Janitor" or "Laborer." | | |
| 67. Have unhardened concrete floors been sealed? | 2 | — |

| | MAX. POINTS | YOUR SCORE |
|---|---|---|
| Sealed floors provide a surface which permits easier, more effective cleaning, protects the floor from wear, and avoids odors and stains because the porous structure is closed. | | |
| 68. Are multi-purpose chemicals in use? | 3 | — |
| Chemicals are available, for example, that clean, deodorize, and disinfect in one operation; or that combine an insecticide with a floor wax. Multi-purpose chemicals eliminate separate duties. | | |
| 69. Are you trapping soil at strategic points? | 4 | — |
| Matting, carpeting, runners, shoe cleaners, mechanical mats and other devices keep soil from being distributed throughout the facility. | | |
| 70. Are adequate litter-control devices provided? | 4 | — |
| An insufficient number of urns and receptacles is an invitation to use the floor as a depository for waste. | | |
| TOTAL | 320 | — |

Percentage score=your total score ............ ÷3.2= ............ %

# Part II

# SUPERVISION

# 19

# Housekeeping Supervision

IT is the responsibility of the housekeeping supervisor to transform his organization's housekeeping and sanitation objectives into actual physical conditions, achieving this under the regulations and limitations as prescribed by management.

In any organization many pressures are exerted on the housekeeping department from a number of sources. For example:

1. Other departments want safe and clean working surroundings.
2. Administrative personnel want attractive surfaces and wish to be free from interference or interruption.
3. Management is concerned with overall cost control.
4. The custodians themselves want more status and cooperation.
5. Management wants to satisfy as many of the housekeeping objectives as possible but is concerned about the time required to do this.

In order to cope with the pressures exerted upon it and to properly fulfill its objectives of all types, the housekeeping department must be properly supervised. Providing sound supervision, both in quality and quantity, is one of the best dollar investments that can be made in a housekeeping department. Particularly in the field of housekeeping and sanitation. the intensity and quality of supervision has an almost direct relationship to the results obtained, up to a point.

The problems in morale, status, relative pay, education, and training makes housekeeping the trade requiring the most intense supervision.

The basic objective of housekeeping supervision is to maintain high productivity from each custodian, within the framework of a well-defined program consonant with management's objectives. In accomplishing this, three subordinate functions are performed:

1. It has been said that 95 percent of our workers depend on the other 5 percent for their planning. Although it is true that individual workers can sometimes bring specific improvements to their work, nevertheless it is true that the basic planning and organizing must be done by others. This the worker expects, and, as a matter of fact, would have no other way.

2. The supervisor is the custodian's most direct representative of management. Identification and communication are provided by the supervisor as the employee's contact with the realities of management.

3. Supervision provides control. People do not work together automatically—they need the cohesive force of leadership and purpose.

## Characteristics of the Supervisor

We know the magnitude of the problems facing us. Let us look at some of the abilities, characteristics, and qualities of the supervisor who will bring to the housekeeping department the greatest possible benefits of good leadership.

The supervisor should be a specialist in obtaining the greatest and most efficient effort from others, rather than in providing it himself. Thus, although the hardest worker is able to lead by example, he may lack the ability to use motivation and control as well as required leadership. There is a certain value in the concept of a good supervisor as an inherently "lazy" person, but the laziness must be expressed as an untiring effort to avoid all *unnecessary* and *inefficient* work.

In *Psychology Applied to Life and Work,*[1] Harry Walter Hepner points out that "both employees and employers have the same mental traits. Their behavior differs chiefly in that they react to different situations. . . . Most of the conflict between employers and employees are not the rumblings significant of a history-making class conflict. Points of friction usually boil down to simple individual grievances."

[1] Copyright 1957, Prentice-Hall, Inc. (3rd Edition).

The good housekeeping supervisor is observant and perceptive. Often supervisors (and workers and management as well) have become inured to slovenliness and soil through long and intimate association. The effective supervisor can react to poor housekeeping in only one way: he determines to improve the condition as quickly as possible. As simple as this may sound, good eyesight, or properly corrected eyesight, is required for proper observation of housekeeping conditions.

The supervisor should consider every occurrence involving his personnel, no matter how bad it may appear at first sight, as an opportunity for a possible improvement. Every termination and every prolonged illness should cause the supervisor to consider what changes this might permit which will benefit the housekeeping operation.

It is a good practice, after any controversial or difficult contract with personnel, to ask the question "What have I learned from this experience?" This deliberate analysis leads us to build a storehouse of valuable human relations material from even the most apparently commonplace occurrences. Such a fund of *studied* experience is invaluable in everyday human relations.

In his work, the supervisor must be persistent and determined. Hannibal, on being told that the crossing of the Alps to attack Rome was impossible, replied, "I will either find a way or make one." Much later, Peary said, "I will find a way or make one to the North Pole." The supervisor should also be prepared, with help, to find a way or make one. The problems that he is beset with may recur again and again, and if necessary, they must be solved again and again. He is often working with people who do not consider their work important, in an atmosphere where the "janitor" is mocked, and some or all levels of management may be indifferent.

He must have the ability to think. This requires concentration on important matters and the self-discipline necessary to restrain himself from becoming overinterested in matters relatively unimportant or trivial, merely as relief. We find that the powers of thought and concentration are improved through deliberate and personal application to important and difficult problems, and through worthwhile reading and conversation.

He must be understanding—or at the very least, tolerant—of nationality, race, sex, age, language difficulties, and limited abilities. This is discussed further under "empathy."

Through a growing knowledgeability of the technology of housekeeping and the techniques of applying its principles, the supervisor will develop self-confidence. Of course, this is not the same as pomp or self-esteem based on power or authority. Norman Vincent Peale tells us, in "The Power of Positive Thinking," that "without a humble but reasonable confidence in your own powers you cannot be successful or happy."

He must be flexible and not so steeped in outmoded attitudes and methods that he cannot comprehend and work with new ideas and methods. Artist-educator Leo Katz, in "Understanding Modern Art,"[2] said that "our rapidly changing conditions necessitate constant adjustment of our ideas. . . . For people carefully trained along the lines of past conceptions, it is usually difficult to keep pace with the new spirit. They have not only to do a great deal of learning, but also an equal amount of unlearning, and it is easier to learn than to unlearn." Beyond this, many people passionately resist new ideas probably because they fear the loss of status or security, or for hidden psychological reasons. One of the limiting factors of the housekeeping program may be the ability of the leader of the housekeeping department to work with new tools and concepts.

The supervisor should have a reasonable memory, or, if he is unable to commit information to personal memory, he should be able to organize data so that they are readily available when needed. The simple check list is one of the easiest, and yet most effective, tools available for this purpose. The supervisor should also make use of a pocket-memory aid. Several organizations publish and distribute these monthly. They contain calendars, memo space, suggestions for better supervision, etc. No supervisor should leave his office without a pencil and note pad, or memory-aid booklet, because he must be prepared to write down observations and ideas as they occur.

He must not fear manual work or soiling his hands or clothing, and he should be prepared to exhibit, on occasion, a frame of mind indicating this, if necessary. Custodians involved in soil removal need to be shown that there is nothing distasteful or derogatory about such work in the mind of their supervisor. Nor must he fear criticism. Thomas A. Edison said: "There is no truer test of a man's qualities for permanent success than the way he takes criticism."

Housekeeping leadership requires a certain amount of scheduling and paper work, which the supervisor must be prepared to handle, and handle well.

He should be able to practice "imagineering," unhampered by common conceptions of what is "possible," "practical," or "dreaming."

He must be able to translate thought and planning, based on past information and imagination, into action. This requires not only the initial action but also a determination to "stick with it."

[2] Copyright , The Delphian Society.

He must have the ability to inspire confidence and cooperation in others through integrity and honesty, humor, and example.

The good supervisor can communicate. He must be able to speak to his personnel in groups to conduct training classes and for the informal exchanges of suggestions and ideas. He should be able to put his ideas and instructions into writing that is easily understood.

If the supervisor does not want his custodians to be clock watchers, he must not be one himself. Beyond this, he must be prepared to vary his hours according to the given situation when required by emergency, absenteeism, special requests, or when a special training session is needed.

He should stand up for his men. For example, he should not stand by and permit the deterioration of the status and organization of his workers by allowing supervision and instructions to come from outside his own department.

## Improving Methods

A housekeeping supervisor should always be looking for better work methods for his custodians. It is true that he will not be as well equipped for this work as are engineering personnel, and it is also true that the housekeeping program should outline the methods to be used. But no program can be complete in all possible details, and furthermore, there is always a better way of doing a given job. These fundamentals can be utilized in bringing about methods improvements:

1. The work should be eliminated if a useful function is not being performed.
2. Two tools performing separate functions should be combined, where possible, into a single tool performing both functions.
3. Two materials performing separate functions should be combined, where possible, into a single material performing both functions.
4. Both tools and materials should be located as closely as possible to the work.
5. Long stretching and deep bending should be avoided, where possible.
6. Arm and hand motions should be made simultaneously in opposite directions where possible.
7. Motions should be circular or continuous rather than jerky or interrupted, where possible. Rhythm should be introduced into work where possible, involving such smooth body motions as swinging, rocking, etc.
8. The weight of a tool should be considered. It should be heavy if its momentum is to be relied on, or light if its direction must be changed regularly. Furthermore, tools must be sized properly for the job.

## Levels of Supervision

There are four basic levels of supervision in housekeeping operations: (1) the housekeeping department manager, (2) assistant managers, (3) foremen, (4) group leaders.

The housekeeping department manager is in overall charge of the sanitation and cleaning function. Alternate titles can include custodial supervisor, general housekeeping foreman, service manager, etc. It is up to the manager to provide the creative effort needed to improve the level of productivity of the custodian. It is he, primarily, who works directly with a consultant toward this end. In many ways the housekeeping department can be compared to a business within a business, in which case the manager can be considered its "president."

The assistant housekeeping department manager (or other applicable title) can be considered the "vice-president of the housekeeping business." This level of supervision is only required in the larger industrial organizations where the number of foremen required is beyond the capacity of one individual to properly supervise and motivate them, along with the other functions which must be performed. A typical example would be a large industrial organization having the department manager on the day shift, with a few foremen reporting directly to him, while having one or more assistant managers on the later shifts, each with several foremen subordinate to him.

The housekeeping foreman is in direct charge of the work being performed. This is the basic level of salaried supervision. The title for the position varies considerably and ranges from "shift foreman" to "assistant supervisor." The latter title may be used where the second echelon of supervision is unnecessary. The extent of his responsibility should always be defined as a specific number of personnel, and usually within a given work area, such as a certain number of buildings, or a subdivision of a single large building.

On the day shift where he is supervising policing operations, one assistant foreman may be in charge of the entire policing operation in all buildings. The effectiveness of the assistant foreman will be almost directly proportional to the amount of time spent in contact with the custodians.

The group leader is a *working* lead man. Where the operating climate of the industrial organization permits and encourages a situation where group leaders will actually perform a fair share of the work and lead primarily by example, the advantages that they can provide are considerable. On the other hand, where the group leader merely becomes a wage-earning supervisor without performing his due share of the work, he should not be tolerated. Here are some of the benefits that the *true* group leader can bring:

1. If group leaders have accumulated considerable seniority, "bumping" within the housekeeping department can be better tolerated because the group leader is able to train workers at the work scene and "ride herd" on them.
2. The use of working group leaders brings an on-the-spot practical direction to housekeeping work which is difficult to obtain otherwise, particularly where the work is spread over a considerable area. Thus, it is through these persons that some of the problems normally associated with training, limited abilities, motivation, etc., can be alleviated.
3. The pyramid of supervision, and especially the group-leader level, provides a method and incentive for advancement for ambitious and deserving workers.
4. Some of the advantages of team cleaning can only be realized when the team consists of a leader and several custodians.
5. Similarly, special projects can be efficiently performed in such a team led by a group leader. It is often desirable to assign the most expensive or complex mechanical equipment used by that team to the group leader for his personal care and use.

Just as relief is required for custodians who may be absent for any number of reasons, so must relief be provided for supervisory personnel. If there are more than a very few supervisors, they will be unable to "cover" for each other, and specific relief may have to be provided for this purpose. Where a number of group leaders are used, it is possible for them to handle the situation in the absence of an assistant foreman for a short time.

Just as training of the custodians becomes an important function, which is discussed in the next chapter, so must subordinate supervisory personnel be trained. Training and general discussion sessions can be combined at times. One supervisor can train the others in some aspect of housekeeping with which he has become particularly proficient.

Although there is a great amount of information available to industrial supervisors in general, actually there is a dearth of material directed to the *cleaning* supervisor. The specific material available to production supervisors, for example, is of only limited use and interest to the housekeeping supervisor—there must be material made available and directed to him.

Chapter 4 pointed out the need for college-level courses, and even degrees, for housekeeping supervisors. Short of this, supervisors would benefit materially from seminars and clinics conducted by professional service engineers and management personnel, either on the premises or at another location.

Insofar as the general principles of supervision are concerned, it is desir able that the supervisor join some national organization in order to receive periodic literature and information concerning methods and techniques. In addition, from a general standpoint, there is a great wealth of literature on this subject, both books and periodicals, which are of real value. Specifically, membership in the Institute of Sanitation Management is desirable.

## The Organization Chart

The organization chart is very effective for clarifying duties and indicating lines of authority. On a single document it provides management, and others as well, information concerning the structure of the housekeeping department, the supervisory work load, the levels of authority, and the position of any individual within the department.

It is desirable to post a large organization chart in the housekeeping department so that it will be in view at all times. Some organizations utilize individual photographs in the organization chart, particularly for those persons having at least one year of service. This does much to bring to the housekeeping crew a feeling of departmental unity. On the smaller charts, individual names can be written in. Where turnover is high, the smaller chart can be covered with acetate and names changed as required.

## How Many Supervisors?

The determination of the make-up and size of the supervisory staff of the housekeeping department depends upon a great many other factors in the program (the interdependence of factors deserves stressing through repetition). For example, such factors include the abilities of the workers, the quality of the supervisors, morale levels, interdepartmental cooperation, physical distribution, etc.

A very definite statement can be made on this point: there must be one person in whom the responsibility and authority for the whole housekeeping function is ultimately vested. In everything but the smallest industrial operations, this individual should have housekeeping as his primary responsibility. In the larger operations it should be his sole responsibility and, as the organization grows in size, he will need supervisory assistance.

The ratio of supervisors to custodians varies considerably. Even a range of one salaried supervisor for each ten to forty workers would not cover every situation found in industry today. The majority of cases should fall in the range of one supervisor for each twelve to twenty custodians. On the wage-earning level, the range is one group leader to each two to eight custodians, occasionally a little more. A relatively high supervisor-custodian ratio indicates management's understanding of the scope and depth of the problems that should be handled at the supervisory level. The case where a housekeeping department is staffed with too many *qualified* supervisors would constitute an extreme rarity.

## Using Empathy

The biggest word in the fields of personnel relations and supervision is *empathy*. The Thorndike-Barnhart Dictionary defines empathy as "the complete understanding of another's feelings, motives, etc." It is mentally putting oneself into the other person's shoes. When a specific problem arises in dealing with personnel, the attempt to visualize the other person's feelings, in the context of his education, background, job position, and experience, will often give a clue as to the proper steps to be taken. This approach is particularly valuable in every case where the employee is trying to explain his problem or grievance.

At times, the reverse of the coin can be used as well. The subordinate may be asked to try to visualize the problems and situations with which the supervisor is faced. The method of asking the subordinate to sit at the desk of the supervisor brings interesting results.

One of the benefits of empathy is that it equips the supervisor to be a good listener. Under such circumstances, a problem loses its theoretical and dull, dry nature and takes on the aspect of a personal problem of considerable significance. The effective supervisor is much more a good listener than a good talker.

You can buy a man's physical presence on your premises for a given period of time—but you cannot buy his spirit or his desire to do a good job. These have to be cultivated in personal relationships.

Each of us, to a greater or lesser extent, considers himself as the center of all internal emotions, thoughts, and responses, as well as the center of all external stimuli, actions, and experiences. Therefore, dealing with people is basically motivated by self-interest. A realistic respect for the powerful influences of fear, hope, and pride clearly points out the value of these principles of supervision:

1. An individual's approach to his work should always be considered or discussed in relationship to the situation, rather than in vague general terms.
2. Criticism should only be made when clearly deserved, and always privately.

3. Commendations should also only be made when clearly deserved, but they should always be made publicly. Probably the five most effective words in the whole field of human relations are "I am proud of you."

4. Show your respect for the individual, and your belief that he can do a somewhat better job than perhaps he himself feels. Place your trust in him to do a better job.

5. Most people experience confusion and anxiety in a changing situation. Be certain to give all the facts as early and as clearly as possible.

6. Temper is a luxury of childhood in which mature adults cannot afford to indulge. A display of temper is a certain signal that something irrecoverable has been lost.

7. It is a compliment to an individual to ask his help. Many an unusual effort has been made in response to this appeal.

## Remembering Names

Nothing sounds sweeter, it has often been said, than the sound of one's own name. No supervisor should ever be guilty of not knowing the names of persons working for him. Where the number of persons involved exceeds 100 and turnover is high, it is understandable that the learning of names might be restricted to employees with a year or more service. Some supervisors go further in their human relations by learning some of the personal details concerning their workers, such as nick-names, hobbies, family details, etc. Naturally, the use of this information should be applied carefully and without familiarity. Along the same lines, sending of a personal birthday note to each employee, even if it is only a few words, helps in its way to weld a housekeeping organization into an effective team.

There is practically no limit to the amount of information that can be stored in the human brain. It is impossible to fill the brain with information. There have been some very interesting individual cases of retentive ability. Mozart, for example, was able to write down, after one hearing, lengthy and complicated liturgical music, note for note and rest for rest. In our own times, former Postmaster General James A. Farley had the uncanny ability to remember every name of the thousands of persons he met. Theodore Nadler, of television quiz show fame, has committed the equivalent of a small library to memory. True, these are unusual feats of memory, but the undeniable fact remains that we all use only a small fraction of the mental resources at our disposal and that the improvement of our memory for names is merely a matter of applying these resources. It is only a matter of making the effort.

Here are some ways to help remember names:

1. Convince yourself of the importance of names. In great part, success or failure is based on our association with our fellow workers and business contacts. Support, understanding, and cooperation are not going to be forthcoming from a nameless person.

2. Use memory association devices, called mnemonics, to help remember.

3. It is desirable to repeat a name as quickly as possible after hearing it in order to embed it in your memory. Try to repeat the name several times within the space of a few hours. Asking a person to spell his name also helps.

4. In a conversation, try and center your interest in the other person, rather than in your own feelings.

5. Finally, remember that the basic requirement is a little effort.

## Discipline

It is up to the housekeeping manager to maintain discipline within his department. This must be done within the framework of company policy, but this policy alone will not provide all the answers.

Exercising discipline protects the department, and the whole organization as well, from personal abuse. Thus, demonstrating to the workers that each and every person must conform to given principles and rules and carry his full share of the load insures good discipline, which is always a factor in high morale. In order for discipline to function effectively, the workers must clearly understand precisely what is expected of them. This is one of the functions of the housekeeping program.

Discipline does not mean trying to keep everybody gloriously happy—nor does it mean brow-beating or threatening the use of extreme measures. What it does mean is firmness in the belief of the justness of the rules and principles which are being followed, and a very great care that their application is fair. The supervisor must make certain that he is prepared to exercise disciplinary measures to every single subordinate, with equal impartiality. If, on reflection, the supervisor feels that he may be unable to do this, he has probably fallen victim to his own favoritism and is damaging the morale of his department.

## Selecting the Supervisor

Where selection of a supervisor for the housekeeping department becomes necessary, either because of an expanded operation, retirement, or for other reasons, it must be very carefully done. Just as the best machine operator does not necessarily make the best shop foreman, neither should a supervisor be selected by choosing the best custodian.

One of the following methods can be used for the filling of the top housekeeping department positions:

1. The promotion of someone from within the department, or someone from another part of the organization, to this position. This is not normally recommended unless the person being considered has unusually good qualifications. This may not occur in the housekeeping department, and other staffs have their own demands and desires for upgrading personnel. Employees being advanced from within to such a position are under the considerable handicap of having to "unlearn" outdated methods and attitudes.

2. Older, more experienced persons may be hired for such a position. Here we have the disadvantages of cost if the person is qualified, of attitude if he has done other types of work which he might consider to carry higher prestige, and again the problem of "unlearning" systems which may have been satisfactory in other locations and times, but perhaps which should not now be used. We must bear in mind, also, the potential number of remaining work years. Many organizations, nevertheless, have achieved excellent results in utilizing retired military personnel for this position, for example.

3. An excellent approach for a large housekeeping department is to hire a young college graduate or perhaps a person having two or three years of college training. There are advantages to this being his first job after graduation. He may be attracted to an organization because of its reputation, and he can be convinced of the importance of housekeeping and sanitation to the company. To such a man the job can become an opportunity for growth and expression. The position may well become a training ground for advancement into other positions within the organization. Certainly, he will be receiving excellent experience in dealing with subordinates and superiors, in methodology, and in liaison with most of the other departments. There is a further advantage that he would carry with him to another position, a realistic regard for the values of housekeeping and sanitation to the organization. These education qualifications may be used:
   a. A college graduate no more than thirty-five years old (less education if necessary because of budget or other circumstances).
   b. Preferably a Bachelor of Industrial Engineering or Industrial Management degree. Other degrees might be acceptable, such as Bachelor of Business Administration, particularly if some specific promotion from this position were contemplated.
   c. Ranking in the lower or middle third of his graduating class.
   d. Personality traits as can be described and measured by psychological tests selected or prepared for this purpose.

## People Need Supervision

The housekeeping department manager carries the direct responsibility for achieving the goals of the housekeeping department. The only way that a number of custodians can be controlled, in order to obtain increased productivity from them, is through close supervision. It is the responsibility of management to provide the resources and climate suitable for the exercise of this supervision. At one time or another, every department or function within an organization makes decisions which affect sanitation. These effects can be beneficial or detrimental. Thus, it is the further responsibility of management to provide the housekeeping department with an organization strong and effective enough to prevent the complete subservience of the housekeeping department to the desires of others.

Custodians need many things in order to give their best efforts. First, though, they need and *want* good supervision. We owe it to ourselves to give it to them.

# 20

# Motivating the Service Worker

OVER 100,000 supervisors are involved in the attempt to motivate 2,000,000 custodians–hundreds of thousands of other service workers are also involved.

It is a formidable problem, not so much because of the numbers, but because of the special difficulties with this type of worker. Since his is a burden activity–a "necessary evil"–not connected directly with the fundamental objective of the organization, he feels, in many cases, that he is a forgotten person. He suffers from poor status and low image due to such factors as humor made at his expense, the soil and dirty equipment connected with his work, the use of people with little or no training, the failure to provide attractive uniforms and the like.

We see all types of personnel performing service activities in terms of their age, sex, background, experience, appearance, etc. We look for the ideal worker in terms of the above factors, and cannot find him. What we do find are excellent service departments using all types of personnel–we must merely conclude that the superior service worker is one who has the advantage of an excellent supervisor and management! Management and supervision, successful in this type of motivation, have discovered that a complete program is required–there are simply no magic one-time answers. A dozen courses must be pursued simultaneously—to attempt one or two activities at a time would literally require years for benefits to be apparent.

A word of caution: any serious attempt at motivation would require a number of changes from current practice. Care must be taken that these do not conflict with various legal requirements, such as those involving the equal pay law, equal opportunity regulations, union and civil service contracts, OSHA regulations, and the like.

The largest source of complaint from service workers, and the cause of most grievances, deals with equitable treatment. Thus, the first consideration in motivation would simply be "fairness". Each worker must be treated the same under the same circumstances. For example, an organization that has a published schedule of disciplinary actions is better off than one where the discipline can be meted out according to the supervisor's whim or state of mind. Another example involves sick leave policy: only where all workers either receive, or are paid for, all sick leave each year is a policy really equitable (most policies benefit the poorer workers and penalize the better workers, since the latter do not use all their sick leave and a portion of it is taken away from them).

The service worker is a human being, and human beings very much desire individual recognition. Here are some of the ways that this can be applied to the service worker effectively:

1. Be sure that the first-line supervisor knows the name of each worker and uses it regularly. But the name must be used in standard format with each foreman either using all first names or all last names, since to mix up first, last, and nicknames would indicate a form of favoritism, even if not intended.

2. Titles are also important as they can be either negative or positive. Titles such as "Porter," "Janitor," "Laborer," or "Mop Boy" are destructive, while titles such as "Service Operator" and "Maintenance Aide" are positive.

3. A monthly recognition and award, which might be entitled "Service Operator of the Month", if properly administered, can be very beneficial. In the larger organization, each supervisor would nominate his worker, with the department head involving management in making the final decision. Weighted quality points can be used in making the selection, perhaps with perfect attendance being a requirement for nomination. The awards should carry publicity (such as in the company newspaper) and a certificate awarded at a meeting.

4. Uniforms can be a positive motivator, if the uniform is attractive and especially when the worker has been involved in the selection of the color, style, and material. Colorful embroidered shoulder patches can show the name of the department, embroidered stars or other devices can show length of service and name badges can identify the worker individually.

5. On being hired, and periodically thereafter, the worker should meet members of management or administration. This is to demonstrate to the employee that he is working at a job that is meaningful to people in higher positions. Management should make itself seen periodically on tours of inspec-

tion, at meetings and award ceremonies.

6. A personal letter to the home of the employee on being hired or transferred to the department is also very meaningful (many workers are employed for dozens of years without ever receiving a personal letter addressed to their home). Letters can also be used to congratulate the employee after one year of service, and each year thereafter, as well as for other purposes. Letters stuffed in the employee's paycheck merely indicate to him that the subject—and perhaps his recognition as well—is not worth another postage stamp.

7. Similarly, birthday cards, or other cards should be sent by the first-line supervisor to the home of the employee (perhaps with a hand written note appended) involving such matters as length of service, the worker's family, his progress, and the like.

No one likes to do a job where he has not been given the proper tools for the task. In terms of equipment we want to be sure that the employee is provided with the proper quantity, size, and type of equipment—sharing should be avoided with another worker on the same shift, as this causes dissention and a considerable loss of time, and makes it difficult to hold a single employee responsible for the upkeep of that particular piece of equipment. The equipment should be attractive, and the investment of another 20% to obtain that is worthwhile. From the standpoint of supplies and chemicals this is no place to cut the worker short! The quality of the materials is too important to his morale to use something which does not work properly. Further, be sure not to exhaust your stock, as this would convince the worker that the job is not important.

In a way, another type of "tool-for-the-job" is the training of the worker. Employees of any kind do not feel good about their work if they cannot perform it efficiently and properly. The basic rules for service worker training are fundamentally simple:

1. Use pictures of your own worker following the standard procedure which shows the work performed easily and safely, using efficient equipment and materials.

2. Involve the first-line supervisor in classroom teaching to augment on-the-job training (a training officer in a large organization might be used for general subjects and for coordinating the programs).

3. The training should be continual and never really come to an end. After all, there is the value of repetition and bringing in of new subjects.

4. The worker should receive recognition for his attendance, including a certificate signed and presented by management.

5. For supervisory service personnel the role play technique is unbeatable as a means for "supervisory practice"

The job assignment for the service worker will be extremely important since it controls the way the worker spends his time:

1. The assignment, as discussed in training, should be based on easy, safe methods, using efficient equipment and materials.

2. The assignment should represent a reasonable day's work—not more and not less.

3. Even more important, work loads should be balanced so that no employee feels that he is asked to do more than another.

4. Be sure to avoid the use of stopwatches in determining work loads—there are many other effective ways of determining time requirements.

5. Short-interval scheduling for service work can be extremely damaging to the morale of the workers. Trying to program each minute of the day suggests that the worker is a machine or a number, rather than a thinking human being. It is better to assign a worker a reasonable day's work, and to get his thinking involved in what daily variations are necessary due to occupancy, weather, availability of equipment, trouble calls and the like. For many types of service work, (area assignments are best) the employee is asked to perform all repetitive work within a geographical location.

6. The area assignments should be supported by personnel who will perform project work (infrequent activities) and will relieve absent area-assigned workers.

7. Rotation of job assignments is desirable for those workers who become bored and who request such a change—on the other hand, it is undesirable for those who enjoy their fixed assignment and want to remain in it.

8. The worker should be in an organization where promotion is possible, although not all workers desire promotion. The use of a working group leader is desirable because of the additional pay, title, and development as a possible future supervisor. The use of group leaders should only be considered if appointment can be made on merit rather than on longevity.

The supervisor and manager alike can have a positive effect on worker motivation simply by setting an example—in walking briskly thoughout a building, in avoiding reading newspapers or magazines in his office, in picking up litter to demonstrate personal interest, in fighting for the rights of the worker when he is verbally abused by others, and by pitching in in a real emergency (but not otherwise).

There is an old saying that people support those activities which they, themselves, help to design, or those things they help to create. Thus, worker participation should be obtained in as many situations as possible, such as his suggestions for changes in building design to promote more economical maintenance, selecting chemicals, materials and equipment, choosing uniforms, determining methods (although his must not be the last word here), etc.

It is clear to most people these days that neither pay nor fringe benefits is a great motivator—these things simply make it possible for us to select from a better labor market and reduce the amount of turnover.

We also generally recognize that the principal function of the supervisor—

and manager alike—is simply "to get the work out". Thus, we try to avoid the conflict of "Attitude versus Performance". It is difficult for a supervisor to act as a psychologist or psychiatrist in evaluating worker attitudes and psychological conditions, but we certainly can become involved in his performance from many standpoints.

Finally, one must not become discouraged when one finds that it is not possible to motivate all employees. The ability to motivate workers would fall into the same distribution curve as shoe sizes and other conditions involving variables. That is, most people fall in the middle, can be motivated and will be motivated with a comprehensive program as described above through the use of effective supervisors. At the thinner end of the curve there might be a few workers who are self-motivators and need no attention whatever except, of course, fair treatment and recognition where it is due. There are also people who cannot be motivated and they should simply be terminated as soon as this fact has become apparent, and as quickly as possible. For those people who complain that it takes two years to terminate someone because of union or civil service requirements, we provide the same recommendation as we do for those people who are considering a comprehensive recognition program: "Get started tomorrow morning—you don't have a day to lose!"

# 21

# Managing the Absentee Problem

NO one needs convincing that absenteeism is a serious problem, not only among building service workers but in general. Over 2½ million workers are absent each Monday and Friday! The annual cost in the United States alone is over 20 billion dollars–the costs to management include wages, social security benefits, overtime premiums and substitutes, unemployment and workmen's compensation, loss in production, replacement and training costs–while the cost to the public is in terms of higher taxes and higher prices.

But the opportunity is big, too. A one percent decrease in the absentee rate for 100 workers would mean a $15,000 annual saving.

The problem is compounded by the fact that not everyone sees this as a problem. Many workers as individuals, and many union and civil service representatives, see absenteeism as an "earned" benefit. Where management has looked the other way during good times, this viewpoint may appear to have been sanctioned.

There are many causes of absenteeism, and even the following list is not complete:

1. Illness. But not all absenteeism recorded as illness was for an actual illness; this category also includes a feigned illness, one that is imagined, one that is perhaps coming on, or an illness in the workers family.

2. Accidents, both on the job and off.
3. Problems with transportation, public or private.
4. Personal activities.
5. Personal problems, such as financial or marital problems.
6. Problems in morale—a worker who does not feel good about his job and the treatment he has been given, simply does not feel like coming to work.
7. Permissiveness - the worker feels that management doesn't really care.
8. Desire for attention (one can see the analogy with an unruly child).
9. Unfair policy, which tends to drive the worker away from his job.

Some unions are now asking for a contract which provides a certain number of "to hell with it" days. That is, if the employee does not feel like coming to work, he can simply say "to hell with it" and be paid for that day!

If you were to study an organization or a department with a history of low absenteeism, you would probably find these factors in effect:

1. The workers have the opportunity to discuss job problems with their immediate supervisors.
2. The immediate supervisor has time—and takes time—to discuss personnel matters.
3. The supervisors have developed a feeling of unity among themselves and that each is handling problems in a basically similar way.
4. Attitudes of confidence and team spirit have been developed.
5. The workers are satisfied with their opportunities for promotion and upgrading.
6. The supervisor recognizes the presence of the worker, and recognizes his work effort.
7. A training program is in effect.
8. Workers feel that they have good working conditions, and adequate supplies and equipment.

### A Program For Attendance Improvement

If you wish to improve your absenteeism rate and thereby overall performance, you must get launched into a careful program of accomplishing this. It will be well worth the time and the money!

The first step is to find out precisely what the situation is:

1. What do the absentee records show?
2. What is official policy concerning absenteeism and disciplinary action ("The official union contract and civil service agreement, if any, with addendums.")
3. Determine to what extent management is willing to enforce its rules and disciplinary procedures.
4. Establish a good system of maintaining attendance records if the existing system is not adequate.

5. Inform employees of the campaign to improve absenteeism, stressing the benefits to them in terms of job security, improved earnings, protection of sick leave benefits in case of illness and equitable treatment.

6. Examine working conditions and institute any necessary improvements.

7. Display an active interest, involving as high a level in management as possible—this display of interest should be continual and not just a "one shot" thing.

8. Discuss absences with individuals, and make it clear that good performance is expected.

9. Take disciplinary action without fail, where called for in company rules.

As part of the program for attendance improvement, it must be borne in mind that the first line supervisor is one of the keys to better attendance, and that management should work with supervisors in improving their ability to handle this situation. Certainly the first step a supervisor can take is to set a personal example in terms of his own absenteeism. Probably the one word that is most meaningful in supervisor-employee relations is the word "fair" and where workers are treated unfairly, or where they feel that favorites are being played, absenteeism is a natural result. Supervisors must understand that workers need to be dealt with equitably, humanely, and with understanding. Especially today, workers want to know "why" and need to be given adequate information. The worker wants to be involved in goal-setting and problem solving, and this should be encouraged where consistent with company policy. But throughout, the supervisor should express a positive expectation with respect to absenteeism.

To go one step higher, the absentee problem really must begin with management. As earlier indicated, management must express its intent with respect to enforcing company rules—or changing the rules if necessary—and management must be involved in the provision of adequate publicity and information concerning employee absence. Any changes in working conditions, supplies or equipment must be authorized by management.

It is best if management actually publishes the disciplinary guidelines for infractions of absenteeism rules as well as other rules and also a clearly worded sick-leave program.

### The Sick-Leave Program

Unquestionably many sick-leave programs encourage absenteeism. In the housekeeping seminars conducted by our firm, we typically make the statement that many sick-leave programs are administered in a way that penalizes the loyal employee while rewarding the poor employee. When a show of hands is asked to determine how many of the audience agree that the description describes their own sick-leave program, typically most of the audience admit that this is the case!

The typical sick-leave program provides a certain number of days available for sickness, but in practice awards the paid time off for almost any reason—or for no reason. Those organizations requiring a doctor's certificate find that this does not solve the problem, because many people are able to obtain such certificates for just a very few dollars. At the end of the year, unused benefits are either taken away (from the loyal employee), or accrued to some long-term benefit (which most workers never expect to use or do not take into consideration).

Why not consider a sick-leave program with the following arrangements, which would treat all employees perfectly equitably:

1. No sick-leave is paid for the first day of absence (some organizations do not pay sick-leave for the first two days of absence). This does not lessen the number of sick-leave days paid—it simply requires the worker to fund his own absence for a very short period time.
2. Two weeks before Christmas (when everyone needs money) unused sick leave should be split, half being paid to the employee in cash, the other half being accrued to a long-term benefit. The long-term accrual should not exceed a specific number of days (such as 90), and once that figure has been reached, then all unused sick-leave is paid in cash.

The above arrangement treats everyone equally and fairly, while discouraging the diseases of "weekend-itis" and "after-payday-itis".

Actually, sick-leave is a crutch for the worker, indicating that he does not know how to manage his own money and therefore it must be saved for when he needs it. A tremendous amount of paper work would be saved if the employee obtained the value of the sick-leave in terms of wages or salary, and then funded his own absence (using insurance, perhaps, for long term purposes). If this were done, the absenteeism rate would drop greatly. But probably management would not be interested in such an approach, thinking that once the value of the sick-leave were given to the worker in terms of additional compensation, then later on a demand would be made to reinstate the sick-leave without affecting the new compensation rate.

### Discipline and Termination Checklists

It is helpful in managing absenteeism programs to make use of two checklists. First, a list should be reviewed before discipline is administered relating to absenteeism. One needs to ask questions of oneself such as whether the rules and an explanation of the rules have been explained to all employees—whether prior verbal and written warnings have been given—are the records available concerning the employees past record of absenteeism–whether management's enforcement of its rules has been consistent–whether or not there are unusual or mitigating circumstances in the specific case and whether or not the action considered is comptable with past action for similar cases.

A second checklist should be used in preparing for termination based on absenteeism. This checklist again should question the policy as to whether or not it is reasonable, and if so whether it has been properly communicated to the employee. One must question his own objectivity in terms of treating the employee as any other employee would be treated. Again, as for discipline, has the employee been properly reprimanded verbally and in writing in the past and thus been prepared for this action. Additional items may be on the checklist, such as providing an opportunity for the employee to resign, consideration of a suspension rather than a termination, etc.

Improving one's absenteeism rate is hardly a question of taking one or two steps to solve the problem. It is true that certain details do improve the situation somewhat–for example, when a supervisor takes phone calls concerning a worker who is absent, rather than a clerk taking the phone call, this tends to decrease absenteeism. When an organization provides a reward–such as trading stamps–for a month of perfect attendance, and the reward is sent to the spouse of the worker, this also improves attendance.

But the general improvement in absenteeism, sustained over a long time, requires action on a number of fronts beginning with management and the development of a realistic and equitable sick-leave policy, the involvement and management of many aspects of the problem, through the first line supervisor in terms of the fundamental principles of fair and equitable treatment to employees. It involves moving on a front consisting of a number of interdependent factors–all of which is worth the time and effort! When we improve our absenteeism rate, we have improved a number of other things as well, basic to the healthful operation of the organization.

# 22

# Salesmanship In Management

AT A housekeeping seminar I once asked the audience of Environmental Service Directors, Executive Housekeepers, and Building Service Supervisors to hold up their hands if there were any salesmen in the group. A few hands went up. I again asked, "will the salespeople in the audience please hold up their hands". On the second try, about a dozen hands went up. After my third question, with no explanation between questions, most of the people in the audience held up their hands, because on reflection, they began to realize that selling is a part of the daily job of the manager and supervisor.

Selling is not a requirement for just "most" managers, but for "every" manager.

Selling is simply convincing others to accept your point of view. One can be a clerk and not be a salesperson, but one cannot be a manager and avoid selling. Thus, you do not have the choice of whether you will be a salesman–you only have the choice of how good a salesman you will be: good, bad, or indifferent. And this decision will inevitably affect your future.

## The "Buyers"

The Housekeeping Manager has many audiences, or "buyers" to whom he must sell if he is to function effectively:

1. Possibly the biggest sale is to one's upper management with the concept of "good housekeeping is good business". Further, we must not forget that our cleaning department is in competition with all the other departments of the organization for the limited amount of money, space, and time available from top management. We would like for funds to be allocated strictly on the basis of justice and need, but very often they are distributed on the basis of the selling job that has been done by a given department. For people operating their own "in-house" Housekeeping Departments, a great selling effort to their management is being conducted by outside organizations such as Contract Cleaning or Contract Management firms–one must competitively conduct his own selling program in order to convince management that one's own in-house department is the most effective way to perform the work.

2. Selling must be regularly done to supervisors within the department, who must be convinced that the actions being taken are, in truth, beneficial to both the organization and its personnel. Of course, we want to avoid a situation where the supervisor transmits instructions from management as if he were an errand boy–we want the supervisor to state to his workers that he himself believes that a given action should be taken and for named benefits.

3. And that brings us to selling to the workers themselves. These days, of course, the proper approach to productivity improvement is through "working smarter". Not only the department head, but every supervisor has the job of "making the sale" to the very people who are doing all the cleaning, with respect to procedures, use of equipment, chemicals, safety precautions, public relations and the like.

4. Yes, the public must be sold so that we may advance the professionalism of building sanitation operations–so that we can develop the image of the custodial worker and so that we can secure cooperation from visitors and other employees alike in our cleaning effort.

The rules of salesmanship are simple, although thousands of books have been written on the subject. And, like public speaking or writing, one cannot become a salesman by reading books about it–one must practice at it. If these three simple rules are followed, most sales will be made:

1. If you want something from someone else, it is obvious that the first step is to ask that person for the thing that you desire. As obvious as this step is, most people ignore it. They complain that they are not getting cooperation from others, but they have not asked those others for their cooperation! People simply do not normally volunteer to do the things that you want them to do.

2. However, a single request for your "customer" to accept your viewpoint is rarely sufficient. Most professional salesmen point out that their average sale is made after six or seven or even more rejections! Thus, it is persistence and repetition that is so important. Let me share with you one of my favorite quotations: "Nothing in the world can take the place of persistence. Talent will not–nothing is more common than unsuccessful men with talent. Genius will not–unrewarded genius is almost a proverb. Education will not–the world is full of educated derelicts. Persistence and determination alone are omnipotent."

3. Obviously, it will be difficult to be a good salesman unless you know and believe in your 'product"; or, in your case, the service that you are providing and its benefits.

4. The last rule is to simply capitalize on human nature: most people instinctively question "what's in it for me?". You buy something from a salesman, not because it will benefit him but because it will benefit you. You purchase an insurance policy not because it makes the salesman feel better, but because you are obtaining protection for your family or your property. The best form of selling is when you are demonstrating to the buyer the benefits of your proposition to him.

As W. Clement Stone said, "If a person has nothing to lose by giving, and you have lots to gain by asking, give him the opportunity to grant you the favor that costs him nothing."

As a profession, selling is almost unique with respect to its possible earnings compared with the education, experience, intelligence, or other factors normally considered fundamental to success. It is a useful—if not indispensable—aspect of every leadership job. Whatever you may do in the future, improving your salesmanship will be beneficial.

With this in mind, although reading a book on the subject will not make you a salesman, perhaps your next book should be on selling, to better prepare you for this.

Oh, yes, you'll have to get used to rejection. Some people find it difficult to sell because they cannot get used to being told "no", sometimes in very forcible terms. But remember that your boss might think that you are not even serious about a topic unless you bring it up a half dozen times! And that "big sale" you make will more than make up for a long list of turn-downs.

In becoming a better salesman, you will find yourself, inevitably, becoming a public speaker, a teacher, a psychologist, and more. You'll become more interested in your subject, you will learn something about body language, timing, and more.

And your future will not only look brighter—it will be brighter.

# 23

# Training the Custodian

OF the roughly half-million personnel involved in housekeeping and sanitation in the United States, only one in ten has had any formal training in cleaning techniques, and probably less than one in twenty has had such training on a regular or periodic basis. This exists, despite the fact that the turnover rate is relatively high in housekeeping operations, and that new methods, materials, and equipment are coming into use more and more rapidly.

Many industrial housekeeping departments of today are amorphous, static organizations which may have remained unchanged over a considerable period of time. In such an arrangement, there is considerable experience in outmoded methods; their perpetuation calls for minimal supervisory ability and performance. Probably training classes have been attempted in such an organization but perhaps have been dropped after a few sessions because they were troublesome and results were not immediately observed. Under such conditions, the training of a new housekeeping worker may have consisted simply of handing him his cleaning tools and putting him on the job with a "buddy" for a few days, after which he is placed on his own. The new employee, in turn, perhaps less than a year later, similarly "teaches" another new worker.

The potential savings available in housekeeping, in direct dollar savings or improved levels of housekeeping, result from the relatively low productivity of housekeeping personnel. This undeveloped productivity cannot be utilized without a serious and continuing training program. Although the housekeeping program should indicate the training program procedure, and although the consultant should actually conduct the initial training courses, it will eventually be the job of the supervisor to continue the training function. Even where the program has been fully implemented and is achieving the basic goals set for it, however, it is still desirable from time to time for training programs to be conducted by outside personnel. This brings a freshness to the approach, excites new interest on the part of the custodians and helps to keep the department up to date. No matter how mechanized the cleaning operation becomes in the years ahead, we will still depend on the individual custodian for the housekeeping work, whether he utilizes machinery or common hand equipment. Less than a generation ago, the suggestion to train the janitors of that day would surely have been met with derision. Today, such training is essential.

The investment in plant and equipment per worker is constantly rising. Management has certainly already realized that the entrusting of expensive production equipment to untrained machinists, or of complex electronic data processors to untrained clerical personnel, is an unsound practice that can be extremely costly. Slowly, this attitude is being developed concerning the training of the custodian.

An untrained custodian in the modern industrial environment can also be costly. Modern finishes for paneling and furniture now so favored by architects can be damaged by the use of the wrong cleaning materials. Wasted motion means that some housekeeping jobs are not going to get done properly or on time. Housekeeping tools and equipment are abused and become unnecessarily short lived. Accidents may occur, and unnecessary fire hazards will exist.

A sound training program carefully conceived and earnestly carried out must be a part of every effective housekeeping program. A training program is beneficial not only to show the worker the best way to perform a given job, but also, and equally important, it has a tremendous morale-building value.

The training program becomes a link connecting past practices with the new program. The soundness of this link has a considerable effect on the overall soundness of the program as a whole. Not only must the program be continuous, but it should also actually be a permanent part of the housekeeping operation. Continuous training and retraining is the only way we can meet the changing conditions that are certain to occur.

## Organizing the Orientation and Training Program

The training program, to be effective, requires careful planning. These steps are involved:

1. There must be a clear definition of the needs and goals of the program. A training program should not be designed to entertain the custodians, but rather to give them the information they need to help them grow in their jobs.
2. Responsibility for carrying out the program should be placed with an individual who, in turn, may delegate responsibility for various phases of the training. Full support must be given to the training leader.
3. The program must be properly organized. Meeting places must be selected, training aids obtained, training methods determined, and arrangements made with participants.
4. The training schedule should be set up at least a month in advance and for a period of a number of weeks or months. Care must be taken so that the dates selected for various meetings will not conflict with other company activities, holidays, vacations, etc. Such a schedule should indicate dates and times, subject matter, participants, and the location.
5. Make certain that the personnel who are actually going to conduct the training classes are qualified to do so. Additional qualification may be gained through reading, or the teachers may actually undergo certain phases of the training themselves in such techniques as public speaking, industrial psychology, training aids, etc. The trainer must know his subject in order to hold the attention and respect of the audience.
6. Training methods should be reevaluated at intervals to make certain the proper subjects are being covered, the custodians are actually learning from the classes, and the best methods are being used.

## Orienting the New Employee

First impressions are very meaningful. They are often never forgotten. This principle can be used to very good advantage if the new custodian reporting to work is properly greeted and oriented.

The new custodian should be greeted by his direct supervisor, with an informal discussion. The supervisor can help to make a new employee feel "at home" and at ease by asking questions which will get the custodian to talk about his former work experiences, military duty, size of his family, birthplace, or things that the two persons have in common, such as some physical characteristic or name similarity.

The company's attitude toward housekeeping, the objectives of the housekeeping department, and the general nature of the housekeeping program are discussed. The importance of the individual custodian's work toward the attaining of overall company objectives is stressed.

A number of industrial organizations provide a general orientation concerning the company, its products, and its place in the community before the worker reports to his department. Therefore, the extent of orientation within the housekeeping department will depend on whether or not a general orientation has been provided. In either event, the employee must understand the general operating rules of the company and the specific rules of the housekeeping department.

The new custodian should be properly introduced to the department manager and other supervisors, as well as to his fellow employees. In the large operation, initial introduction should be made only to those custodians in one supervisory section so as to avoid confusion. The introduction should include an identification of each employee's duties and how they affect the new custodian. He should be taken on a familiarization tour of the plant. In the very large plant, he should be shown a map of the entire facility and given an actual tour of the buildings in which he will work. The tour should include:

1. Familiarization of the premises to prevent becoming lost and to put aside fears of the unknown.
2. Observations of custodians performing their cleaning tasks.
3. Pointing out examples of both good and bad housekeeping techniques and results.
4. Housekeeping facilities, such as the central storage area and closets, cabinets, location of water taps and drains, etc.
5. Personnel facilities such as rest rooms and locker rooms, water fountains, etc.
6. First-aid facilities including first-aid cabinets, dispensary, eye-wash fountains, fire blankets, etc.
7. Food service and vending areas.
8. Fire warning and extinguishing equipment.

The new custodian should be shown his work station, or the route of his work tour. He should be given the program data and other literature affecting his work.

The new custodian is now ready for his on-the-job training.

## On-the-Job Training

Most workers want to receive specific work instructions. It is only when the worker knows exactly what is expected of him that he can do his best and feel secure.

The training of the worker in more than one job is often desirable, but the capacities of the individual must be considered. A person hired as a relief worker would naturally have to be trained for all the jobs that he might be called on to perform. In this regard, in order to improve flexibility, employees can be rotated from one job to another at periodic intervals for on-the-job training under group leaders or other supervisors.

Where turnover is high, the supervisors will find themselves spending a great deal of time with on-the-job training if adequate instructions are to be given. This situation can be improved through the use of group leaders or a temporary specialist trainer, if properly qualified.

On-the-job training should not be considered complete after a given time interval but should be continued until the individual has demonstrated suitable understanding and proficiency.

Let us consider the four basic steps in the teaching process:

1. PREPARATION. Put the worker at ease. Explain the benefits of the methods to be shown him and the fact that you expect that he will not get it exactly right without some practice. Discuss possible hazards of the work, safety practices, and the materials and equipment to be used. Be sure that the equipment is in good condition and that the materials are uncontaminated or unspoiled. Be sure the custodian's working facilities are properly arranged and outfitted. Keys should be made available, or a procedure set up for unlocking doors that are normally kept locked. Written instruction sheets should be reviewed.

2. DEMONSTRATION. Explain the operation to the worker. Proceed one step at a time and only as quickly as you feel that the information is being absorbed. Give him an actual demonstration of the work, moving slowly and methodically. If the operation is difficult, first give a demonstration at slow speed, then follow it with a demonstration at normal speed. The "before-and-after" method can be used to demonstrate the results of good housekeeping technique. For example, clean one half of a lobby or rest room and compare it with the half not cleaned. At all stages in the training, give the custodian an opportunity to ask questions. Answer his questions carefully and with patience.

3. APPLICATION. When he feels he is ready, let the custodian demonstrate the job back to you. He should also be able to tell you *why* the various steps are being taken. Encourage him to look at written instruction sheets if he feels this is necessary. Again, try and keep the custodian at ease by using patience and

understanding. Do not stand too close. Compliment and encourage him. If he becomes confused, give him another complete demonstration, or take a break for refreshments. When you are both satisfied, congratulate him on learning this new skill and let him perform the work alone and unassisted.

4. INSPECTION. Check the work often, giving the custodian plenty of opportunity to ask questions. He will want to know from you whether satisfactory progress is being made. Do not hesitate to tell him what needs correction, but first compliment the part of the work that has been done properly. And remember, training should never involve criticisms. If the worker has not properly learned the job, it is best to blame yourself for incomplete motivation or instruction.

## Arranging a Training Class

Training classes pay considerable dividends in improved productivity. The cost of training classes is primarily in the supervisory time necessary to organize and conduct the classes, the time of the workers actually spent away from their work because of the classes, and also may include other costs. Unless suitable arrangements are made for a training meeting, time will be lost, and retention will be reduced.

Here are the steps needed for a complete arrangement:

1. Arrange the proper time and date with the individual responsible for coordinating meeting rooms or auditoriums. Provide the coordinator with a complete schedule of future meetings for his benefit. Check back on each proposed meeting the day before to make certain that all arrangements have been taken care of.
2. Check to be sure that proper announcements have been made and that all custodians who are to attend have been notified. It is also desirable to invite representatives from other departments who have an interest in cleaning, such as quality-control personnel, maintenance personnel, etc.
3. If the meeting place is not a familiar location to all, simple directions should be given as to its location, or the group should go to that place as a body.
4. Physical conditions of the meeting place should be as well prepared as possible, including comfortable heating or air conditioning, sufficient lighting, sound and noise control, etc.
5. Be certain that there will be enough seats for everyone and that they are properly arranged.
6. For those who will address the audience, there may be a need for a podium or lecturn. Of course, there should be a table, chairs, and a water carafe.
7. If the meeting is to be held to coincide with a break period, refreshments may be suitable. This would also be true if a meeting were lengthy, or held after usual working hours. Be sure that all materials and equipment to be used are on hand and in proper working condition.

## Training Aids

Training aids promote learning because they supplement and relieve verbal instruction and information with material that appeals to the senses. Pedagogy is definitely improved through increased retention, interest, and appeal when the subject matter can be seen, felt, smelled, or heard.

There is seldom a danger of too many training aids. Here are some useful training aids:

1. Flip charts can be used to emphasize key points in a presentation. The best flip charts contain only a few words in large print, are colorful, and contain illustrative cartoons, which are sometimes humorous.
2. Posters may be used to convey key points or morals and can remain on view throughout the session.
3. Sample products and equipment can be passed around or examined.
4. Chalk boards or sketch pads can be used for drawings or diagrams.
5. Various types of literature may be made available or passed around, such as program instruction sheets, product labels, trade magazines, manufacturers' literature, etc. Printed material may also be shown on a screen with a suitable projector.
6. Laboratory-type demonstrations are very effective. These might include demonstrations of surface tension, chemical hazards, emulsification, spontaneous combustion, etc.
7. The most useful training aid is the motion picture. A good second choice is a strip film with phonograph record.

By all means, visual aids, particularly motion pictures and strip films, should be tested out before the meeting. Here are some points to watch:

1. Locate the projector and screen where they will not obstruct vision and can be seen by the entire audience.
2. Designate an individual to control the darkening of the room by closing blinds or pulling shades and turning out lights. These should all be checked out in advance.
3. The film should be wound onto the projector and actually run for one or two minutes to make certain that everything is in proper order. This, of course, will require suitable current and extension cords.
4. It is best to let an experienced person handle the actual projection. He should be supplied with extra bulbs, fuses, or other parts of the projector that may fail during the showing.

## The First Training Class

The conduct of the first training class is extremely important. It should be the vehicle for orienting all cleaning personnel to the housekeeping program and will set the tone and pattern for all classes to follow. Do not let it be "the big one that got away!"

Although subsequent training classes may be broken up into groups of only fifteen to twenty persons, for greater effectiveness the first meeting should contain the entire housekeeping force, if at all possible. This will tend to emphasize the team effort viewpoint and will help to build personal pride in the housekeeping department.

The basic purposes of the orientation meeting are to *inform, reassure,* and *create enthusiasm*.

INFORM. Begin the meeting with an honest description of its purposes. Make whatever introductions are necessary. Point out that housekeeping is much more important to the success of the company than most people realize and that management's realization of this fact has led to an engineered program especially designed for the housekeeping department. Point out examples where planning and organization by qualified specialists, depending heavily on wide experience with persons just like those in the audience, bring a number of benefits to the housekeeping department, to their fellow workers in other departments, and to management alike.

REASSURE. Point out that the training classes particularly, as well as the program in general, have been set up as a benefit to the custodians so that they may work *smarter* rather than *harder*. This can be done through the use of timesaving equipment and materials, by using ways of cleaning that are less tiring and more effective, by arranging the work so that some things are not done unnecessarily or too often, and by receiving cooperation from others. Point out that the program will be carefully paced and that improvements will be made carefully and without friction or confusion. If management has agreed to this point, guarantee that the program will not result in the loss of any jobs for anyone in the audience. Be sure that there is plenty of opportunity for questions on these points.

CREATE ENTHUSIASM. Show that it is through the custodians themselves, and their spirit of interest and cooperation, that the housekeeping department will achieve its goals and finally get the recognition and cooperation that it deserves. Describe how the program works, and the fact that the housekeeping department, in working with the help of the program, depends on the custodian for advice and suggestions, since he is the man "on the firing line" and most directly involved with the problems of cleaning.

Wind up the meeting with a short training film such as "The Importance of Sanitation" or some sample demonstrations of how the most simple jobs might be done easier and more effectively. Finally, ask for questions or suggestions, or if there is any part of what has been said that requires repeating or clarifying.

## Training Classes

If the custodians are to learn improved methods and later put them into use, the training classes must be conducted in a way that encourages learning. A friendly, informal atmosphere and approach will help to remove tension, suspicion, and anxiety. Before the meeting actually begins, talk to some of the individual custodians about their families or hobbies or compliment them about truly good work or attitude. Begin the meeting with a good, clean joke that relates to cleaning, holding meetings, or the like. At times like these, audiences react particularly well when the speaker tells a joke on himself somehow dealing with the subject matter. Do not belittle the importance of the meeting, but keep it informal.

By all means, introduce guests from outside the company or from other departments. This is also a good opportunity to introduce new members of the housekeeping department or to make announcements such as forthcoming retirements, holidays, etc.

If the leader wishes, an invocation is certainly in order. This may be given by a custodian who is an official in his church or may be rotated among volunteers. Individual participation is desirable.

Training classes should always be opened and coordinated by the housekeeping department manager. The actual presentations may be given by him or by others, but no individual should speak continuously for more than fifteen or twenty minutes. In addition to the housekeeping department manager, talks and demonstrations may be given by:

Management representatives
Housekeeping department supervisors
Custodians with special skills
Supplier representatives
Consultants
Guests from other departments (methods, safety, medical, and industrial relations, etc.)
Teachers from local universities

Training classes should be limited to an hour (forty-five minutes without training aids) for these reasons:

1. The span of attention of the typical custodian is limited. This is true particularly when the course departs from tangibles such as equipment or materials

and goes into abstractions such as cooperation and morale. Here the ability of the custodian to concentrate weakens.

2. In any kind of training, retentive value increases when training sessions are shorter and more frequent rather than being longer and more rarely held.

3. Morale is improved, as well, with shorter classes held more often.

4. It is possible for many of the workers, with a little extra effort, to "catch up" on their work after the loss of an hour. When more time than this is lost, it becomes very difficult to make it up.

5. Training classes should be anticipated by the workers. If they are overly long they will tend to be boring and oppressive.

"The Art of Persuasive Talking," by Surles and Stanburg, contains ideas which help the speaker to obtain response and action from his audience.

Meetings should be held at regular intervals, such as every two weeks in the early stages, and later on each month. Ample notice should be given, and the selection of time and location should create the least possible hardship for the attendants.

The number of housekeeping training subjects is impressively large. The more important subjects, such as safety, floor maintenance, and the care of equipment, may require more than one session and may be repeated after a time.

Once a subject has been covered in a training class or demonstration, it cannot be merely put aside as "completed." The average person retains only a percentage—often as low as 25 percent or less—of what he is told or shown on any given occasion. Training subjects will have to be repeated, the frequency and total number of repetitions depending upon the complexity of the subject, whether or not the subject is interesting or dull, the skill of the trainer, the ability of the audience to learn, etc. It has often been said that in order to teach people something you must:

Tell them what you are going to tell them
Tell them
Tell them what you have told them
Remind them of what you told them!

All classroom training should be followed up by on-the-job training and inspection.

The purchase of training certificates is well worth their small cost. These certificates may be awarded after twelve classes without absence, and after the taking of a suitable qualification test.

All training sessions should provide a period of time at the end for discussion, questions, and final remarks.

# Certificate of Achievement

_______________________________

*has participated in*

**A SEMINAR IN**

**HOUSEKEEPING MANAGEMENT & SUPERVISION**

*and is accordingly awarded this*

*certificate of achievement.*

**Presented by:**
**Service Engineering Associates, Inc.**
**Housekeeping Consultants**
**Atlanta, Georgia 30319**

_______________________________
Service Engineering Associates, Inc.

_______________________________

An attendance certificate helps to promote the professionalism of a training program.

# 24

# More on Training

"TELL me about your housekeeping training program," suppose someone asked you.

We find that many housekeeping department managers give one or more of the following answers to that question:

"We have so many things to do, I can't take time out for training."
"My turnover is so high, what's the use?"
"I tried it once, and couldn't see any results, so we stopped."
"My people don't like them."

And yet, despite these answers, there is a logic in the consideration of training that is inescapable. Let's examine this syllogism:

1. If we are going to obtain any improvement, it is necessary that we make change; obviously, without change, things must stay the way they are.
2. To bring about a change requires a knowledge of the nature of the change on the part of those who are responsible for making the transition.
3. Except for the few people who are self-taught, the operative worker must gain this knowledge from another party; of course,

the transmission of information from one party to another can be called teaching or training.

4. Thus, leaving out the intermediate steps, improvement requires training.

We can only really begin to evaluate custodial training when we accept the fact that it is an unavoidable part of any program of quality or cost improvement in cleaning.

But what about the man who said that he tried it once, didn't see any benefits, and then stopped? One of the most common mistakes in the conducting of training classes is the expectation of immediate benefits. There are four reasons we should not expect change to happen right away:

1. Maids and custodians have a natural "let's wait and see what happens" attitude, and will remain rather aloof and reserved until they are confident that the training classes are not in the nature of a short-term experiment, but are more on the nature of a serious, permanent activity.
2. The skilled teacher realizes the fundamental importance of repetition in the learning process. It is a matter of the forgetting rate: so much is forgotten so quickly, that it requires a number of repetitions to make it a matter of ready memory.
3. In many cases, the training involves techniques that differ or reverse activities that may have been done for a number of years by the trainee, over thousands of repetitions. One has a "selling job" to do to convince the listener of the desirability of change.
4. Much of the value of a training program is in terms of worker morale, which is difficult to assess, especially in the early stages.

And, the one man who says he doesn't have time to give training? He is creating a vicious circle for himself, and, as the pressures grow, he will have less and less time to give the training that becomes more and more important, since it is so fundamental to achieving a higher productivity—which is really what he needs to gain more time! He would be better off to break the cycle now, to make time for the training classes, arranging them in such a way that they least interfere with the necessary work. After all, we are only talking about an average of some five minutes a day for classroom training!

The man who says he can't train because of his high turnover, I'm afraid, may well have things backwards, because that is just the time when training is most essential. His problem may well be the method of scheduling and assignment, to permit people to do simplified, repetitive jobs that require a small amount of training, but train he must in order to be able to utilize personnel for even a short period of time while avoiding damage to surfaces and equipment, eliminating safety and fire hazards, and creating conditions that have to be corrected later.

Training of cleaning employees should begin when they are first hired. From this moment, habits, practices and working rules should be put into effect. Orientation, or initial training, is a one-time opportunity, and first impressions are very lasting. It would be desirable to prepare a booklet or notebook containing fundamental information on company policies, rules and procedures to show or present to the new worker. He should then be introduced to his fellow employees, department heads with whom he will come in contact, as well as people he may need to go to for help. A familiarization tour is in order, with special emphasis on fire protection devices and personal facilities, as well as on the location of maintenance items. Naturally, he should be shown his work area and the duties which he is to perform.

In addition to the on-the-job training, which should be continual, we should plan a classroom training series very carefully. But no housekeeping training program should be started unless there is a serious intent to continue it for a reasonable length of time—at least a year. It would be better not to have *any* training classes than to have one or two and then abandon the project, since this would seem to prove to workers that management either considers the whole situation hopeless or of little importance.

The watchword in training is *selling;* we must stress the personal benefit to the worker. Simply, we want to "work smarter, not harder." The worker, basically, is a human being who is primarily interested in himself and the things that happen to him during his working day. He is much more interested in the fact that a new technique will cause him less fatigue than he is that his employer will make another few thousand dollars profit; he wants to protect himself from injury more than he wants to make his boss happy; he wants people to compliment his work more than he wants to give his organization a better image. Quite plainly, he wants to know "what's in it for me?"—and your changes will indeed do something for him as well as for your organization if they are to work properly.

Here are some hints for developing an effective housekeeping training program:

1. Establish the needs of the workers and the goals of the training. These will differ for various types of organizations.
2. Standardize on techniques, materials and equipment. There's no point in training employees in a system that is inefficient or undeveloped.
3. Place responsibility for the program with an individual who may then delegate responsibility for various phases of training.
4. Set the calendar at least a month in advance (try to use the same week day of each month) so everyone involved will know when and where training is to take place. The schedule should indicate dates, times, subject matter, program duration, participants and location. By all means, training should be conducted on organization time.
5. Subjects and general content should be laid out far in advance, as should meeting place be selected, training aids obtained, training methods determined, and participants enlisted.
6. Be sure instructors are well qualified. Teachers may gain additional qualification through reading or undertaking training themselves in public speaking, leadership psychology, training aids, etc. A worker particularly skilled in using a certain piece of equipment could, for example, assist with a demonstration. Suppliers can aid in training and actually provide training tools.
7. Invite persons who do not perform cleaning as a regular job, but who have some contact with your program, to be guests at certain meetings where they are involved, such as a safety engineer or personnel director.
8. Re-evaluate training methods periodically to be sure proper subjects are covered and that trainees are actually learning the best methods from the classes.
9. Choose the audience for each class separately; don't include people who will have absolutely no interest in the subject; their boredom will infect others.

10. Use visual training aids. People learn more and retain more when they can see as well as hear. Let the aids show your own personnel at work.
11. Appeal to the individual wherever possible; show him how learning the new technique or using the new equipment will benefit *him.*
12. Keep the sessions short. Cleaning personnel normally do not have a span of attention longer than an hour; it is recommended that classes do not exceed 45 minutes without training aids.
13. Keep it light and informal; loosen up the audience with humor; refreshments will help keep the training class a pleasant event.
14. The use of attention-getting devices gives you a better chance of a topic being remembered. One can use hidden items which are suddenly exposed; an unexpected result of a demonstration; an unusual noise or movement; an unusual slide, such as one upside down; or an unusual prop.
15. Enthusiasm will be the most valuable tool of the person conducting the training sessions; it will create a spirit of interest and cooperation, without which learning rarely takes place.

Traditionally, the cleaning worker has less education and receives lower pay than any other employee in his building or department. Frequently, though conscientious and hard-working, he has neither the skill nor the competence required of the job and little or no status, recognition, or chance for advancement. Since higher caliber, skilled workers are, for the most part, neither available nor interested in cleaning work, this is your housekeeping labor market. These are the people, with their typical defects and limitations, who must be trained and developed to meet their job requirements in an efficient manner.

Developing personnel to put modern cleaning technology to work is no easy job. But, as every cleaning administrator realizes, the effectiveness and productivity of his staff rests with the success or failure of his training program. Further, the training program is a great morale builder, since the worker feels that finally someone is interested in him and what he is doing. When combined with good leadership, an equitable wage system, provision of worthwhile equipment and materials, organization-wide publicity and recognition, selection of titles, and incentive and bonus systems, a comprehensive training program—carefully conceived and earnestly carried out—will result in top-quality, high productivity cleaning at reduced cost. Custodians and maids cannot be expected to fulfill the objective of their work unless they are aware of that objective, identify their needs with it, know what is expected of them, and how they are to perform.

Despite all these things, we are faced with the sobering reality that probably less than one out of twenty of the nearly one million cleaning workers in the United States has received adequate training for his job.

Now, what about *your* training program? Have you got one? If not, why not start today!

# 25

# The Body At Work

THE brain is a great asset in operative work, as it finds means to improve procedures, equipment, materials, scheduling–and so many other factors, in order to get more done with less effort.

But other than operations performed mechanically, it is the body that actually performs the work.

Proper use of that body (which means using one's head, again!) protects the worker from injury, strain, and undue fatigue. (Fatigue is a natural aspect of every type of work, and we shouldn't feel concerned or guilty that we are managing people who become fatigued–but we want to reduce the fatigue to reasonable proportions).

Let's review some of the more basic aspects of the use of the bones, ligaments, tendons, and muscles, often called "body mechanics", or, more technically, kinesiology.

### Work Avoidance

As in so many human activities, the questioning attitude may save a great part or all of the physical effort. Consider these questions:

1. Can the load be shared? Example: Two workers, rather than one, lift a box.

2. Can the load be reduced? Example: Buy supplies in one-gallon jugs rather than fifty five-gallon drums.

3. Can the work be performed mechanically? Example: Rather than manually handling buckets of water and swinging a mop, utilizing an automatic scrubbing machine.

4. Can the load be moved in another way? Example: A load can be moved on rollers, rather than lifted.

5. Should someone else do the work? Example: A person more physically suited might move a load, rather than someone who is slight or frail (although this may appear to conflict with the equality of work, it is a very practical choice in some cases).

6. Can the load be handled in parts? Example: Rather than lifting a case of six one-gallon containers, remove two, one in each hand.

7. Can the force of gravity be used to advantage? Example: Slide a box down a plank rather than carrying it.

8. Is a specific device available to perform the work? Example: Use an automatic stair-climbing machine rather than manual labor.

### The Use Of The Muscles

The basic thing to remember in body mechanics, concerning muscle use, is to use the largest muscle possible to do the work. The smaller muscles, such as those of the back or arm, are much more susceptible to strain or injury than the large muscles, such as those in the leg. A simple example is with the use of the dust mop: in cleaning a corridor, swinging the mop from side to side not only creates a great deal of dust, and fatigues the worker, but it is possible to cause a strain of the muscles. By placing the handle of the mop stick on the hip, or locking the elbows against the body so that the hands don't move, one can walk straight ahead, using primarily the large muscles of the leg.

The same would be true in lifting: one needs to lift primarily with the large leg muscles by squatting down and bringing the load near the body, rather than attempting to lift with the small back and arm muscles by bending over and picking up a weight.

One needs to master the "control" of mechanical equipment, rather than attempting to use it by "force". While I was conducting a training class on the use of floor machines for a group of women, one of the ladies in the audience remarked that she was fearful of operating a machine that weighed over a hundred pounds and had a motor larger than one horsepower. I remarked that I once taught a young 18 year old lady, who weighed less than one hundred pounds herself, to operate a machine weighing over two tons with a motor one hundred times larger than that on the floor machine! I then explained that I was speaking of teaching my young daughter to operate our automobile.

Body mechanics in this and other aspects unfortunately is a regular subject of training in too few custodial departments.

In considering the wonderful machine that is the human body in its general terms, we observe two things: First, the body is bilaterally symmetrical (that is, the right side is generally a mirror-image of the left side), and secondly, the appendages (hands, arms, etc.) work in radials, so that the hand pivots around the wrist, the forearm around the elbow, etc. This helps us understand that work is simplified when the hands, for example, work in opposition and in circular motions rather than straight or reciprocal motions. A simple example of this is the erasing of a chalk board: the most difficult way is for the worker to use one hand and eraser in straight-line patterns–it is much easier (and quicker) to use two erasers working in circular patterns, the right hand moving, say, counterclockwise and the left hand clockwise.

## The Rules Of The Game

Here are some guidelines for improving safety and reducing fatigue in body activities:

1. Wear protective devices, such as gloves and safety shoes when handling loads.
2. Make sure the path is clear, without obstructions or projections, before attempting to move something heavy.
3. Similarly, be sure that when moving the object, one's vision is not obstructed.
4. When two people share a load, it should be decided in advance who will be group leader, so that only one person is giving signals.
5. Where two people are involved, it would be better if they were of similar size.
6. Study the load; when lifting it, be sure something is not knocked over, or when putting it down, be sure that something else does not fall.
7. Give the load adequate physical support: attempting to use too small a moving device, or a rack that is not strong enough, could be serious.
8. Wherever possible, push rather than pull.
9. Where possible, move a weight on wheels or rollers rather than by sliding (wouldn't this save a lot of damaged floors!).
10. Use a device where available, or where it can be purchased, such as a desk-moving device, or file cabinet mover.
11. Avoid lifting and moving on slippery floors, which may be caused by excessive dust, water, paper clips, etc.
12. When carrying a can or box, keep the load close to the body: the further away the load is from the backbone, the less mechanical advantage one has.
13. When lifting above the shoulders, use a sturdy platform so as not to unduly extend the arms and strain the back.

14. Do not use the top step of a step ladder, since the legs are not braced.

15. Be sure the feet of a straight ladder are secured (the author once had a painful and embarrassing injury of this type, the ladder simply slid back on a wet floor, and dropped him ten feet).

16. Take a secure grip on the load before the full weight is carried, and try not to change the grip while the load is being carried or moved.

17. For those containers that are moved regularly, consider attaching handles or other devices that simplify lifting.

18. Avoid twisting the body when carrying or moving a heavy load.

19. In lifting loads from the floor or a low position, be sure to bend the knees, not the back.

20. When placing a load on a horizontal surface, put it down on the edge of the surface, and slide it back, to avoid holding the load out from the body.

21. Move slowly and gradually—avoid rapid or jerky motions (remember inertia: starting something into motion is much more difficult than keeping it in motion).

22. If a low load to be lifted is small enough, it will help to straddle it with the legs and then follow the correct lifting procedure (again, this brings the load closer to the backbone).

23. If two people are moving an object on a stairway, if going up, the lower person should always face the load and the upper person may face the load or face away from it—but going down, both persons should face the load.

24. In carrying long objects, such as pipes or ladders, it is best to use two people even if they are not heavy, to prevent striking someone in the face, or damaging surfaces.

25. Remember that lifting for a heavy person is more dangerous than for a person of normal weight, for two reasons: the load is further from the backbone, and there is a greater strain on the heart.

26. A habit of good posture—walking or sitting with the backbone nearly straight—improves health in general, while better conditioning the body for physical exercise.

27. People with chronic back problems, of course, need to get professional medical advice—but some things help almost everyone, such as sleeping on a firm bed, taking regular exercise, keeping the knees bent when sitting, etc. The best and safest exercises appear to be walking and swimming.

28. To avoid accidents, remove any grease, oil or other slippery material from the object to be carried, as well as from the hands or gloves.

29. Do not attempt to move or lift a load that appears beyond your safe capacity, or that is in any way hazardous.

## "TAKE IT EASY"

A professional dancer makes difficult steps look simple. A virtuoso violinist makes the playing of the violin appear as if it were child's play and a professional custodian makes all the hard cleaning work look easy.

Actually, effective and safe cleaning activities not only appear to be easy, they are easy.

The sanitation worker who is struggling with the floor machine, who is trying to mop the corridor in a single pass, who is flinging the dust mop around the floor is straining, using a lot of energy, and getting little work done–if not actually creating more work to be done later. It is the worker who controls the machine, swings the mop in a comfortable pattern, pushes the dust mop in a straight line that gets the job done and with no discomfort or injury to himself.

And he knows that three out of ten accidents which cause a temporary injury are the result of moving objects in the wrong way.

Look easy, look safe –be easy, be safe.

# 26

# Safety and Fire Prevention

INDUSTRIAL accidents take an enormous annual toll in human misery, business expense, and social losses. Safety engineers point out that most of these accidents are preventable. A large percentage of industrial accidents are related to *housekeeping* practices.

The National Safety Council reports that the direct costs of injuries resulting from around 2 *million* nonvehicular accidents amounts to several *billion* dollars. Each year it is estimated that accidents cost the United States *a half million* man-years in lost time.

The total extent of the cost of any given accident, to the individual, industry, and society, is surprisingly high. Money is spent, as the result of an accident, many months or even years after the occurrence. Such costs and expenses represent a complete and total waste.

1. Payment of compensation claims.
2. Payment of noncompensated medical expenses.
3. The lost income of the injured employee.
4. Lost time of fellow workers who stop to help the injured worker or merely to watch.

5. Loss of time of supervisors, staff personnel, and management in investigating the accident, helping the injured worker, training new workers, etc.
6. Damaged equipment or materials directly involved in the accident.
7. Impaired quality on the part of other workers because of emotional involvement.
8. Related accidents by other workers because of emotional concern.
9. Decreased production rate due to diverted interest or upset of workers.
10. Lost production because of machine stoppage.
11. Decreased effectiveness of the injured employee after his return to work.
12. Lost business or good will resulting from a decreased production rate, poor quality, etc.
13. Legal expenses and costs of litigation.
14. The general loss to society and industry because of reduced industrial skill, potential, and capacity and loss in earning power.

From the above list it can be seen that accident prevention can be a far more profitable investment than is usually realized.

Some supervisors, who are not familiar with the principles of insurance, consider compensation insurance costs as a fixed expense. Generally speaking, however, most liability contracts include a feature called the *experience rating*. This is an incentive feature which lowers premiums as the result of a low-accident rate, but also increases the premiums following a high-accident rate.

The Walsh-Healey Public Contract Act affects all industries and all businesses involving contracts of $10,000 or more with the government. This Act recognizes the importance of industrial safety and the fundamental need to protect the worker, which, in turn, must benefit the industry. The Act states that: "no part of such contract will be performed nor will any of the materials . . . furnished under said contract be manufactured . . . under working conditions which are unsanitary or hazardous or dangerous to the health and safety of employees. . . ."

Any serious safety or fire prevention campaign must directly involve housekeeping. In its *Bulletin No. 1237,* the Bureau of Labor Statistics of the United States Labor Department analyzes fifty-seven different accidents in boiler-shop and foundry-type operations. Of these, approximately 15 percent were *directly* traced to poor housekeeping practices. Again and again in the bulletin, the analyst pinpoints poor maintenance and housekeeping as the specific causes of accidents. It repeatedly calls for an adequate inspection and maintenance program.

Actually, either directly or indirectly, poor housekeeping practices probably contribute to well over half of all industrial accidents. Some safety engineers feel the figure is as high as three-fourths!

Not only must a housekeeping program document contain specific instruction sheets summarizing safe housekeeping practices, but also each section of the program should have safety "built in" in methods, materials, etc. Training and promotion relating to housekeeping safety must be continuous.

## Safety Assistance

Most industrial firms have, or should have, a safety director, safety engineer, or individual who has been assigned overall safety responsibility and authority. Such personnel generally jump at the opportunity to work with the housekeeping department in improving general safety practices and conditions. The entire housekeeping program document, for example, should be reviewed by the safety engineer.

The duties of the safety engineer, or the safety coordinator who handles this work as a part-time function in the smaller plant, should include:

1. Conducting monthly safety meetings.
2. Keeping accident records and making accident reports.
3. Calculating safety records and making safety awards.
4. Interviewing workers who have had accidents.
5. Coordinating safety publicity.

Union officials are usually very cooperative where improvements are considered in the interrelated fields of safety and housekeeping. Their assistance and suggestions should be encouraged.

Industrial relations personnel will also provide the same positive response. They will assist in promotion and publicity campaigns, the details of which have already been discussed.

Personnel outside either the safety department or housekeeping department may be given specific safety duties to perform. For example, either internal plant protection personnel or hired guards may be provided a check list of potential unsafe conditions which are to be reported in writing.

Safety and fire prevention should be promoted, similar to the promotion of housekeeping as described in Chap. 26. Awareness must be developed. For example, The National Safety Council points out that "the worth of the safety poster has been demonstrated so often that there can be no question of its value in any industry program. Invariably, it proves an effective medium for attracting the employees' attention in getting across a safety message that will stick."

Good advice and direct assistance in the safety and fire prevention program are gladly furnished merely by asking workmen's compensation in-

*Posters Courtesy National Safety Council*

surance companies and fire insurance companies. They are prepared to furnish posters, show films, conduct training classes, provide promotional literature, and make inspections and reports. Professional groups, such as The National Safety Council, National Board of Fire Underwriters, and the National Fire Protection Association, are anxious to be of help. The Government Printing Office also has valuable material on safety and fire protection, such as *Safety and Health Standards* prepared by the Department of Labor, Wage and Hour and Public Contract Division.

The previous page indicates some of the safety posters, related to housekeeping, available from The National Safety Council.

## Safety Regulations

Safety should be promoted positively through its universal benefits. It is necessary to supplement the informal activity, however, with a formal statement of the company's safety rules and regulations.

Accident prevention consists merely of performing every operation, whether large or small, the correct way. It merely puts into practice the intelligent and common-sense methods of getting the job done. No industrial activity is so important that an unsafe condition or act should be permitted in its performance. Management's basic responsibilities with respect to safety are:

1. To provide equipment which has been properly investigated for safety, and equipped with necessary guards, shields, and other safety devices.
2. To arrange all processes and operations in a safe fashion.
3. To train and motivate supervisors and employees in safe operating practices.
4. To provide regular safety inspections and condition reports.
5. To investigate all accidents in order to prevent recurrence.

The worker has his own basic safety responsibilities:

1. To use the safety equipment provided.
2. To be aware of and follow safety rules and regulations.
3. To be on the lookout for hazardous conditions, and report them.
4. To caution fellow workers about unsafe conditions or practices.

Below is a typical list of safety regulations for housekeeping personnel. Many of the items are applicable to other personnel as well. Special regulations concerning floor, materials, and equipment safety are treated later in the chapter.

1. In wet weather, dangerous falls can be avoided by using rubber mats or runners near entrance areas. This will also prevent the tracking of mud and water onto clean floors. Put out and maintain mats at doors assigned to you.

2. Poor vision can cause nasty accidents. Report necessary light bulb or fluorescent tube replacements in critical areas such as stair landings, near doors, at aisle intersections, etc.

3. Watch out for vehicles; assume that the drivers do not see you and get out of the way.

4. Smoke in designated areas only.

5. Be sure to report any accident to your supervisor.

6. When lifting heavy objects, keep your knees bent and the back straight. Be sure you know the proper method of lifting. Get help when necessary.

7. Horseplay can result in serious injury. Avoid horseplay on company property.

8. Be familiar with and use emergency treatment measures where necessary. For serious injury, get first aid or medical help. The injured person should be moved only if unavoidable.

9. Apply artificial respiration if breathing has stopped. In case of chemical splashes, wash off immediately with plenty of water.

10. Report safety hazards and recommended improvements to your supervisor.

11. When entering a tank, pit, or other enclosed space, obtain inspection and approval by your supervisor, who must make certain that proper ventilation is provided, respirators and other protective equipment are used, and assistance is available.

12. Be sure to pay attention to all plant safety signs and notices.

13. Familiarize yourself with the potential safety hazards of your work area.

14. Stay out from under hanging loads.

15. Run through the plant only in an emergency.

16. On stairs, watch your step and use the handrail.

17. Carefully step down from platforms, stacks of material, or other high places, rather than jumping.

18. Use a brush to clean your clothes, rather than compressed air. Flying chips can cause eye injuries.

19. In emptying waste receptacles, dump the material out rather than reaching in. This will avoid cuts.

20. Disconnect fans and other electrical equipment before cleaning them. Pull on the plug rather than the wire.

21. It is dangerous to take chances with safety. In any questionable situation, always check with your supervisor. He will thank you for it.

## Materials Safety

These suggested regulations apply to the handling and use of cleaning chemicals:

1. Pour and use solvents in well-ventilated places. Avoid the use of gasoline for cleaning anything. Replace the cap on all solvent containers after each use.

2. If any liquid gets in your eyes, even just dirty water, flood the eyes with plenty of tap water immediately.

3. Avoid using carbon tetrachloride for cleaning purposes. This chemical has been directly responsible for a number of deaths through inhalation of its fumes.

4. Keep the protective equipment which you have been provided in good

condition and wear it whenever it is needed. This includes face shields, goggles, hoods, gloves, aprons, respirators, etc.

5. Use safety shoes when moving drums, working with heavy equipment, or in areas where objects might fall.

6. Inflammable materials must be stored in accordance with fire regulations.

7. Where strong chemicals are used, keep the skin covered for protection against chemical irritation.

8. Dry the hands and face carefully after washing to further prevent skin irritation.

9. If it is necessary to dilute an acid, be sure to pour the acid into the water, rather than vice-versa.

The housekeeping department supervisors can improve the safety of handling cleaning materials in these ways:

1. Poisonous materials such as arsenical weed killers, and various types of rodenticides and insecticides, should be very carefully marked and controlled and should be used by designated personnel only.

2. Avoid the hazards of broken glass by buying materials in plastic containers wherever possible.

3. The use of acids for descaling bowls should be done only on a periodic basis, and then only by specially designated, trained, and equipped personnel.

4. Conditions should be provided to permit safe storage. It should not be necessary to lift heavy loads, stack materials too high, etc.

5. Harsh caustic or acidic products should be replaced by neutral or milder products wherever possible.

For a more detailed treatment of safety precautions in the handling of chemical materials, see "Chemical Safety Supervision" by Joseph Guelich.

## Equipment Safety

These rules pertain to the use of housekeeping equipment and the general use and cleaning of other types of equipment:

1. Watch how you carry your equipment, mops, brooms, etc., so that no one is injured in the eye or otherwise injured by handles sticking out, etc.

2. In areas having low ceilings, special care should be taken to avoid hitting sprinkler heads with mop and broom handles in order to avoid the severe water damage that a sprinkler leak would cause.

3. When cleaning stairs, be careful where buckets and equipment are placed so as not to cause someone to fall.

4. Keep your equipment and maintenance area neat and in good order.

5. Keep equipment tightened up. If a particular bolt or nut keeps working loose, obtain a special lock washer or fastener.

6. Be sure wiring is in good repair and not frayed, knotted, or cut.

7. Electrical equipment should be grounded when in use, particularly when there is water on the floor.

8. Ladders should be kept in good repair and properly used.

9. Tools should be inspected regularly to make certain they are in safe operating condition. Personnel washing outside windows must have their safety devices inspected daily by their supervisor.

10. Fans and other electrical equipment must be disconnected before cleaning.

11. Be sure that switches on equipment are in the "off" position before plugging in the wire.

12. Keep greasy or oily cloths or dust-mop heads in closed metal containers.

13. Floor-machine handles should be in the lowered position before starting the motor.

14. Floor-machine brushes and attachment plates should be manually secured to the machine before the motor is turned on.

15. Keep equipment out of aisles and traffic lanes to avoid tripping hazards.

16. "Pay out" only the length of electrical wiring needed, rather than an excessive amount.

17. Loose clothing, neckties, rings, or other items should never be worn around moving machinery.

18. In driving a vehicle, obey all the plant traffic rules.

19. Use machinery only if authorized.

20. Use the right tools and equipment for a job. Do not use a knife blade for a screwdriver or a brush block for a hammer, for example.

21. Tools and equipment must not be thrown from one worker to another.

22. In using scrapers or other sharp tools, apply the force in a direction away from the body.

23. Wear a helmet when your cleaning job must take you under construction or maintenance work.

24. When riding on a vehicle, stay seated.

Management should provide tools and equipment which will reinforce these rules and also will, in themselves, avoid safety problems.

## Floor Safety

Most rules concerning floor safety are directed to preventing falls, which is normally considered the most serious type of accident exposure. Falls cause most lost-time accidents and occur more often than any other type of accident. These rules will help to avoid falls and other painful accidents related to floors:

1. Use "Wet Floor" caution signs freely whenever floors are being wet cleaned, waxed, sealed, or stripped. Signs should be placed so that they will be visible from all avenues of approach.

2. Keep floors stripped according to schedule to prevent build-up of brittle wax, which causes slipperiness.

3. Rain, snow, or other liquids should be mopped or vacuumed immediately to dry the floor.

4. Oil drippings or grease spots should be removed by vacuuming, scraping, or mopping, or by using a nonflammable oil absorbent.

5. Floors should be dry cleaned regularly to remove slipping hazards.

6. Heavily trafficked areas should be policed, in addition to the basic maintenance, in order to remove slipping hazards such as paper clips, toothpicks, etc.

7. Solvents must not be used in the maintenance of composition floors. Their chemical action will cause softening and roughening of the tiles.

8. Natural soaps should be avoided in floor cleaning, as the film remaining may become slippery. If natural soaps are used, they must be rinsed carefully.

9. When working on a wet floor, walk carefully and take shorter steps than usual.

10. When it is necessary to apply liquids to a floor during usual business hours, the floor should be maintained in sections, and each section should be dried carefully before proceeding to the next.

11. Be sure that mats and runners lay flat. Wrinkles or turned up corners cause tripping accidents.

12. Report defective flooring, loose handrailings, bad stair treads, and dangerous projections from walls to your supervisor so he can arrange for their repair

13. If any furniture must be moved, be sure this is done before the floors become wet. Attempting to move heavy objects on a slippery floor can cause back strain or injuries due to falls.

14. Be sure that all drawers or doors are closed before moving furniture.

There are a number of steps that *management* can take to improve floor safety:

1. Floors and stairways should be designed to have an antislip surface insofar as possible.

2. Existing slippery surfaces may be corrected. Particularly bad slip problems at critical concrete areas can be lessened by etching the concrete with a mild muriatic acid solution. This should be done very carefully as overetching will cause the floor to become rough and more difficult to clean. Etching should always be done under the direction of a supervisor. Adhesive carborundum strips may be used on stair treads or ramps.

3. Floors should be loaded within their capacity limits. Sagging will cause unevenness, which may lead to tripping.

4. Worn floors should be repaired quickly. This includes loose or curled tiles, cupped wood, holes and cracks, etc.

5. Carpets should be kept in good repair. A nonslip pad should be used under small rugs.

6. A general rule should be put into effect that the person causing a spillage should be responsible for either removing it himself or, for a serious problem, obtaining help.

7. Adequate lighting should be provided to prevent persons walking into obstructions, slipping on foreign articles, or being struck. A good time to check for lighting problems is at night.

8. Arrangements for ice removal should be carefully set up and checked on. Ice removal materials and equipment should always be available for the worst possible condition.

9. "Stiletto" heels on women's shoes have caused a large number of accidents. A number of concerns, particularly in Britain, have forbidden women to wear them on company property. Other concerns furnish plastic caps that cover the heel and make it wider. Where such heels are worn, perforated or rubber matting should be removed and replaced with cocoa mats or molded matting.

10. Floor finishes, waxes, and coatings should be purchased which have adequate slip-resistant qualities.

## The Floor Finish Problem

Most persons associate the shine on a floor with slipperiness. This is such a strong association that people actually slip more often on a floor that is thought to be slippery than on a floor that is thought to be unslippery, although the two surfaces may be absolutely identical in their coefficients of friction! Actually, shiny floors are not necessarily slippery and slippery floors are not necessarily shiny.

Contrary to some belief, high-grade floor maintenance may be achieved by using waxes and other floor-surfacing materials without increasing the slipping hazard. In some cases, the slipping hazard is actually reduced.

Properly compounded and approved floor-coating materials may be used with assurance that they will not increase the slipping hazard, particularly if proper application techniques are used. Investigations by independent organizations have shown that the slipping hazard of floor finishes has been highly overestimated. Studies of the National Safety Council and Underwriters' Laboratory have both indicated that many manufacturers are marketing floor waxes and floor-surfacing materials that provide a higher coefficient of friction than the bare floor itself.

The basic cause of floor slipperiness, then, is not the application of wax where a quality product is used. The Underwriters' (UL) Label is a worthwhile guide when selecting waxes, as this approval indicates a wax that falls within their minimum slip-resistance standards. Many of the newer polymeric floor coatings, which contain little or no natural wax, have particularly good slip-resistant qualities. Where a good floor coating material is used, most falls will be caused by wet floors, foreign objects, bad physical repair, poor lighting, and other such hazards.

## Fire Prevention

Industrial fires can be considered as a special class of accidents. Although an accident which involves a breach of safety normally only directly involves a few persons, a major fire can—and half the time does—put a company out of business. Fires, like safety accidents, can be prevented.

Rubbish and waste materials of various kinds contribute to a large number of fires. Storage of any useless material, from rags, rubbish, waste paper, and wood to obsolete flammable liquids should be discouraged. Storage in out-of-way places such as attics, basement areas, under stairs, and in other hard-to-use locations should be avoided if possible.

Proper and regular disposal of waste products are basic and elementary rules of good fire prevention and housekeeping. Metal receptacles, preferably covered, should be utilized in critical areas, for many of the plastic receptacles on the market today actually contribute fuel to fires. Tests have been conducted on several types of these units with the result that they actually ignite in a short period of time and are totally consumed by the rubbish burning inside.

The maintenance of a high standard of cleanliness and order is one of the first and basic elements of good fire prevention.

Management cannot consider that adequate fire prevention steps are being taken unless the following steps are followed:

1. Sprinkler protection should be provided in all necessary areas.
2. Watchman service should be provided, consisting of well-trained, alert individuals.
3. Fire extinguishers of suitable type and size, in accordance with National Board of Fire Underwriters Pamphlet No. 10, should be provided and posted in well-marked locations throughout the plant.
4. Smoking should be positively restricted throughout the plant and prohibited in certain hazardous areas. Smoking rules should be completely enforced and smoking should be permitted in designated areas separated from combustible material, utilizing the best of housekeeping.
5. All employees in the plant should be trained in fire-fighting practices and in the use of the fire protection first-aid extinguishers throughout the plant.
6. Employees should be properly trained for job hazards in handling flammable materials and other hazardous operations.
7. A fire brigade, properly trained, organized, and equipped, utilizing periodic drills, should be installed. The local fire department should be encouraged to inspect the plant periodically. The plant fire brigade should be trained with help from the fire department and the insurers.

8. Weekly self-inspections of the plant should be made by trained personnel, covering fire-protection equipment, electrical equipment, and housekeeping. Reports should go to management for review and correction, and copies should be kept for the insurer's representative.

9. Proper special protection from industrial hazards, such as welding and flammable liquids handling, should be set up in conjunction with the insurers in order to minimize the possibility of serious flash fires shutting down the plant.

10. Cooperation should be assured with the fire protection engineers who visit the plant from the insurance company.

11. Many cleaning compounds, polishes, floor oils, insecticides, etc., are hazardous because of a tendency to heat spontaneously or because they contain flammable solvents. Such materials should be handled carefully, and, before any products are used, their properties should be identified so that proper precautions may be taken in handling.

The number of fires caused by poor housekeeping in well-protected industrial plants is both surprising and unnecessary. Fire protection engineers say that "a clean plant is a fire-safe plant."

### Occupational Safety and Health Administration

In 1970 an act was passed to protect employees of organizations involved in interstate commerce from unsafe or unhealthful working conditions. The definition of who is involved in interstate commerce is so broad as to include most organizations. The Act, known as the Williams-Steiger Occupational Safety and Health Act, otherwise OSHA, has gone through numerous interpretations and changes, and will probably continue to do so for some time.

The Act covers many types of operations and maintenance, including custodial. Building service management and supervision are well advised to be aware of the provisions of this Act, which contains requirements ranging from the avoidance of slippery floors to the visibility of electric cords on floor machines. Somewhat overdone, as many government activities are, OSHA regulations nevertheless do represent a benefit to workers, unfortunately with an additional possible cost to management. At least the Act points another finger at the importance of proper cleaning activities.

# 27

# When Salesmen Call

YOU'RE sitting at your desk, about to receive a sanitary supply salesman who wants to sell you a can of soap, a floor machine, and other cleaning supplies. You have already seen two salesmen earlier in the day, and this one will make the tenth one this week.

The first question that comes to your mind is, "Another one?"

Yes, another one. The woods are full of them! There always will be because it is easy to get into the sanitary supply business, and it can be quite profitable. If you, yourself, wish to go into this business, it only requires the printing of a letterhead and business card, and the contacting of some jobbing house which will print labels with your name on them to put on chemicals they will ship to whomever you name. You can mark up the item 100 per cent. If you hire other people to sell for you, you can pay up to a 25 per cent commission—sometimes more.

There are a number of large companies in the business, but there are also thousands of small local jobbing outfits, averaging only two or three persons to the company. If everyone were selling the same formula materials, the decision would be easy: price and delivery.

But sanitation chemicals are a blind item. The way they *look* and *feel* and *smell* has no relationship at all to their actual performance. People buy a detergent, for example, because it smells good, has a nice color, foams a lot, and is viscous—but none of these things has anything to do with how well that product will clean.

If that were not enough, there are other problems facing you in selecting housekeeping chemicals. Because of the nature of the detergency process, there is no such thing as an all-purpose cleaner. A detergent for one situation will not work in another because of the nature of the soil or the surface to be cleaned—or even the condition of your tap water.

Improper selection can do more than just waste a few dollars in initial cost. It can also:

1. Increase labor costs in time and effort
2. Damage your building surfaces
3. Be corrosive to the skin, and cause rashes and irritations
4. Produce noxious fumes
5. Introduce legal problems due to slips or stains
6. Demoralize the workers who have to use these chemicals

People have tried various ways to overcome these problems. One method is to test samples submitted, and select the best performer. This is not practical because it is extremely expensive to test all the hundreds of chemicals which will be brought to your attention. Furthermore, many tests—such as for waxes—take several months for the results to become clear. A test in the summer might give different results from a test in the winter!

Some people purchase by specification, but this is only useful for large purchases, and where the specifications can be carefully coupled with actual performance characteristics so that the cheapest product need not be purchased.

Other people use "approved listings," but such listings do little more than force the quality down to an almost unacceptable level, and eliminate the desirable quality products which are, in the long run, the best investment. Further, these listings exclude those companies which do not believe in the need to pay fees to have their products approved. The approving organizations exercise no control over the testing and maintenance of the quality of the chemicals involved. Yet, formula changes in the sanitary chemicals industry are frequent.

What, then, is the user of these materials to do?

Probably the answer for you is to "buy" the supplier *and* his representative rather than the individual product. There are a large number of reputable (this does not necessarily mean large) manufacturers and suppliers of sanitation chemicals. And there are plenty of others who are not quite in that category. It is desirable for the buyer to select at least two but not over three such reputable companies, and deal with them exclusively, avoiding a vast expenditure of time in dealing with many jobbers and minor suppliers. Furthermore, when you are dealing with only a few suppliers, you are able to give each of them a reasonable quantity of business. This permits the salesman involved enough income from your account to be able to spend time in giving you service, such as in providing demonstrations and training classes, inspection, and minor repairs of equipment, assisting you in solving cleaning problems, and the like. It is particularly desirable to deal with chemical companies in interstate commerce. A recent government survey showed that only about 20 per cent of local manufacturers of disinfectants produced materials which were effective *when used as directed!*

This concept can be expanded further by making an annual purchase commitment to your two or three quality suppliers, so they merely replenish your stock without the need for additional purchase orders. To the numerous other salesmen calling, your account is of little interest to them until such time as you renegotiate the annual purchase commitment. At this time you can arrange to have supplier salesmen call on you during a two-week period rather than a fifty-two-week period.

In dealing with reputable companies, if a piece of equipment, such as a floor machine, is purchased and is not satisfactory, or if a new chemical does not work out properly, the supplier is expected to take back the items and issue full credit.

Which sales representative do you select, finally? This depends on (1) your personal evaluation of their sincerity and desire to serve your needs, (2) their reputation as individuals and the reputation of the company they represent, (3) whether they do business in interstate commerce, and (4) the amount of time various representatives have invested with you.

Try not to waste his time—remember it's the only real commodity he has to sell.

Why not decide on your course of action before the next one walks in the door?

# 28 Public Relations

CONSCIOUSLY or not, your housekeeping department is deeply involved in public relations. The principal purpose of any maintenance effort is to provide clean, comfortable and pleasant surroundings for several groups: building tenants and employees, visitors, and often, the community at large.

Your maintenance staff can increase its good image and earn respect from these groups if they:

1) Direct extra attention to lobbies and entrances. Employees and visitors form an instant impression the moment they enter the building. Make sure it is a pleasant one. Give special attention to lobby floors and don't overlook walls and glass doors. Smudges and fingerprints at eye level are particularly offensive. Give them a quick spot check at least once daily with a spray bottle of detergent and a sponge.

2) Concentrate on thorough cleaning and sanitation of all restroom facilities every day. Keep a spray bottle and sponge available and spot clean restroom walls frequently. Provide enough paper towel receptacles to prevent floor litter. Two receptacles side by side are better than one where it's needed and a second out of reach.

3) Keep mirrors and sinks spotlessly clean.

4) Watch carefully for litter and spillage in work areas around desks and vending machines.

## Visitors

Place even greater emphasis on the points above to make a favorable impression on visitors. Clean walls, floors, doors and counters in reception rooms have tremendous effect on your building's prestige and should rate at the top of the cleanliness standard. Do not overlook parking lots, entrys and other outdoor areas.

Use "Caution" signs to warn visitors of areas being cleaned during the day. This tells people you are concerned about their safety. It also puts cleaning operations on a level with other company activities and boosts custodial morale.

## Personal Relations

The director of the housekeeping staff must form good relationships with the people who work for him and with people further up the chain of command. Here are suggestions to generate good will in both directions:

1) Soften criticism with praise. Give criticism in a friendly and helpful way. Show the worker how to do a job and don't be impatient if he doesn't remember the first time or two. Keep showing him until he gets it right. Be patient and helpful; you'll earn his good will and get as much cooperation as he's capable of giving.

It is absolutely imperative to give orders in a courteous, friendly manner. Sharp speech, careless criticism or abusive remarks destroy subordinate morale and may well limit supervisory usefulness.

2) Praise an employee as often as you can. Most supervisors fail to commend workers often enough. Don't take the attitude that "Joe doesn't do anything worth praising." Inspect his work area and FIND things that merit praise. An employee who doesn't do anything worth praising may need to be replaced, not just ignored.

3) Always use subordinates' names when speaking to them. Names are the most important thing in personal relations. By using his name, you instantly capture the worker's attention and create a favorable attitude. Don't say, "Hey you," or "Come here." Say, "Join me please, Joe." Names are vital. Know them, pronounce them correctly, spell them right and use them every chance you get.

4) Keep subordinates informed. A department bulletin board is an

excellent way to do this. A good place for it might be near the time clock. Don't assume there isn't enough news. MAKE news out of personal events. Sample notes you can put on your bulletin board regularly are:

"Compliments to the third floor crew for their excellent work cleaning the corridors."

"Best wishes to Mary Smith who has just become engaged to Bill Brown. The wedding date is October 2nd."

"Mr. Jones, our executive vice president, complimented us the other day on the high level of cleanliness in the executive suite. Good work!"

A good supervisor can engineer some compliments from top executives toward certain crews or members of the housekeeping team. A compliment is still a compliment, even if it is deliberately developed for the benefit of subordinates.

Other items for a bulletin board include amusing cartoons, especially if they relate to housekeeping; short verses, slogans, and posters.

5) Give housekeeping awards. Many companies now have a "Custodian of the Month" program. A $5.00 prize goes to the custodian or maid who does the best work during that month or makes the most helpful suggestions. Money should always be supplemented by a badge bearing the title the worker has won. Recognition is often more important than money. Try to see that this award is passed around and that poor workers are encouraged to compete for it too.

## Up the Line

A housekeeping executive or foreman must keep building management informed. This communication should include:

1. Projects under way
2. Projects planned for the the future
3. Projects recently completed
4. Special problems to be solved
5. Ways the manager can help

A busy manager has no way of knowing what is going on in the housekeeping department unless he is told. Do this in a standard way. Use a monthly report, concise but complete. Put into writing even matters which have been discussed personally. This gives both management and supervision a running record of maintenance staff activities. It lets the building manager know which personnel deserve promotion or praise. It keeps him aware of problems and best of all, it lets him know how much work the maintenance staff is doing.

## You and the Community

Don't overlook the relationship between your housekeeping department and the community. Not only does the community judge the company by the appearance of its buildings, but also by the conduct of its employees. Set standards of personal appearance for custodians equal to those for other employees. Too often custodians are given no feeling of pride in their work. Consequently, they may travel to their jobs in dirty, ragged or unkempt clothes. They will not wear dress-up clothes on the job, of course, but a dressing room or shower room at work makes it possible for them to appear on the street in clean dress. Uniforms for custodial help can do wonders for morale and improve the attitude of other employees toward them. Cleaning workers encouraged to be neatly dressed both on and off the job soon raise their standard for the building itself.

Efforts to improve worker spirit do more than improve work output. Remember, highly critical employees can damage the company image. Some people may judge the entire operation by one disgruntled employee who circulates criticism among his friends. He may also make it difficult for you to get good employees to work for you.

# 29

# The Housekeeping Worker

JUST as the body must have food to live and grow, so must the business organization have profits to live and grow. In order to earn these profits, the business depends completely on its personnel who, after all, must be considered the principal asset of any company.

The fundamental approach to achieving our housekeeping goals must be through the people doing the work. The success of the housekeeping program will depend on the ability of the custodians to work efficiently and effectively.

In the previous chapter, we considered the supervisor's approach to improving productivity. In the next chapter, we discuss the benefits of training. In this chapter, we consider the motivation and control of the worker himself.

## Mr. Custodian

Although in hospitals, hotels, and sometimes in office buildings, women may outnumber men in the housekeeping department, in industry quite the reverse is true. The typical industry performs the bulk of its cleaning operations using custodians and employs maids primarily for policing operations in personnel areas during the day. Depending on the labor market and the situation within a given company, some organizations use maids in the lighter-duty cleaning of office areas as well. Thus, in the typical operation, custodians outnumber maids at least ten to one, and there are a number of operations where maids are not utilized at all.

We must remember in dealing with the custodian that each worker is different. Each has his own special abilities and limitations. The individual has his personal, physical, and mental needs. And every man has his private and unique set of problems. Although a number of principles apply to custodians in general, we should always temper them with the individual approach.

Habits are hard to break; many workers have preconceived ideas and opinions. In implementing a housekeeping program, we could not expect every custodian to agree to, or be happy with, every change. In *Time Study and Motion Economy,*[1] Robert Lee Morrow says, "Almost any change is received with distrust and frequently condemned without even a trial. Changes disrupt the established habits of people and temporarily make them uncomfortable. Often the most obviously beneficial improvements are resisted." Most of this resistance is generally based on fear. Such resistance is often expressed "psychosomatically" through complaints concerning products or equipment, trivial grievances, etc., and are best prevented through proper orientation and training.

The admiral in Gilbert and Sullivan's *H. M. S. Pinafore* claims that he became the ruler of the Queen's Navy because of the care with which he polished the handle of the big front door! Of course, this is exaggerated, but most people do agree with Thomas Edison that "there is no substitute for hard work." Many individuals, and organizations as well, are putting a great deal of hard work into the housekeeping operation, however, without achieving worthwhile results. More than hard work is required. It is easy to claim that low productivity is caused by loafing on the job. We should face the hard fact that, rather, it should be charged to ineffectual planning or implementation. Thus, in the opening sentence of the clever and instructive book *Parkinson's Law,*[2] Professor Parkinson says that "Work expands so as to fill the time available for its completion." Give a man a day in which to perform a five-hour job, and he will use, as a rule, the entire day for its completion.

The work should be arranged in order to avoid monotony. Where there is monotony, the slightest excuse becomes a reason for stopping work. Beyond the organization of the work to this end, rest periods often serve to break unavoidable monotony and, as carefully documented studies have

[1] Copyright, The Ronald Press Company.
[2] By C. Northcote Parkinson. Copyright, Houghton Mifflin Company.

shown, actually normally increase the day's production. Forming workers into groups or teams also helps to avoid monotony and, for such simple tasks as are often found in housekeeping, it has been found that light conversation does not decrease efficiency, and may even improve it. This is particularly true where the group leadership principle is followed.

Boredom is also avoided by not placing people in jobs beneath their intelligence and education.

It sometimes appears that an almost endless series of questionnaires is conducted in an effort to find out what it is the employee wants most from his job. Surprising to many is the fact that the things that most workers consider important and want to have as a part of their job are also the things that management would like for them to have. It is up to all of us to transform such a meeting of the minds into actual practice. Unquestionably, management and labor can best achieve their objectives where they can be shown to coincide. In good housekeeping, the coincidence is complete.

Let us provide the conditions that will inspire and sustain individual initiative and effort from every worker!

## Building Morale and Status

Almost invariably, one of the biggest stumbling blocks in the establishment of an effective housekeeping and sanitation program is the low morale of housekeeping personnel. This condition generally is the result of a combination of these factors:

1. Lack of status.
2. Failure to obtain recognition from management, fellow employees, and the general public.
3. Little or no chance for advancement within the housekeeping department.
4. The absence of sufficient incentive and motivation.

Naturally, poor morale leads to low efficiency, abuse of materials and equipment, resistance to learning improved methods, and slovenliness. It is extremely difficult to effect meaningful improvements in housekeeping in the face of such attitudes.

The morale problem has been almost universally prevalent in housekeeping departments, arising from the stereotyped image of the "janitor" and "janitress" in the public mind, the very nature of the work itself, the lack of publicity concerning the beneficial aspects of housekeeping and sanitation, and the natural tendency of management to focus emphasis and attention on more direct activities.

"If management is to protect and enhance its prerogatives of leadership," says Michael J. Jucius, in *Personnel Management,*[3] "it will do well to take the time and effort to improve the morale of its supposed followers." Let us consider some of the ways that this can be done for housekeeping workers:

1. People need causes, in addition to good leadership, in order to be motivated to a high activity level. This was recognized by Napoleon when he pronounced his cause as the liberation of Europe from monarchy. The results achieved by his armies were entirely disproportionate to their size. Housekeeping personnel actually are entrusted with the custody of the physical facility and the safety and health of *all* the workers. This is a trust of tremendous importance. Let us establish in the mind of the custodian the cause of *guardian of health and property,* developing it to the fullest extent possible in all of our personnel relations. Such a cause can help to fill the vacuum created in so many lives through failure to set, and work toward, a personal goal.

2. In a similar vein, probably the most important factor in the determination of whether a worker will provide a good output is his regard for the nature of the work. This is not only on the basis of what he himself feels about the subject, but also about what his fellow workers say. Although most workers do not want overly responsible jobs, nevertheless they want jobs that have some meaning. Such meaning should be publicized at meetings, with posters, and with every other conceivable means. As a bonus effect, this aim is usually achieved in part by the overall publicity program.

3. Some organizations claim pride in the fact that most personnel are unaware of the existence of the housekeeping crew! True, the custodial staff must perform its activities without interfering with the work of others—but the staff can hardly develop the attitudes required for a greater effort if they are to remain unknown. After all, "man does not live by bread alone." Through publicity, company-wide recognition should be given to the housekeeping worker. Such a course of action, naturally, would be desirable for the employees of every department within the organization, but there is a particular need for this in housekeeping because of the low repute generally accorded this department. "One of the deepest drives of human nature is the desire to be appreciated," said psychologist William James.

4. The provision of uniforms is very effective in morale building and departmental recognition and identification. Where uniforms cannot be provided, even brassards (or arm bands) help. Uniforms of a light blue color seem to have the best acceptance. Some companies embroider the name of the custodian over the left-hand shirt pocket, which permits identification for either commendation or complaint.

5. An incentive campaign may be especially helpful. For example, if company policy permits, a small bonus or certificate may be given once each month to the worker who has made the greatest contribution to cleanliness during that month. The prize winner can be publicized in the house publication and his name posted in a place of honor, or it can remain strictly a departmental project.

[3] Copyright, Richard D. Irwin, Publishers.

6. Morale building is an aspect of a number of other activities carried on in the complete housekeeping program, such as orientation and training meetings, selection of titles, providing worthwhile materials and equipment, maintenance of custodial facilities, an equitable wage system, and good leadership.

It may be desirable to conduct a morale survey of housekeeping personnel. If the survey is carefully conducted by personnel outside the housekeeping department, with responses recorded anonymously, the results can quite easily be predicted if there has been no positive move to gain the morale-building effects of a comprehensive program.

Large organizations, after having received the basic benefits of the housekeeping program, can go still further through the use of a worker-qualifications system. Such a system combines the approaches of training, identification, employee awareness, and may also include a form of rating for compensation. Such a program identifies the custodian as a trained, qualified specialist:

1. New employees are hired as "housekeeping trainees."
2. After six months of service, the employee is given the opportunity of taking a written (or oral) examination and practical performance test. If he passes this test, the employee will receive the rating of "Qualified Sanitation Specialist, Third Grade." He should be provided a certificate of achievement (suitable for framing) and a distinctive shoulder patch bearing a single red star.
3. Similarly, advancement into the grades of "Qualified Sanitation Specialist, Second Grade" (two stars) and "Qualified Specialist, First Grade" (three stars) will be made on the satisfactory completion of tests at one-year intervals. Certificates and shoulder patches will be provided on each of these occasions.
4. If the employee fails the test, he may repeat it after a period of six months, during which time he will undergo his training.
5. Personnel already assigned as group leaders should be permitted to take these tests without the need for the intermediate waiting periods. All new group leaders must have passed all three tests.
6. If it is desirable for foremen or supervisors to wear shoulder patches for greater identification with their department, they should be permitted to do so without the need for qualification tests.
7. Still further effectiveness may be given to the program by tieing it in with wage incentives, if this is not contrary to the organization's operating policy. Each qualification could carry with it an increase of five cents to ten cents per hour, for example.
8. The program may be set up as either mandatory or optional.

Each new employee of the housekeeping department should be welcomed in person, as well as with a personal letter, from the housekeeping department manager. Such a letter can accompany the copies of the portions of the program new personnel receive when they begin their training.

## Selecting the Custodian

The housekeeping department can never achieve the status it needs for most effective performance (not only considering the individual worker, but also the department as a whole) unless applicants are carefully screened before hiring and only competent workers are employed. If the housekeeping department is forced into the position of hiring incompetent workers through pressure of other departments, or for charitable reasons, it will have great difficulty in performing its true function so as to give the organization the greatest possible benefits. The housekeeping department will be judged by some persons on the basis of its most limited worker. And so may the custodians themselves judge the importance of their work to the company, to themselves, and to the public. On the other hand, care must be taken not to hire persons who would become bored with the work.

A review of the personnel aspects of the program with the personnel department should be made to insure recruiting, interviewing, hiring, and indoctrination practices that will provide workers properly prepared to play their part in a dynamic and progressive housekeeping program. Unfortunately, some housekeeping departments have become hosts to incompetent people who have sifted down through the various production departments. These people may be mentally limited, suffering from ill health, or may have serious physical limitations. The housekeeping department seems to become the natural resting place for the compensation case arising from an accident in the production department. Such a procedure should be seriously reconsidered because of its catastrophic effect on the morale of the housekeeping department. Injury cases should be retained in the production departments, where work suitable to their limitations are sometimes even more readily found than in the housekeeping department. This would further provide an incentive to reduce accidents.

The method of recruiting and hiring employees for the housekeeping department, or of transferring personnel into this department from other parts of the plant, is generally quite clearly defined in most plants. Where rules permit, these points should be followed:

1. A thorough physical examination should be given prospective employees in order to eliminate the hiring of disease carriers. If this cannot be provided for all

housekeeping workers, it should at least be given to custodians who will work in rest rooms or medical and dietary facilities.

2. Personnel should be retired from the housekeeping department at age sixty-five if unsuited to the work. As company rules and government retirement laws are changed, the age limit may be lowered.

3. Emergencies do not justify the hiring of an incompetent worker. With a consideration of the total labor market, we should wait for an employee who generally fits the predetermined qualifications.

4. Where wages are not set by labor agreement, some housekeeping departments utilize part-time workers, as well as full-time workers also having other jobs, as they can be obtained at a low pay rate. These should be eliminated from the housekeeping department since the attitude of such persons is usually poor, and they are usually fatigued even before beginning work. The additional amount of pay required to obtain personnel without other jobs will be economical in the long run.

5. Prejudices on the part of supervisors can have a serious effect on the quality of personnel, and therefore the results obtained from the housekeeping effort. If housekeeping department supervisors make their own selections of personnel, prejudice will result in a diminished field of acceptable applicants. Where the selection is made by another department, prejudice in the housekeeping department will increase turnover and, through lower morale, decrease the effectiveness of remaining workers. In a large housekeeping department, the total absence of workers from a local minority group should be investigated carefully.

6. In the preparation of labor agreements, both the union and management should be encouraged to consider the effect on all the workers of utilizing the housekeeping department as a bumping station for all the other departments. The housekeeping department should have its own seniority "family tree." At the very least, the agreement should protect the group leader position in the housekeeping department by insistence on its being a *working* job, by permitting only persons well trained in housekeeping techniques to occupy such a position, and by preventing their displacement by bumping.

### Job Descriptions

Job rating sheets should be set up for housekeeping workers. Normally it is not necessary to prepare a separate specification for each worker. A breakdown such as the following will usually suffice: (1) group leader, (2) custodian, (3) window, overhead, or other hazardous type area cleaner.

A standard specification cannot be drawn up for industry as a whole, although it would be desirable to have standard ranges throughout industry. The specification must consider the local labor market and the specific needs of the organization.

The typical job-rating specification sheet will cover a series of a dozen or so characteristics which may be described either verbally, or by using numbers or letters in a standard code. A typical list of such factors would be:

1. Education
2. Experience
3. Mental ability or aptitude test ranges
4. Manual skill or manual dexterity test ranges
5. Initiative and ingenuity
6. Physical requirements
7. Materials and equipment to be used
8. Difficulty of the work
9. Hazards of the work
10. Comfort of the work
11. Responsibility for damage or loss
12. Responsibility for wastage
13. Responsibility for safety

Of course, care should be exercised so as not to set the standards too high, even in a very good labor market, because of the problem of boredom.

Job descriptions should always be prepared with the housekeeping duty as a full-time function. Only in the smallest industries should custodians be permitted to perform other functions, such as guard service. Where there is a split in duties, the worker normally lacks interest and fails to develop a proprietary attitude concerning either duty. In addition, the make-ready and put-away times (which are, of course, nonproductive) are increased considerably when custodians turn from performing one type of task to another.

The job descriptions should be based on a personnel assignment that assigns specific jobs to the individuals so that responsibilities do not overlap. This provides a system whereby any deficiencies may be traced to the individual, and whereby the worker may take pride in the results of his own personal efforts.

## Testing

The use of sanitation tests for custodians can provide several advantages:

1. The tests are a support for the training program, as they measure the retention of information by the trainees.
2. The tests can be used as guides in selecting group leaders or other supervisory personnel.
3. Tests can be used to help implement bonus or incentive systems.
4. They can be utilized to screen prospective applicants or personnel "bumping down" from other departments.
5. The tests help to boost custodial morale by creating interest, if they are not used punitively.

6. They can be used as the basis for a worker qualification system, as previously described.

Certainly there is no question of a company's moral or legal right to use such tests on a basis that is optional for the employees. Where the tests are to be mandatory, such as a requirement to bid for a job in the housekeeping department, it also appears that this is perfectly correct. Arbitrator Harry J. Dworkin recently ruled:

> The use of tests, fairly and objectively administered, would appear to be desirable and in the interests of the employees and the company. A solid testing procedure should serve to allay any suspicion of favoritism or discrimination. Any written or oral test should be reasonably related to the duties required of a particular job. Since the responsibility of determining the relative rights of applicants is primarily vested in management, it may employ proper tests in arriving at a fair decision.

## Discipline

Housekeeping personnel, like other personnel, should know exactly where they stand with respect to breaking company rules. Some companies believe the punishments for breaking company rules should be posted in a conspicuous place, such as a locker room or near the time clock. At any rate, the personnel should be made aware of what these rules are. A typical company might use the following rules:

These offenses are usually met with immediate discharge:

1. Commission of an important crime on or off the premises.
2. Deliberate damage to company property.
3. Falsification of documents, such as reports, applications, etc.
4. Insubordination to a superior, or abuse of authority to an inferior.
5. Strike action that is either unauthorized or illegal (counsel should be sought on all labor relations problems).
6. Subversion.
7. Theft or pilferage, positively proven.
8. Time-card falsification.
9. Working for competitors.

These offenses may be met on the first offense with a lay-off of a suitable period of time without pay, and on the second offense by a discharge:

1. Brandishing a weapon.
2. Fighting on the premises.
3. Gambling on the premises.
4. Sleeping while on duty.

These offenses usually are met first with a warning, a lay-off on second offense, and discharge on third offense:

1. Drinking on the premises.
2. Garnishment. In this case, the discharge is sometimes preceded by two warnings rather than a warning and a lay-off.
3. Horseplay.
4. Peddling of unauthorized items.
5. Flagrant safety violations.
6. Smoking in restricted areas.

In the following cases, the first two offenses may be met with warnings, followed by lay-off on the third offense, and discharge on the fourth:

1. Unexcused absence from work.
2. Carelessness or indifference.
3. Chronic absenteeism or tardiness.

## Wages

Whenever industrial personnel are surveyed concerning those aspects of their work which they consider to be of the greatest importance, it is a surprise to some that wages are not placed at the top of the list. This is apparently because the typical worker is generally satisfied with the wages he receives, but not as relatively satisfied with other phases of the work, such as security or supervision.

Generally speaking, wages are an overall company problem and prerogative about which neither the housekeeping department nor its consultant can have any real effect apart from making recommendations. Of course, any inequities in wages between the housekeeping department and other departments, for work requiring equal skills and abilities, should be eliminated.

Wage rates for custodial workers are normally prepared for the housekeeping department and stated as figures which may not be exceeded for starting rates, raise increments, or maximum wages.

The group or individual responsible for determining these wage rates has to consider such factors as:

1. Labor contracts
2. The current minimum wage
3. Quality and intensity of supervision
4. Cost of living in the given area at that time
5. Fringe benefits offered or available
6. Job requirements
7. Hours
8. The local labor market

9. Financial condition of the organization
10. Wage structure within the organization
11. Wages for similar jobs offered in the same type of institution or industry
12. Turnover history

The housekeeping department manager can only make limited recommendations within the above framework.

Many organizations, treating housekeeping as a necessary evil rather than as a vehicle to further its fundamental objectives, make it a practice to hire housekeeping personnel at the least figure that will bring enough applicants to fill the open positions. This can create problems of real significance.

The survey of wages being offered locally or in the same general region by similar organizations can be useful. Care should be exercised to interpret the responses on the basis of results achieved, the personnel turnover rate, etc.

Wage statistics prepared by local and federal organizations are also useful. It must be remembered that the statistics indicate averages only for those organizations which contributed to the study. Their chief value is that they indicate the wage which would be paid by one of those organizations, and against which another organization may have to compete for a qualified worker.

## Employee Rating

An employee-rating report is similar in certain respects to the job description, as both are concerned with various job factors. The basic difference is that the job description is primarily used as a selection device, whereas the rating report is used to grade worker performance.

To be meaningful, any kind of grading must be numerical in order that comparisons may be drawn and rates of improvement noted.

The proper steps to set up a rating form are:

1. Devising series of characteristics applicable to all workers in all positions.
2. Preparing a series of descriptions indicating the various degrees of excellence of each function.
3. The assigning of numerical values to each of these descriptions.
4. The summation of values to obtain a total grade.
5. The relation of total grades and total grade ranges to overall worker performance.

Experience shows that the *whole* story cannot be shown in this way, and the numerical rating must be supplemented by notations and remarks of a general nature.

Such a list of general comments might concern answers to such questions as these:

1. Is the employee performing the task best suited to his ability? If not, what kind of work should he be doing?
2. What are his most desirable traits?
3. Where does he need to improve himself?
4. To your knowledge, what is he doing to improve himself?
5. What is his attitude and record concerning safety?
6. Any other comments?

A refinement of the above procedure is to weigh each factor depending on its relative total importance, rather than assuming that each is equally important. Thus, the top grade in one factor may be five points whereas the top grade in another factor may be twelve points, for example.

Through the rating report, each employee's ability and fitness in his present occupation, or consideration for promotion, may be appraised with a reasonable degree of accuracy and uniformity. The rating requires the appraisal of an employee in terms of his *actual performance*. It is essential, therefore, that snap judgment be replaced by careful analysis. In order to make ratings carefully, we should:

1. Use our independent judgment.
2. Disregard general impressions of the employee and concentrate on one factor at a time.
3. Study carefully the definitions given for each factor and the specifications for each degree.
4. Remember instances that are typical of his work and way of acting. We should not be influenced by unusual cases which are not typical.
5. Make our rating with the utmost care and thought and be sure that it represents a fair and square opinion. We should not allow personal feelings to govern a rating.

Generally speaking, a rating report is prepared by a direct supervisor and reviewed by his superior. Supervisors may be asked to prepare self-rating reports in some plants, which will later be compared with reports prepared by another individual.

In order for the ratings to be of the greatest use, they should be as accurate as possible. A regular system of preparing notations concerning individual workers will improve accuracy, as these can be referred to as the rating report is made. Such information can come from supervisors' daily notes, commendations, complaints, suggestions, etc.

# Part III
# OPERATIONS

# 30

# Soil Prevention

IN ALL forms of maintenance activities an ounce of prevention is worth a pound of cure, as the old saying goes. In custodial work we are concerned with time, which typically represents over 90% of a building cleaning budget. We might alter our saying to claim that an hour spent in soil prevention might save ten hours later in removing that soil.

That this is so is not always readily apparent. For one of our clients a consulting recommendation that we made concerning entrance matting was questioned. We answered the question by putting down a single length of matting 4 feet wide by 12 feet long. In the next twenty-four hours it collected twelve pounds of dirt, or one pound for each linear foot of matting! Without the matting, that soil would have been distributed throughout the building, not only on the floors but on all surfaces, as floor soil tends to become airborne after a time.

This serious and important subject can have humorous overtones. In the seminars which we conduct for custodial supervisors and managers, we often discuss this subject. We were given one report that a custodial worker objected to entrance gratings because the catch-pan underneath kept filling up with dirt which he had to remove! (Apparently it did not occur to him that it is easier to shovel dirt out of a pan by the kilogram than to clean it up by using a mop or other manual tool to remove it by the microgram.) Another story concerned a custodian responsible for the lobby of a building—he was assigned to keep carpet matting in place. One day his supervisor, not seeing the matting at the entrance, asked the custodian where it was. The worker replied that he had rolled it up and put it in storage because, since it was raining, the matting would get dirty! Without the matting, of course, it was the lobby floor which was getting dirty and slippery.

## The Building Approach

Each building entrance should serve as a barrier for reducing the volume of soil brought in by foot traffic (the same would be true inside a building, where areas change from a heavily soiled situation to that such as from a manufacturing area to an administrative area, or a shop area to a cafeteria).

As many actions as possible should be taken to prevent soil—a good place to begin is the walkway to a building. If the walkway is canted or tilted to one side, or crowned so that it is higher in the middle than on the sides (the latter is more expensive to pour and finish) then water and mud would be less apt to collect on the walkway and be tracked into the building. Also, a rough surface would be helpful in trapping soil before it is walked into the building.

A canopy over the walkway is helpful in preventing the fall of rain or snow directly onto the walk. Snow melting devices are also useful and can prevent accidents as well.

## The Entry

If there is an exterior stair landing, this might be an ideal place for a grating, especially if it is protected by a canopy. Since one of the problems with a grating is removing accumulated water (gratings should not be connected to plumbing systems, as they tend to foul them up), it is easy to provide a drain from the grating to the outside through the landing.

In many cases the space between the two sets of entry doors is the best location for the grating, and this preferably should be drained to the outside as well. The grating should be as large as possible, preferably covering the full area of the entryway. The grating bars should run perpendicular to the line of traffic—the bars should not be spaced more than 1/10" apart to prevent any possibility of narrow heels being caught in the spaces. Some types of grating are provided with plastic ribs, which act as squeegees, to remove as much soil as possible from the soles of the shoes. Naturally, the grating must be recessed to present a level flush with the floor so that there is no tripping hazard. Immediately behind the grating, so as to provide a two-stage soil prevention system, should be some form of matting to trap that small quantity of soil and moisture which the grating did not receive. Ideally, this would be a type of carpet runner or removable carpet squares. The matting should preferably be in a recess, again to avoid tripping accidents. Attention should be given to keeping this matting clean and replaced when necessary, and it should have good absorption characteristics to retain moisture in large quantities. Such a mat should be large enough, and properly located, so that each person passing over it will take at least two steps (preferably three) across the mat with each foot. This would require a mat to be at least 12 feet and preferably 16 feet long. Be sure that the mat is wide enough so that a pedestrian will not step off it on his way into the building.

There are many forms of matting. Rubber mats, even with holes in them, do little good as there is no place for the soil to go and it accumulates on the surface to be walked off again onto the other floors. A type of matting which resembles the floor machine pad, made of synthetic materials, can be very beneficial and has a long wear life. However, it does not remove quite as much of the finer soil and moisture as the carpet-type material. Where any of these matting devices cannot be recessed, then a beveled edge or "nosing" should be attached to prevent accidents.

For the maintenance of the carpet-type mat an upright vacuum with a pile-lifting device should be used periodically, supplemented by shampooings. The polyester open-weave mats must be hosed off periodically. By all means avoid link mats which are difficult to clean and cause tripping accidents, and cocoa mats which are almost impossible to clean.

Soil prevention at entrances is required for all types of floors—terrazzo, vinyl asbestos, and carpet. If an entire carpeted floor is used as a soil entrapment device, it will wear too quickly.

### Special Matting

Many types of special matting are made. Here is a sampling of the wide variety available:

1. A "tacky mat" consists of a stack of plastic film with a sticky side up, for all the world like a large stack of jumbo cellophane tape! This type of mat has the advantage of being able to remove great quantities of very fine dry soil and is useful for critical areas such as industrial super-clean rooms. When one surface becomes "loaded" it is simply peeled off, exposing the next layer.

2. A "sponge mat" is a device sometimes used at the entrance to hospital operating rooms or research facilities involving bacteria, where the pedestrian steps into a stainless steel pan containing a sponge which is impregnated with a disinfectant (such as a synthetic phenol). The reason for using this device is that wet disinfectant is most effective in bacteria control—for less severe conditions

a simple soil remover is sufficient since most bacteria are attached to small soil particles.

3. The "mechanical mat" is a device consisting of parallel metal bars between which are nylon bristles—when one steps on the mat (or crosses through an electric eye) the mat is actuated and the bristles begin to vibrate, thus literally scrubbing the bottom of the shoes, with the soil falling between the grating into a catch-pan beneath. This is a rather expensive device requiring considerable installation time—ideally, it should be installed with new construction or renovation. It is very efficient when quite a bit of soil must be removed in a small space (such as the entry from a factory producing products using carbon black, chemical dust, or the like) before entering an administrative area.

4. A special form of matting might be considered a "removable grating" because this mat consists of slats with a carpet or plastic surface spaced to form a grating, the whole thing being constructed in such a way that it can be rolled up and removed for cleaning. The floor underneath must be protected so that the soil falling between the spaces does not grind the floor with foot action on the slats.

> NOTE. All matting should be removed and cleaned regularly, since if placed on a finished floor it can permanently damage the floor by the action of moisture and bacteria beneath it. This is especially true of matting with rubber or plastic backing.

5. A special form of matting involves the carpeting of selected areas to prevent the transfer of soil from one floor to another. The best examples of this include the carpeting of elevator floors and stair landings. The tending of these mattings is simplified if both the interior of the elevator and the stair landing contain an electrical outlet (as a matter of fact, the entire subject of soil prevention is one best approached from a new construction or renovation standpoint, although a great many things can be done after the fact).

6. The laminar airflow entrance can be considered a special type of matting. In this case a flow of air from above passes over the pedestrian who walks across a grating and the air is pulled into the grating beneath. This provides a cleaning action not only for the shoes but also for removing soil from the clothing as well. This is a device used at entrances to prestige buildings, industrial clean rooms, some department stores, aerospace activities, etc.

### Other Soil Prevention

Other approaches to soil prevention involve removal of soil or prevention of soil from other than the shoes:

1. Waste and ash receptacles should be placed or mounted wherever people are likely to discard waste or cigarettes. One should not wait for cigarette burns to appear on the floor to decide to install ashtrays or urns! Normally, both a cigarette urn and a waste receptacle (which indicates that a combination fixture is most desirable) would be needed wherever there is a change of activity, such as at a doorway, in an elevator lobby, in the elevator itself, on a stair landing,

at a water fountain, by each telephone, between every two urinals, etc. Wall-mounted urns and receptacles have the advantage that they do not impede floor-cleaning activities and are less apt to become damaged or to stain the floors.

2. The use of plastic waste liners should be encouraged in all receptacles where wet waste, food, or drink is likely to be deposited. The liners avoid the costly washing of the receptacles, they eliminate odors and vermin problems, and maintain the cleanliness of the receptacles with a minimum of cost. Liners need to be replaced daily in health care and research facilities, but in administrative-type areas they should be replaced only when necessary (typically about once per week, often less frequently).

3. Spills often occur when drinks dispensed in cups are carried from one area to another. The frequency of such spills is decreased whenever the fill level of the cup is lowered by using larger cups—lids to cups should also be provided when the drink will be taken to another location. Personnel should be informed of these actions.

4. Soap dispensers should be mounted so that any dripping is into the bowl of the sink, not on the floor, the faucet or on the handle.

5. Waste receptacles should be large enough so that they do not become over-full, with a resulting waste on the floor, especially in restrooms. Further, unless push-plates to waste receptacles (where they exist) are kept clean, people will refuse to use them and will throw their waste on the floor.

6. The use of roll-type paper toweling tends to create less litter than folded-type toweling. In the latter case, toweling may fall to the floor (particularly where too small a size is being used)—and people tend to pull out two or three towels, often dropping one. The roll-type toweling dispenses the paper in a single sheet (which is also less expensive).

7. The use of high-quality floor sealers and finishes can significantly improve the maintainability of an area, while reducing airborne dust. Spills mop up more easily—the floor is more resistant to heel marks, scuffs, and grit—and the floor can be maintained at a high level of appearance with less frequent scrubbing and buffing than if inferior materials, or none at all, are used. The use of sealer on a fairly rough floor such as concrete also provides a surface which can be maintained with simpler equipment. For example, a client in the garment industry had an entire carload of clothing rejected by a military service because the clothing was covered with a fine film of dust, not apparent to the naked eye. It was discovered that the plant factory had a concrete floor surface and the wheels of the carts were grinding the surface into dust, which then became airborne and sifted onto the fabric. The sealing of this floor with the proper solvent-type sealer not only reduced the dusting (which also comes from the pores of the concrete) but created a surface which could be cleaned with a treated dustmop rather than having to use more sophisticated and expensive equipment.

# 31

# Graffiti Control

WHEN was the last time you wrote on a wall, and what did you say?

Such a question would upset a number of people, who might think that they have never written on a wall, and wouldn't want to be classed with those who have. Yet, just about everyone has written on a wall at one time or other (yes, including the author, who still vividly remembers being caught at it by his grade school principal). There seems to be such an instinct for wall-writing that we must discuss the subject in terms of its "control" rather than its "elimination".

Several books have been written on graffiti as an art form, and it has even been suggested that art historians of the future might look back on our times as the most creative flowering of this important art form! (I shudder to think that this might be our society's most important legacy).

Personally, the graffiti I most enjoyed was in an Atlanta rest room, where someone had written "anyone who would write on a rest room wall must be mentally ill." Perhaps the writer of that gem felt that he must include himself as being disturbed since he was in the act of writing that observation. Yes, it must be true that some people who write on walls are psychotics, but most people who write on walls are perfectly sane.

To understand how we might better control graffiti, we need to know more of the psychology of the wall writer. Dr. Jean Rosenbaum says in "Practical Psychiatry" that "the person who lacks self-esteem or feels psychologically threatened may write obscene or insulting words on public surfaces for everyone to see. By doing so he relieves his frustrations and is expressing his hostility toward a special person or even toward the entire world."

But this is not the only reason that people mark surfaces (the term graffiti, of course, can include writing on surfaces other than walls, such as on furniture, automobiles, etc.) Let's summarize what motivates an individual to do such writing:

1. An expression of general hostility, as described by Dr. Rosenbaum.
2. More specifically, an expression of resistance to authority or conditions posed by authority. Thus, we see much more graffiti where living conditions are crowded, such as in housing complexes and on subways.
3. Part of the crowding syndrome is an expression of individuality, where one does not want to end up as a number, but wants to retain his unique personality. This desire can only be applauded, but other means should be sought to achieve it.
4. Sometimes graffiti is simple advertising, such as an exhortation to vote for a candidate in an election.
5. As an aspect of teen-age or later rebellion, some writers want to shock the establishment, just as children want to shock their parents. The more shocked one becomes, the more successful is the person who caused it.

Graffiti can be quite durable. The first art to be discovered were drawings on the walls of caves found in French and Spanish mountains that were made some 15,000 years ago (only by definition can we separate art from graffiti in this and many other cases). Then, as now, the graffiti writer desires permanence and wide exposure. He wants his writing to be seen by everyone, and forever.

A special problem confronting subway systems, is "gang writing", where clubs of youths are literally founded for the purpose of marking subway cars and terminal walls. One 16-year old Columbia College freshman was interviewed by a New York newspaper reporter. He said, "We had business cards printed, and had meetings every Friday. We had a waiting list of about 150 kids who wanted to join. We made contact with all of the other graffiti groups. It's a kind of recognition thing." The reaction to some of these activities has been vigorous. Various local laws have been passed concerning being caught with an open spray-paint can. But the punitive laws don't seem to help, because the judges don't seem to want to prosecute (perhaps because of comparing this crime with other much more serious crimes that might be substituted in its absence).

## Graffiti Prevention

The use of graffiti is an aspect of human nature which cannot be completely prevented, but it can be reduced considerably by providing surfaces that are more difficult to mark:

1. Carpeting an elevator wall or an entrance door yields a surface that is extremely difficult to mark but is easily maintained and can furthermore be very attractive.
2. A restroom stall of painted metal can be waxed using the same metal interlock polymer that one might use on the floor, since it is so difficult to write on a waxed surface.
3. A tough vinyl wall covering, not completely white and without deep or sharp fissures, can be helpful, especially if it is impregnated with a chemical soil retardant.
4. Laminated plastics such as, "Formica" and "Micarta" are very useful and much easier to maintain than stainless steel or any other surface.
5. Concrete block walls and the joints between them should be covered with epoxy resin which presents a glass-like surface difficult to mark and easy to clean. This is even better than glazed ceramic tile, since the grout between the tiles is so easily damaged. Whenever ceramic materials are used, however, the grouting used should not be white but of a colored material.
6. Where ceilings are reachable, they should not be made of mineral materials, but of perforated metal pan which can be both washed and painted.

## Graffiti Reduction

With this background of the psychological causes of graffiti, we see that it cannot be prevented completely, but it is certainly possible to reduce graffiti in these ways:

1. Rapid removal is by far the most effective means of graffiti control. We remember that the perpetrator desires broad exposure and permanence to his writings–if it can be demonstrated to him that what he has written will not remain longer than 24 hours, then he may well give up. Rapid removal also eliminates the "growth" of drawings or poetry, since it does not provide time for people to embelish on what has already been written.
2. Good illumination is the next best approach to graffiti reduction. The human tendency is to do bad things in dark places, and a higher lighting level tends to dampen the graffiti spirit. One may balk at this approach in the light of energy conservation, but perhaps there are better areas to save lighting. Further, the illumination level can be intensified by using more light-reflective paints, keeping cleaner light fixtures, etc., without actually increasing the wattage.
3. Confronting the graffiti writer with surfaces that are difficult to mark can help to limit the problem, although a confirmed writer might only accept

this as a challenge, and his marking might become even more serious. Surfaces difficult to write on might include an epoxy resin coating on concrete block walls (by all means including the grout, which may be marked with pencil or crayon), plastic laminate rather than natural wood surfaces, enamel rather than flat paint, knurled rather than flat metals, etc.

4. A special case of anti-graffiti surface is graffiti wallpaper, where the pattern on the paper or vinyl surface is practically solid graffiti, the entire area being covered with words and pictures in various colors, so that any additional graffiti simply becomes lost to view!

5. Some people feel that if you can't beat them, join them—they provide writing surfaces in rest room stalls and on walls, by hanging note pads, chalk boards, and the like. An especially successful approach to this is a "graffiti board" in a public corridor, such as in a public school. People are invited to write or mark on the board, which is cleaned off and repainted periodically.

6. A more creative approach may be helpful, and this has often been seen to be the case on plywood walls surrounding construction sites, where artists are literally invited to paint the surface. The same approach might be used with subway cars, with contests and a good deal of publicity and even awards for the best-painted subway car (perhaps a spin-off advantage would be to have the cars painted free!)

7. It is possible that arrest and prosecution can have a beneficial effect, and perhaps the least that can be achieved from this is to have the person who commits the act and causes the damage as the one who ends up paying for it, or is required to remove it personally.

8. An interesting case was seen in the fine arts building of a university campus, where the students were invited to paint and decorate the interior of the building—following this, the students themselves made quite certain that neither those in the building or visitors to the building ever marked one of "their" surfaces.

9. The least effective way—perhaps it has no effect at all—of controlling graffiti is through the use of signs which say "please do not write on the wall". This is taken by some graffiti writers as an invitation to write on the sign!

The removal and correction of graffiti markings can be an expensive and time-consuming activity. This might involve cleaning—perhaps more than one cleaning for effective removal—the use of poultices to draw foreign matter from porous surfaces, repainting, and sometimes even completely replacing the surface.

Various approaches are taken toward this problem. In one public school system, the arts classes are called on to paint over graffiti with contemporary designs. In some industries, rest room doors and even partitions have been removed where graffiti has become a problem very expensive to cope with. This removal might conflict with various legal requirements, but sometimes these laws are "bent" with the claim that the doors or partitions have simply been removed temporarily for repair! Some college housing authorities, as well as public apartment buildings, assess a penalty for damage, including graffiti, which is deducted from an advance deposit.

There are many similarities between graffiti and vandalism, not only in the psychological causes but in the prevention and control. Usually, where action is taken to limit graffiti, it has a beneficial effect on vandalism, and vice versa.

# 32

# Rest Room Care

OF all the areas possible to be constructed in a building, the only required type, by law, is the rest room—not only must there be rest rooms, but in sufficient quantity and with enough fixtures to serve the needs of the numbers of people who might use the building.

And the legal requirements go further: the rest room must be maintained in such a way as to be safe and healthful under OSHA and other regulations.

The care of rest rooms provides a microcosm of the entire custodial field, as it involves specific equipment, chemicals, training, procedures, as well as a number of special requirements such as those involving health and hygiene.

Let us begin with supplies:

> A wide variety of hand cleaners are available, but for a single item the most acceptable as well as economical is a liquid soap dispensed in a foam or lather form.
>
> For toweling, the roll type paper towel where the length is controlled through the turn of a crank or the push of a button provides very good economy, and gives much less litter than folded towels. Electric hand dryers consume too much energy and are not hygienically desirable, as germs are blown from the dried skin into the air.
>
> Roll tissue is much more desirable than folded tissue; and a double dispenser is required to eliminate complaint calls.

The waste receptacle should be of large enough size to accomodate the waste, and more than one installed if necessary. If a push plate is used, it must be kept clean so that the receptacle will be properly used.

A brief review of some rest room design features would be in order:

Urinals should be of the flooded open throat type, so as to avoid stoppages and odor problems connected with strainers.

There should be a floor drain in the proper location to permit the use of spray cleaning and to handle overflows.

Fixtures and partitions should be wall or ceiling hung to keep floors clear for cleaning.

Floors should be of ceramic material with dark sealed grout; resilient materials should never be used.

Proper ventilation should be supplied.

Many custodial managers and supervisors—not to mention the cleaning personnel themselves—are frustrated with the problem of graffiti in rest rooms—more of this wall-writing appears in rest room facilities than in all other areas of a building combined. Some of the salient aspects of controlling (it is almost impossible to eliminate) graffiti are these:

The principal factor in control is rapid removal; the rule should be that no graffiti remains as long as 24 hours.

Better lighting means less graffiti, and the lighting should be arranged to illuminate the inside of the stalls as well as the balance of the rest room (merely keeping the light fixtures clean would help a lot).

Surfaces should be used which are more difficult to write on while being easier to clean, such as epoxy resin on concrete block, or waxing metal partitions.

The control of odors in a rest room represents a problem that has several possible approaches, some of which are helpful and some not:

The use of masking chemicals is very rarely advisable, and the time and money spent on these should be more devoted to cleaning and disinfection. Attempt to avoid the use of drip fluid, deodorant discs, and similar items.

If a deodorant is to be used, it should be of the chlorophyll or formaldehyde type, perhaps injected into the room with a timer or through a wicking device.

The proper control of odors is through removal of bacteria, which are the basic cause of other than transient odors. This is provided with good cleaning technique, the use of effective disinfectants, and the sealing of surfaces to permit their proper cleaning.

Ventilation is important for the transient odors. This can be improved in many cases by more open grilles, cutting off the bottom of doors, putting a larger grill in a door, or removing the door altogether if a proper blind is provided.

The complex subject of disinfectants need not be fully discussed here; let's briefly consider only the four most commonly used and their application to rest room care:

The disinfectant most useful is the quaternary ammonium type, which has the advantage of superior odor control and is easy on the skin. It cannot be used successfully, however, where the soil is too oily or greasy.

The next best choice is the phenolic disinfectant, which can be used in oily soils and may be a somewhat better germ killer, but is harder on the skin.

The iodine-type disinfectant is too hard on the skin for this purpose, since it contains acid.

Pine oil disinfectant, although highly touted on television, is a very poor germ killer, and is irritating to the skin.

Other chemicals and materials should be selected with care for rest room service:

For scouring or stain removal, the proper material is the lotion-type, which resembles milk of magnesia. The powdered cleansers contain particles that are too large and sharp for most uses, and cause a good deal of damage by scratching. No scouring powder should be used if it can be avoided, but rather use a liquid germicidal detergent for daily care based on one of the above disinfectants.

Acid bowl descalers should be used for mineral build-up removal periodically, and these materials should never be used on a daily basis. It may be desirable to handle acid descaling on a project basis by specific personnel.

A special chemical is useful for removing scum and scale from shower stalls and other surfaces and, like bowl cleaner, should be available for this periodic use—some cleaning procedures substitute high-pressure cleaning for this, or combine the two methods.

Although toilet seat covers are taken by many to be a benefit, if not a sign of courtesy to the public, by others they are taken to be an indication that the restroom is not properly being cared for—this is in much the same way that people view the use of deodorant blocks and drip fluid. A properly-cared for rest room facility does not require paper seat covers, where the facility is both cleaned and policed at the proper frequency.

Silicone-based stain retardants are injected into flushing water either from a device suspended in the flush tank, or through a mechanism installed on a pipe and fed through a hole drilled in the pipe. It is true that the coating on the fixture surface is smoother, and does not tend to collect foreign material as easily as the more irregular surface of the fixture. However, if a fixture is properly cleaned, the soil is properly removed. Further, the use of the soil retardant may lead to the failure to clean and disinfect the fixture properly, because of excessive reliance on the retardant. Where fixtures are cleaned at the right frequency and with the proper technique, investment in this system is not required.

The worker must have the proper tools with which to do his job—the right tools for rest room cleaning will depend on the method to be used, the size of the rest room, and many other factors. Equipment to be considered would include:

The custodial cart makes it possible to transport minor equipment items, supplies, chemicals, paper goods and the like from a work area to the rest room. In a large rest room this may be brought inside, otherwise stationed at the doorway.

Mopping equipment is important, including the type of mop, bucket and wringer. The mop should be of the type with an easily removable head for laundering, the bucket should be of plastic or stainless steel, and the wringer should be the downward-pressure variety.

Floor machines are required for periodic scrubbing, and use either wet vacuums, or a squeegee to remove the water to a floor drain.

Pressure-sprayers are useful for removing scale build-up in showers and for general cleaning where the rest room is designed to permit this efficient operation.

Manual tools are also required, such as a plastic pail, bowl mop, sponges and cloths, etc.

The question often arises concerning the cleaning of a rest room by a member of the opposite sex. This is being done successfully by many thousands of cleaners in many types of facilities. The first requirement is to be sure that the room is not in use, and this can be done by calling into the room, and even flicking off the lights momentarily to see if there is any response—after that, a sign should be placed to show that the room is being cleaned, and the door blocked by the custodial cart.

Although a given procedure for a specific type of rest room would have to be devised for that individual case, a set of instructions prepared for a client and his specific case might be of interest:

1. These instructions apply to rest room facilities, locker and athletic areas which are equipped with toilet facilities, and medical and laboratory areas where basins are provided for wash-up. The fixtures in these areas should be thoroughly and properly cleaned daily.

2. If the cleaner disinfectant solution is not already mixed in a plastic spray bottle, this must be done. During the normal rest room cleaning operation, a bucket of cleaner disinfectant solution will be used to mop the floor. The plastic spray bottle (and small plastic pail) can be filled from this bucket.

3. The door to the rest room should be propped open, or locked, and a sign posted or some other indication given that cleaning is in progress and that the rest room is not to be used.

4. Normally the first fixtures to be cleaned in a rest room are the basins. These should be cleaned after the mirrors, lights over the mirror, towel cabinets, soap dispensers, and other items in the area of the basins.

5. Turn on the faucets to rinse the interior of the basin.

6. Close the drain to make sure that the sponge or cloth will not catch on the drain plug. Spray the basins liberally with cleaner disinfectant from the plastic spray bottle and wipe with a damp sponge. Wipe the hardware also.

7. Open the drain and spray cleaner disinfectant into it. Wipe the drain plug and the rim of the drain with the sponge or cloth.

8. Spray some cleaner disinfectant into the overflow outlet. Turn on the water and rinse the bowl.

9. Wipe the hardware dry with a clean cloth or paper towel to prevent spotting.

10. Wipe the skirts or sides of the basin with a damp cloth or sponge or, if they are heavily soiled, spray them with cleaner disinfectant and wipe them.

11. The bottoms of the basins and the pipes and valves under them normally do not need to be cleaned every time the basin is cleaned. However, these areas should not be neglected, and should be wiped with a damp sponge approximately once for every five or six times that the basins are cleaned.

12. Whenever the sponge being used begins to show a build-up of soil, it should be rinsed thoroughly in one of the basins.

13. In most rest room cleaning operations, the fixtures which are cleaned after the basins are the commodes and the partitions. Begin the cleaning by flushing the fixture.

14. Spray the top of the toilet seat with a liberal amount of cleaner disinfectant solution and wipe it thoroughly with a damp sponge (or dip the sponge in the pail of cleaner disinfectant and wipe the seat).

15. Wipe the seat thoroughly dry with a clean cloth.

16. Raise the seat and spray the underside with cleaner disinfectant and wipe it with the damp sponge. It should also be dried with a clean cloth. Spray and wipe the entire exterior of the fixture.

17. Spray a liberal amount of cleaner disinfectant into the bowl. Using the bowl mop, thoroughly scrub all surfaces above the water level, paying particular attention to the underside of the flushing ring.

18. Flush the fixture and follow the water line as it decreases with the bowl mop, scurbbing in a circular motion. After the inside of the bowl has been thoroughly cleaned, flush the fixture one more time to insure that all chemicals have been removed from the surface.

19. Because of build-up, due mainly to minerals in the water, it will be neces-

sary to use an acid type bowl cleaner to clean the interior of the commode on occasions. Extreme care should be taken when using this product. Any contact with skin or eyes should be avoided. To use the acid type bowl cleaner, first wet the bowl mop and then pour the bowl cleaner onto the mop. Always hold the mop over the bowl while applying bowl cleaner to it.

20. Scrub vigorously under the flushing ring and at the water level to remove all build-up, rust and scale. Flush the fixture and follow the water level down with the bowl mop, vigorously scrubbing in a circular motion. Flush the fixture one more time to remove acid from all surfaces.

21. After cleaning each commode, damp wipe all pipes, valves and connections.

22. Before leaving the stall, damp wipe all partitions, including the tops. They may be either sprayed with cleaner disinfectant and wiped with a damp sponge, or a sponge may be dipped in a pail of cleaner disinfectant solution and then used to wipe all surfaces of the partitions. Any writing or drawing on the partitions should be removed as thoroughly as possible.

23. After cleaning the commodes and partitions, urinals should be cleaned with a method similar to that used for commodes. First, the fixture should be flushed and any waste which does flush through should be removed.

24. The wall areas above and beside the urinal should be sprayed with cleaner disinfectant and wiped with a damp sponge.

25. Spray and damp wipe all exterior surfaces of the urinal, including the base and areas underneath the urinal.

26. Spray the interior of the urinal with a liberal amount of cleaner disinfectant solution. Beginning at the top of the urinal, scrub thoroughly with the bowl mop, particularly the underside of the upper flushing ring.

27. Flush the fixture and follow the decreasing water level down with the bowl mop, scrubbing vigorougly. Flush the fixture one final time to remove any remaining cleaning solution or residue.

28. As with the commodes, it is necessary periodically to clean the urinals with an acid type bowl cleaner. Begin by flushing the fixture and wetting the bowl mop. While holding the bowl mop over the urinal, pour the bowl cleaner onto the mop.

29. Scrub the interior of the urinal thoroughly with the mop, especially around the edges and under the upper flushing rim. Scrub until all build-up, rust and scale is removed.

30. Flush the fixture and scrub the lower part with the bowl mop as the water level decreases. Flush the fixture one final time to remove any remaining cleaning colution or residue.

31. If there are partitions between the urinals, these should be sprayed with cleaner disinfectant and wiped with a damp sponge (or, wiped with a sponge which has been soaked in the pail of cleaner disinfectant).

32. Pipes, valves, and other hardware connected to the urinal should be damp wiped with cleaner disinfectant and then wiped dry with a clean cloth or paper towel to prevent spotting.

33. In a rest room, no traffic should be allowed until all rest room cleaning has been completed.

34. For clean-up, spray the bowl mop liberally with cleaner disinfectant solution and rinse it. Spray and rinse all sponges and cloths. Wring the sponges and cloths as dry as possible and store them so that they will be easily recognized by their color as equipment which has been used to clean rest room fixtures. If the cleaner disinfectant solution is to be left in the plastic spray bottle for later use, be sure that the bottle is properly labeled before it is stored.

# 33

# Custodial Facilities

WHEN a man is given a job to do, it is wise to provide him with reasonable facilities, equipment, and materials for the work. Only in this way can acceptable work be produced efficiently. We must make certain that this universal principle is also applied to the operations of the housekeeping department.

A painting crew, for example, could not be expected to work effectively or safely unless they were properly supplied with a suitable preparation area and paint locker. They need equipment such as ladders and scaffolding, spray guns and accessories, brushes and rollers of all types, tarpaulins and shields. They require materials such as pigments, vehicles, thinners, varnishes, stains, etc. It is reasonable that the custodian must similarly be equipped with adequate working and storage facilities, equipment, and materials.

## Purchasing

The purchasing of housekeeping equipment and supplies requires information of several types:

1. Specifications, including tolerance limits

2. Determination of stock items, as opposed to items which are to be ordered specifically for a given project
3. Minimum points for stock items
4. Ordering quantities
5. Packaging
6. Vendor

Each person in any way involved with housekeeping has a different approach to such factors. The housekeeping department manager, for example, wants to be certain of having sufficient materials on hand at all times so that his custodians will never be idle through lack of materials. The buyer or purchasing agent, on the other hand, is interested in holding inventory to a minimum while obtaining the best possible price. There is no question that substantial gains can be made through modern purchasing methods, but the buying of housekeeping equipment and materials will benefit from adequate liaison of the purchasing department with the housekeeping department. Only in this way can we avoid the dilemma of requiring plenty of stock but having no money tied up in inventory! The cost of *not* having housekeeping materials on hand can be rather expensive, and the purchasing department should be familiar with the problems involved.

In some organizations, the housekeeping department does its own purchasing of supplies. This has the advantages of satisfying immediate needs, the quick handling of paper work, as well as providing a realistic value analysis of the material in question without overly stressing unit cost. On the other hand, often there is lost the benefits that the purchasing department can provide by obtaining quantity prices, contracts, discounts, etc. The selection of material types, concentrations, and other characteristics should be made by the housekeeping department; but, in general, the actual purchasing in the larger organizations should be done by the purchasing department.

The same forms and controls used in the purchasing of other materials should be utilized for the purchases of housekeeping products and equipment in order to gain the advantages of standardization, using either manually posted or electronic data-processing systems, depending on the size of the operation. The purchasing records are normally broken down on a product basis, indicating the product name, its description or use, source, packaging, and price. Columnar data are posted about requisition date, source, and quantity, the quantity actually ordered, purchase order number, date ordered, date received, and amount received.

## Distribution

Distribution of supplies can be handled either on a "pick-up" or "delivery" basis. On the pick-up system, the custodian presents the list of his requirements at the central storage area and returns with these materials to his work station. In the smaller operations, the custodian may actually remove the materials himself from the central storage area, or, even more desirable, one of the custodians may be assigned this work as a part-time duty. If group leaders are used in the housekeeping operation, one of them should handle this function.

The delivery system is a much better means of supply distribution. Thus, each custodial closet or cabinet is properly stocked periodically against a supply list posted on the door. The custodian using the closet always has sufficient materials and equipment with which to perform his job and never needs to spend time obtaining them. In an extensive operation, a special vehicle may be required for this purpose. In some cases, the driver of the vehicle can also perform the function of stock clerk, but in the largest departments both personnel may be required.

In the larger plants, a full-time supply room attendant or stock clerk would perform these functions:

1. Maintain perpetual inventory and product-consumption information.
2. Make minor repairs and lubricate mechanical equipment.
3. Launder and treat dust mops and dust cloths.
4. Repackage bulk materials into smaller containers.
5. Fill orders for the housekeeping department, as well as other departments using cleaning materials.
6. Dilute concentrated detergents, liquid hand soaps, etc.
7. Prepare requisitions against the purchasing department for the housekeeping department manager or a designated assistant.
8. Receive materials into the housekeeping department.
9. Keep the central storage area clean and properly organized.
10. Perform other functions as assigned, such as special pick-ups or deliveries of housekeeping materials.

## Inventory Control

A well-organized inventory control procedure can provide these important benefits:

1. Provide assurance that housekeepers will have materials and equipment on hand with which to perform their work properly.

2. Prevent an overly large inventory that ties up excessive money and space.
3. Minimize time and effort in the requisitioning and ordering process.
4. Uncover wastage, failure to use the proper product, or pilferage, by pinpointing consumption in various areas.
5. Provide a basis for comparing efficiency.

Here are some ways in which inventory can be controlled:

1. Set up a simple record-keeping system and follow it. Use a card index file or simple looseleaf notebook. Make a card or separate page for each product purchased. For example, a separate stock page would be used for floor cleaner or all-purpose detergent, disinfectant, floor wax, sweeping compound, dust-mop treatment, floor seal, liquid hand soap, bowl cleaner, etc. As for equipment, typical items that would require separate pages might be dust mops, wet mops, push brooms, wax applicators, mop buckets, utility trucks, floor machines, scrubbing pads, etc. To keep track of all of these different items, the record sheet should indicate: (1) name of the product, (2) supplier, (3) minimum stock, (4) quantity to order each time, (5) average use per month. There should be a space where the date that new orders for the items were placed can be recorded as well as the date shipment was received. The stock sheet should also show what quantities of the item have been taken out of the central storage area and distributed to various custodial storerooms on different floors or different areas of the building.

2. Know the correct methods of using each product and correct rate of usage. This prevents waste and incorrect use. The larger organization should develop consumption standards for repetitive jobs so that consumption for new work as well as continuing work can be estimated. We have a right to expect guidance on product use and work methods without charge from suppliers. The material should be purchased from manufacturers who can provide suitable data on the correct frequencies and probable consumption rates of products.

3. One of the biggest sources of waste in housekeeping supplies is the tendency of an untrained worker to overuse a product. He will use a full cup of detergent per pail of water where the instructions call for one-half cup. He will use the most concentrated recommended dilution for the lightest duty jobs, or will add another cup "for good measure." The only reason for the custodian overusing the products is his belief that it will make the work easier. Only when he is convinced that it is *not* to his advantage to overuse materials will such wastage stop. Consumption control records alone will not stop it because the custodian can utilize less water in order to increase the concentration. Proper training in use-dilutions will explain and demonstrate the *extra* work caused by overuse of detergents, for example, because of the formation of too much suds, the need for excessive rinsing, and the formation of a soap film which collects dirt.

4. The custodian should be trained to read labels or instruction sheets and to follow the directions carefully. When the correct methods are used, as well as the proper use-dilutions, the materials will be serving the user at highest efficiency.

5. Utilize use-control aids to provide accurate concentration. A graduated plastic measuring cup is safe and easy to use and eliminates guessing in measuring. The cups can be marked with a fill-to-line where they are to be used for a

single product. Proportioning or simple metering devices, attached to water taps, predetermine the concentration and mix automatically as water is drawn. The use of such equipment gives the custodial closet a laboratorylike atmosphere and contributes to the custodians' sense of status, while performing their basic economic function.

6. Spillages and leakages can devour a significant part of the material dollar. They can be minimized by having small containers filled from drums as a function of central storage and by the adequate distribution of funnels to all custodial closets. When drums are stored on racks, they must be supplied with faucets which do not leak. A bucket should also be placed under each faucet to catch any minor dripping. Drum pumps may be used with the drums standing vertically; this avoids all possibility of leakage and dispenses measured amounts of the material.

7. Certain concentrated materials (such as liquid hand soaps) should be diluted on arrival at the central storage area and marked "ready for use." This provides an accurate dilution and eliminates the task of making a dilution at the use point, thus conserving time as well as materials which may be lost through spillage or failure to dilute. If a container has been purchased which requires dilution with an equal amount of water, a simple procedure is to take an empty container of the same size and fill it half-full of water. The same container is then filled full from the container of concentrated material. The half-full container of concentrated material is then filled full with water (the diluent may be mineral spirits or some other solvent in the case of insecticides, etc.). Water under pressure should never be run into a concentrated detergent as this causes a great amount of foaming and will waste a good deal of time. The water should be poured through a large funnel, to which a hose is attached that reaches to the bottom of the container.

8. The sanitary-supply industry seems to be moving toward "controlled packaging" as a response to user needs. When powdered materials are packaged in relatively small units, such as in five-pound cartons rather than in fiber drums containing several hundred pounds, then, on the basis of consumption standards, a given number of cartons can be issued to perform a certain job, thus avoiding wastage. The economics of using such controlled packages will be clearer in the years ahead, as packaging costs relative to housekeeping wages are reduced through mechanization and volume selling by the larger suppliers. Pilferage is more difficult to control with such packages. At this time, controlled packaging is primarily restricted to hand-cleaning powders, and dishwashing and laundry detergents.

9. Misuse should be avoided through the careful marking on arrival of all containers with the nomenclature used in the housekeeping department. Each storage area should have a given position for every type of material. Color-coding containers help when dealing with illiterate workers or with a high turnover situation. The misuse of a product not only wastes the material that is used, but also often requires an additional product of another type for its removal. More important is the damage to surfaces or injury to workers which might occur.

10. A simple but complete inventory control system will provide a record of

materials distributed and pinpoint consumption at the use point. Thus, pilferage, appropriation by other departments, and wastage can be discovered and the causes corrected. The formal system on paper should be supplemented with a sight check periodically, such as twice monthly.

11. Stock shrinkage has become quite a serious problem in industry. Housekeeping materials are particularly subject to pilferage and personal use because of their universal applicability in business as well as in the home. A good record system, as mentioned above, can help to prevent stock shrinkage. All custodial closets, cabinets, and other supply areas should be locked when not in use. The key should be retained only by the custodian using the closet, his supervisor, and the housekeeping department manager. If an individual is assigned to distribute supplies to all closets, he must also have a key, of course. If materials are issued on a "pick-up" basis, empty containers or worn-out supplies should be turned in to justify replacement. On a "delivery" basis, the absence of containers or minor equipment should be reported.

12. Breakage can be avoided by the purchase of materials in plastic containers rather than in glass containers.

13. Some materials are lost through aging or spoiling. This can be prevented by sound purchasing methods so that supplies do not remain on hand for too long a period of time. The housekeeping department should have a one- to three-months' supply of housekeeping materials on hand, depending on the type of material and consumption rate. Orders for general replacement of housekeeping materials should be made monthly. Older materials should be used first.

14. Adequate and properly arranged storage facilities will prevent damage and loss due to wetting, overheating, freezing, soiling, or contamination.

## Central Storage

The central housekeeping storage area is the heart of the inventory control system, and its operation should be under the direct supervision of the housekeeping department manager or a designated supervisor.

The size and arrangement of the central storage area must be determined functionally. Although some useful estimates have been published which indicate the size of the sanitary-storage facilities as a percentage of the total area to be maintained, these are useful only as generalizations and can lead to real problems if directly utilized. The housekeeping department does not want to suffer the inefficiencies of operating with insufficient facilities. On the other hand, neither does it want to tie up more of the organization's money in these facilities than is warranted. The size and layout of the storage area, therefore, must be determined on the basis of the needs it must serve. For long-term economy, the layout must be flexible enough to permit rearrangement or expansion based on future needs.

The central storage area, preferably, should be located as nearly at the

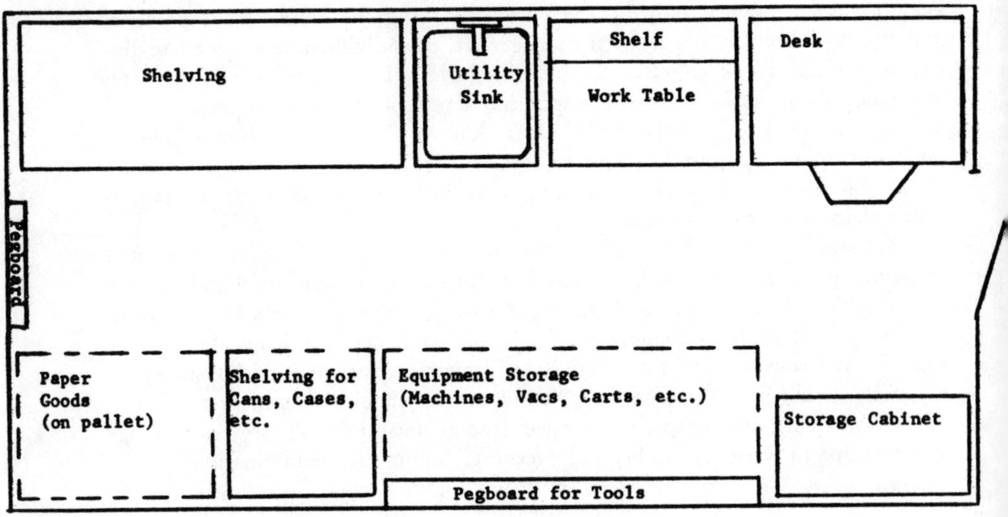


A central supply and work area is a must for other than the smallest custodial departments. The drawing depicts a minimum arrangement.

center of the housekeeping activity as possible, and adjacent to the housekeeping department offices. Where a separate central storage area cannot be set up adjacent to the department offices, a section of the general maintenance supply area may be designated for this purpose. If this is done, the subarea should be separately fenced off and locked.

Some companies provide separate buildings for maintenance stores and operations. This is done in new construction where the cost inside the main production facility is too great or where the location might interfere with operational flexibility. It may also occur in a renovation or relocation project where more area is required for production activities within the main plant building. Such a building might include supplies and work areas for painters, welders, riggers, plumbers, electricians, and other trades. A subdivision of such a building would be very useful for the housekeeping department stores and offices as well.

The location of equipment and materials in the central housekeeping storage area should be carefully organized and properly marked. Each drum pallet or large piece of equipment should have a fixed location with an appropriate identification sign. Each shelf for storage of smaller items should be similarly marked. Finally, there should be posted a complete list of the minimum materials and equipment to be kept on hand in this area. Paper goods (tissue, towels, etc.) should be kept enclosed either in their original containers or in closed cabinets provided for this purpose in order to prevent soiling.

Obsolete or otherwise unusable materials should be removed from the storeroom in order to provide more shelf space, better appearance, and to avoid the possibility of errors.

The ideal central storage area requires:

1. A location or stall for each drum or barrel. If the containers are stored vertically, a drum pump should be supplied. If they are stored horizontally, drum racks will be required as well as a method of placing the drums on the racks.
2. Shelving for small containers, case goods, and hand tools. Adjustable steel shelving is most desirable for this purpose.
3. Tool holders for hanging mops, brooms, and other items with handles.
4. A pegboard for equipment attachments such as hoses, wands, etc.
5. A work table for the inspection, lubrication, and minor repair of equipment. This table may also be used to spread out and treat dust mops and dust cloths.
6. Fireproof metal containers for treated cloths and dust mop heads. One container should be designated for soiled items, and the other for clean and freshly treated items.
7. Adequate lighting. Electrical outlets must accommodate all the various types of equipment for testing purposes.

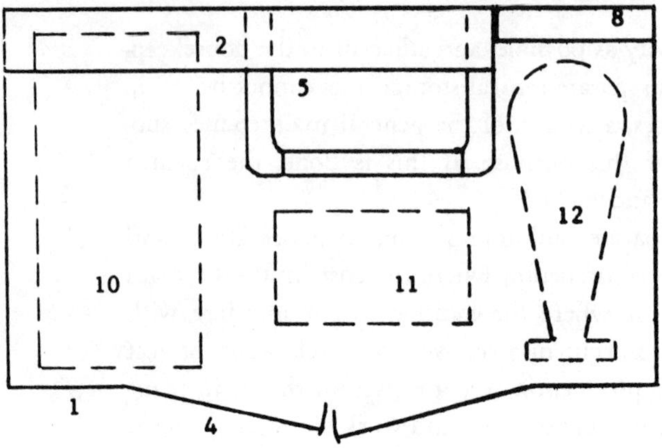


Plan View

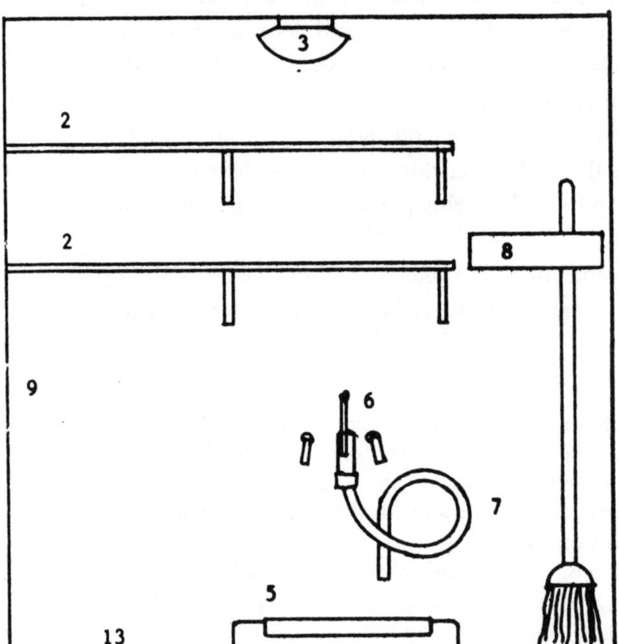


Front View

LEGEND

1 **Dimensions: 8' long, 4½' deep (36 square feet)**

2 **Shelving 10" deep, with bracket supports**

3 **100-watt lamp, with door-hinge switch**

4 **Two 30" doors, pierced for ventilation, and lockable**

5 **Utility floor sink, with stainless steel lip cover. Note off-center**

6 **Bibb faucet with support hanger**

7 **4-foot length of hose**

8 **Tool holder**

9 **Walls ceramic to 4', painted enamel (including ceiling) above 4'**

10 **Location for custodial cart or waste hamper**

11 **Location for 2-bucket (or 3-bucket) mopping outfit**

12 **Location for floor machine or vacuum**

13 **Floor -- concrete, ceramic or terrazzo (not resilient**

A minimum custodial closet, where space is not available for an efficient layout.

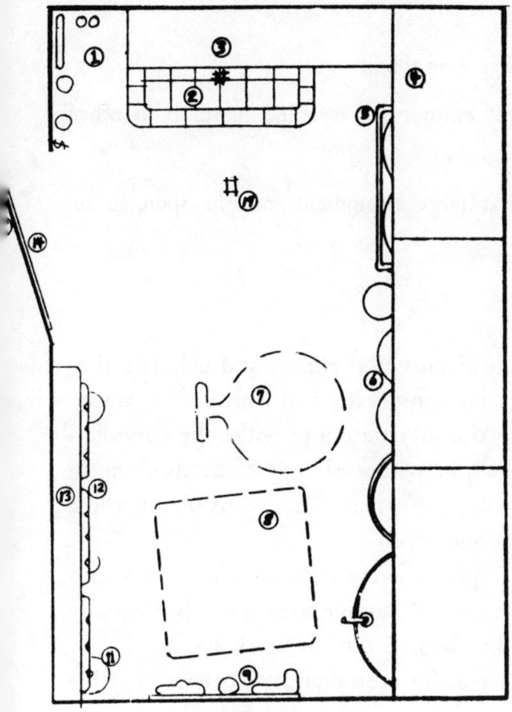

PLAN VIEW

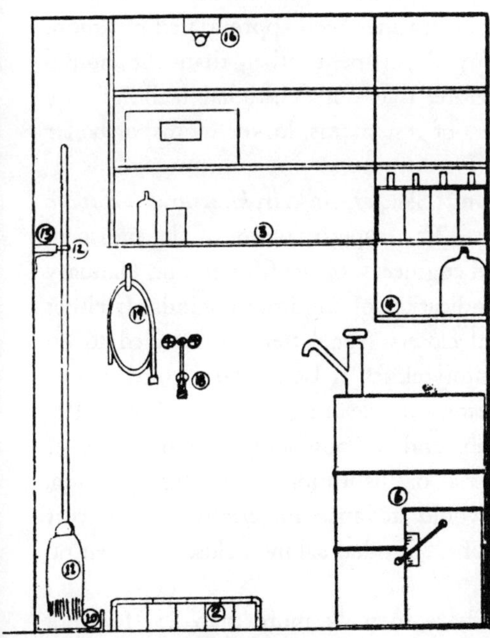

SECTION

# CUSTODIAL CLOSET (MINIMUM)

MINIMUM DIMENSIONS: 6' WIDE, 9' LONG, AND 8' HIGH.

CLOSET TO BE USED ONLY BY DESIGNATED PERSONNEL, AND IS TO BE LOCKED WHEN NOT IN USE.

IF CLOSET IS COVERED, INSTALL FAN PROVIDING 20 AIR CHANGES PER HOUR.

1. STORAGE AREA FOR HOSES, EXTENSION WANDS, PIPES, ETC.
2. BUILT IN CERAMIC TILE, FLOOR SINK, WITH DRAIN. (SECOND CHOICE: WALL MOUNTED UTILITY SINK)
3. SHELVES OVER UTILITY SINK, 9" DEEP, 12" SPACING
4. STORAGE SHELVING OVER FLOOR STOCK, 18" DEEP, 12" SPACING
5. MOPPING OUTFIT IN STORED POSITION
6. FLOOR STOCK (DRUMS, CANS, ETC.)
7. FLOOR MACHINE IN STORED POSITION
8. VACUUM IN STORED POSITION
9. ACCESSORIES, FITTINGS, AND TOOLS MOUNTED ON PEGBOARD
10. ALUMINUM OR CERAMIC DRIP TRAY
11. MOP IN STORED POSITION
12. TRI-GRIP TOOL HOLDERS
13. 4" SPACER TO KEEP MOPS AWAY FROM WALL
14. BULLETIN BOARD CONTAINING INSTRUCTIONS, SCHEDULES, ETC.
15. 30" WIDE DOOR WITH LOUVER. LOCATION OF DOOR INTERCHANGEABLE WITH ACCESSORIES PEGBOARD IF NECESSARY BECAUSE OF ORIENTATION OF AREA.
16. CEILING LIGHT PROVIDING MINIMUM 40 FOOT-CANDLES. LIGHT SHOULD BE SHIELDED TO PREVENT BREAKAGE.
17. FLOOR OF CERAMIC TILE OR CONCRETE, WITH FLOOR DRAIN IF POSSIBLE
18. BIBB (THREADED) FAUCET, WITH BRACE
19. LENGTH OF HOSE, FOR WASHING EQUIPMENT

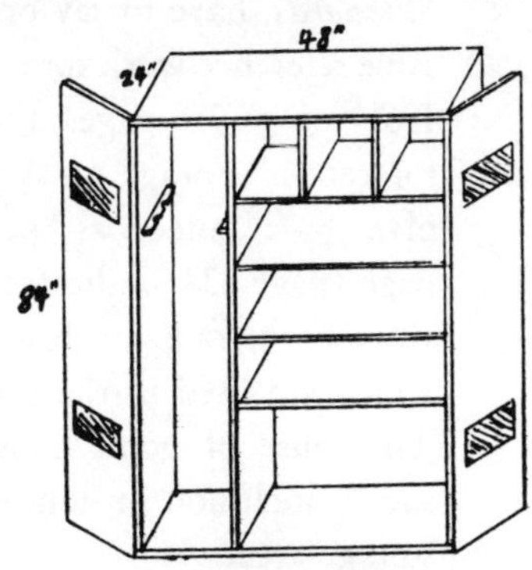


CUSTODIAL CABINET

TO BE USED WHERE THERE IS INSUFFICIENT SPACE TO INSTALL A CUSTODIAL CLOSET

A custodial closet suitable for two workers.

8. A utility sink for the cleaning of equipment, washing hands, and other uses.

9. Adequate ventilation and heating.

10. A floor drain is desirable so that large equipment may be sponged or hosed off while resting on the floor.

## Closets and Cabinets

The proper provision and location of custodial closets and cabinets provide one of the greatest opportunities for conservation of time of the housekeeping personnel. A properly equipped utility station provides the custodian with the materials and equipment which he will need to perform his cleaning work, avoiding the need for him to walk a great distance to obtain these supplies or spend time checking them out.

Not only do insufficient facilities require the custodian to travel back and forth between his work area and the central storage area, but also, equally important, it gives him an excuse to be absent from his work station.

In the past, the selection of utility areas for custodians has been a negative procedure. That is, the housekeeping department was given areas which could not be used by others. Even where specific custodial closets have been designed in new construction, they have at times been appropriated by "more important" departments—that is, any department other than the housekeeping department. We find, therefore, that our "custodial facilities" are often areas under stairways, in corners of rest rooms, locker rooms, or boiler rooms, or piping access spaces.

It is significant what Ramsey and Sleeper, in *Architectural Graphic Standards,* have to say on this subject. This superb volume is the indispensable reference work for architects and engineers. In the fifth edition (January 1961) of its 758 pages, there is no indication of minimum standards either for central storage areas or custodial closets (the latter are referred to on plans by architects as "j.c.," for "janitors closet"). Less than a third of one page (page 428) indicates the dimensions of "cleaning equipment" such as a washboard, radiator brush, wash tub, and a household vacuum, none of which are used in the typical industrial or institutional cleaning operation. The cause of good housekeeping would advance materially by the very simple inclusion in standards books of a custodial cabinet, closet and central storage area.

In general, no custodian should have to walk more than 250 feet nor have to go to another floor level to obtain his supplies and equipment. It is desirable to supplement the role of the closet or cabinet in bringing the tools to the work by the use of mobile carts specifically designed for this purpose. These are discussed in the chapter dealing with manual equipment. The effectiveness of the utility areas can be further supplemented by the strategic location or water taps and drains throughout the plant.

Where room cannot be found for a walk-in closet, or where the area is too small to warrant the use of such a closet, a cabinet made of wood or steel may be used. The simplest approach to this is to use "combination cabinets" sold by most office equipment supply firms, although a special cabinet made for the purpose would be more desirable.

It is interesting that hospitals have found that the custodial closets can be a very serious source of cross-infection if not properly set up and maintained. Where these areas are not given proper attention, the damp corners and dirty equipment become a breeding place for bacteria. The materials and equipment stored in the supply closet are taken to every area of the facility, including such sensitive areas as rest rooms, dietary areas, and medical services. Thus, the proper organization and maintenance of the supply closets becomes essential in a sound industrial hygiene program. Furthermore, the condition of custodial closets is usually an indication of morale and attitude toward the job. The regular inspection of these facilities is important.

The International Sanitary Supply Association has produced a film, "Easy Does It," which points out, among other things, the importance of proper custodial closet organization. The film is done in a clever, humorous vein, and, although directed to hospitals, it is also a good training tool for industry.

The requirements for good layout and operation of an ideal custodial closet are:

1. Custodial closets must comprehend the storage in the work area of modern mechanical housekeeping equipment. The "janitor closet" of yesteryear, which was suitable only for the storage of a mop bucket and certain hand tools, is of little use today.
2. Just as in central housekeeping storage, the custodial closet should be carefully organized and properly marked, so that there is a given location for each item, making the absence of any item immediately apparent.
3. Each custodial closet and cabinet should be assigned a number which is clearly marked on the door, for simple reference.
4. The door should be kept locked, and the contents of the closet or cabinet made the responsibility of those persons utilizing it.
5. Color coding is useful in identifying equipment and supplies with specific custodial closets or work areas. The closets may be referred to by color, rather than by number. For example, all equipment from a "green" closet should be marked with a spot or a line of green, using paint or colored tape.
6. A supply list should be made for each closet to insure its proper supplying

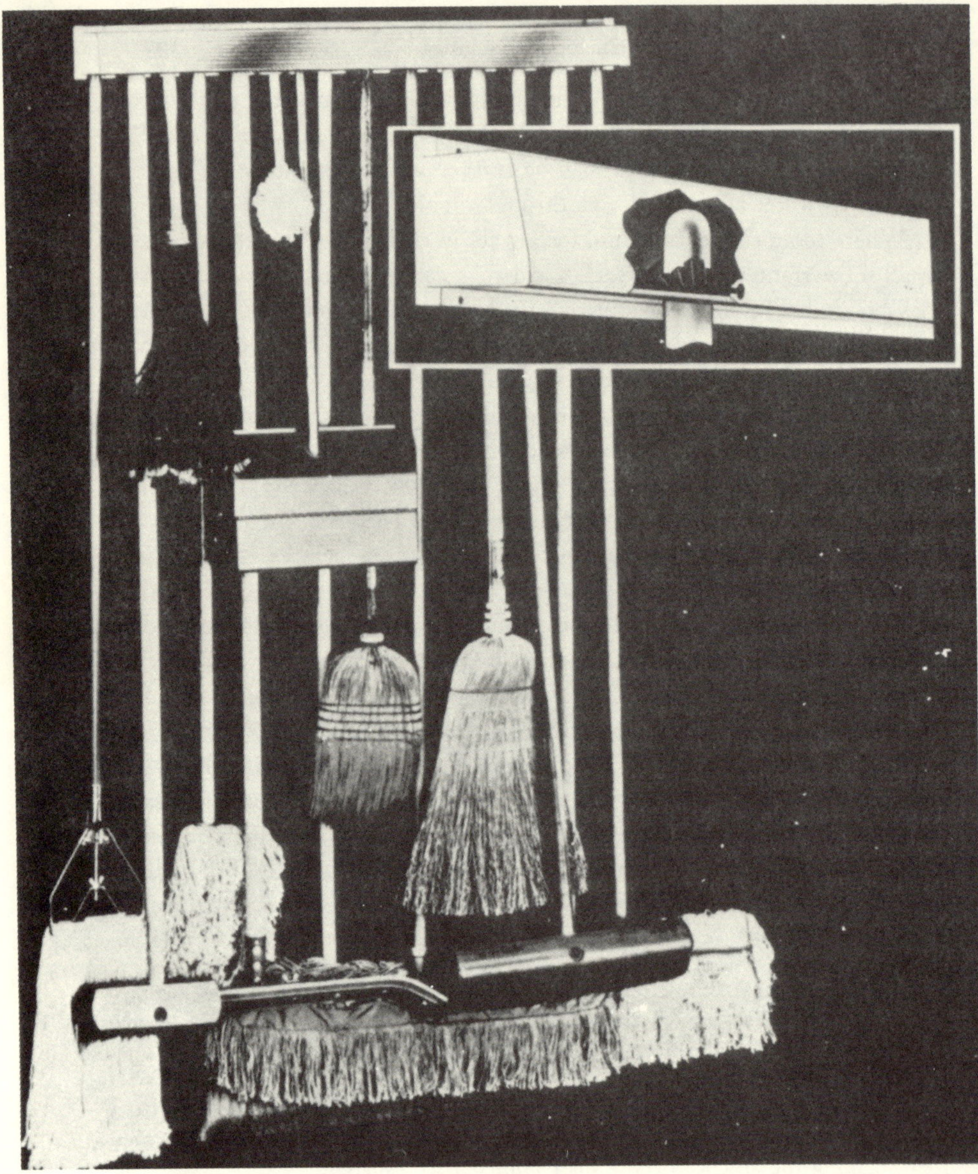

Tool holders help in the organization of a custodial closet (Walton-March, Inc.)

and equipping. All equipment which cannot be stored in custodial closets *must* be returned to the central storage area.

## Containers

Housekeeping products are now packaged in a wide variety of containers. The choice of container sometimes has a direct effect in aiding or hindering the custodian in his work.

These are the basic container types:

1. The most common bulk container is the closed-head steel drum for shipping liquid products. The typical drum has a varnished inner coating to prevent rust by the action of water vapor. Special baked-on phenolic resin liners are available for mildly corrosive chemicals. Mild acids will require a plastic lining and plastic plugs. Strong acids should not be shipped in this type of container as denting may cause lining failure which will result in leakage.

For many years the standard bulk container has been the 55-gallon drum. A drum of wax or detergent will normally weigh about 515 pounds. Sixty-five gallon drums are also available from certain suppliers. Even though these may in some cases offer a slightly lower unit cost than the 55-gallon drum, they should not be used. Generally weighing some 610 pounds, they are awkward and dangerous to handle.

Smaller drum sizes are available. Thirty-five or 30-gallon drums are known as "half drums," while 20- or 15-gallon drums are known as "quarter drums."

In all these cases, even though a lower unit price is obtained because of the quantity purchased, the hazard of using heavy containers far outweighs the economy. There are many organizations which have suffered regular lost-time accidents due to handling drums, who have not only eliminated these accidents when purchasing in smaller containers, but also have lowered their total chemicals budget by avoiding waste and over-consumption.

2. The middle range of containers for liquids are cans and pails, which normally do not require the use of a faucet. The most common sizes are 12, 10, 7, 6, and 5 gallons. The 5- and 6-gallon cans predominate. Most of these containers are fitted with a metal or plastic pouring spout. Tilting racks are available for these containers, but they have not gained wide acceptance. Probably the best combination of unit price and ease of handling is available in the 6-gallon container. These containers are also available with the special corrosion-resistant linings used in the drums.

3. The smallest containers for liquids are cans, jugs, and bottles. These are available in a wide variety of styles, types, and materials, ranging in size from 2-gallon cans to tiny glass ampoules. Recent improvements in the packaging industry have made plastic containers available. Because of the greater safety in the handling of plastic containers, glass containers should not be accepted as a container for housekeeping chemicals except for those materials not compatible with plastic, such as certain types of solvents.

4. A number of chemical products are available in aerosol containers, but these are extremely expensive per unit quantity consumed, and also there are ecological problems concerning the propellents and disposal of the cans. Typically, a pistol-grip small plastic spray bottle gives good performance at low price. Aerosol cans, however, may be considered for infrequent insecticide use and for nitrogen propellent in order to freeze and remove chewing gum.

5. Fiber drums are primarily used for packaging powdered chemicals. These containers are measured in gallons of volume and range from 70 or more gallons to just a few gallons. In the lower-size range they are known as fiber pails or tubs. Fiber containers must be stored so that they will not become wet, as this will cause most products to cake and thereby become useless.

Some fiber containers, with a plastic inner bag, are used for the shipment of liquids. This has met a good deal of resistance because when such containers are wet by accident they may collapse. They may also be punctured relatively easily by drum trucks having sharp forks.

6. Open head steel drums, as well as fiber drums with plastic liners, are used for the shipment of paste or jelly-type soaps. Some of these drums contain built-in solutionizing attachments, where water is run into the drum and then drawn off as it dissolves a certain amount of the soap. This type of container is not often seen now, as paste soaps are falling into disuse, and also because of the variability of the concentration and the fact that complete solution may not take place.

7. Special packages are also available for specific uses. Multiwall bags are used for inexpensive bulk materials such as sweeping compounds or oil absorbents. Chipboard cartons are used for packaging small amounts of powdered materials. Fiber or plastic shaker-top containers or metal or plastic cans and cups are used for powdered and paste-type cleansers and polishers.

Carboys of plastic or glass are used for shipping acids and other highly corrosive materials. These are generally shipped in an overpackage of steel, fiber, or wood. Some of these combinations require a returnable container arrangement which is costly in freight and handling costs. The combination of plastic carboy and fiber or wirebound wood overpackage, in the size range of 10 to 15 gallons, provides a suitable container that is economical enough for a single-trip use.

The future will unquestionably see many improvements in the packaging of sanitary chemical specialties. Some probable examples are:

1. Water soluble packages of PVA (poly-vinyl-alcohol) or other similar material. With such containers, a premeasured amount of detergent, for example, can be thrown into a measured amount of water, with the package. Package and contents dissolve quickly.

2. Eventually, it is likely that a "pill" will be developed which can be thrown into a pail of cold water in order to heat it to a temperature suitable for cleaning purposes. This would be particularly useful in industrial plants where hot-water piping is limited.

3. Tapering containers of metal, plastic, and fiber will be used because they may be "nested" when empty, thus requiring a small space for storage.

4. More containers will be designed with built-on applicator systems or devices.

5. The development of packaging materials and fabrication techniques will provide lighter, yet stronger, containers. More corrosion-resistance will be embodied, resulting in longer shelf life.

6. Collapsible containers, even of steel and plastic, will probably be developed.

# 34 Powered Equipment

THE American Economic Foundation pointed out recently that the greatest problem now facing the American free enterprise system is our burden of obsolete machinery and equipment. In the past, we have been able to overcome our high wage scales in the world market through superior productivity. This productivity has been based on mechanization, and on methods collectively referred to as "know-how." A number of other countries are now investing in capital equipment at a considerably higher percentage of gross national product than we are. This should bring them a productivity increase at a faster rate than ours.

Our position in the world market must be protected through improved methods and greater mechanization. Housekeeping, and other indirect costs as well, must play its role in this quest for productivity.

Very few firms have properly mechanized the housekeeping department. This is true even of those companies which have shown remarkable advances in the mechanization and automation of their production processes. The roughly 100 million dollars spent annually on housekeeping equipment is certainly less than half of what should be spent. Although it is true that improvements in housekeeping will normally only be gained over a period of a number of months, the acquisition of correct equipment can bring startling results in a very short period of time.

As a word of caution, just as the appropriation of more billions of dollars will not give us space leadership in itself, neither will the provision of housekeeping equipment provide overall housekeeping efficiency in itself. Mechanical equipment thrown into a disorganized housekeeping operation soon becomes ineffective or completely inoperative. Without training and supervision, for example, it is not unusual to see work being done manually while a brand-new piece of equipment rests in the storeroom.

## Equipment Benefits

Custodial personnel must be equipped with sufficient labor-saving equipment to enable them to carry out their respective tasks efficiently. This is consonant with the principle that management has the responsibility to properly equip each worker for the job for which he is held responsible.

Housekeeping equipment, if properly utilized, often pays for itself in less than a year's time. Although this justification for purchasing new equipment is adequate, there are other advantages in their acquisition:

1. Properly selected equipment generally does a job better and more economically than can be done manually. Both a mop and a vacuum will dry a floor, but the vacuum will remove the water faster and leave the floor cleaner, for example.
2. Fatigue is reduced through the use of power equipment, raising the general productivity of the custodian even when not using the equipment in question.
3. Equipment, particularly of the mechanical type, is a status symbol. The purchase of modern housekeeping equipment becomes a tremendous morale builder, not only improving the attitude of the custodian himself, but also enhancing the prestige and status of the housekeeping department as a whole. This aspect can be further developed through the selection of attractive equipment (streamlined, chrome plated, etc.) and the marking of individually assigned equipment with the custodian's name.
4. Housekeeping equipment promotes safety. The use of an extension wand and pipe tool on a dry vacuum, as an example, avoids the need for ladders and scaffolding. Floors which are dried more quickly and thoroughly with a wet vacuum are less likely to cause an accident from slipping.

## The Equipment Program

A complete program of housekeeping mechanization will include all these factors:

1. A survey of needs based on an evaluation of the work to be done in the

various maintenance areas. This survey must not be on the basis of the number of machines alone, but should also include all other factors. A large number of small floor machines, as a case in point, does not necessarily indicate a high level of mechanization. For areas where larger floor machines may be utilized, the smaller ones may only be operating at 50 percent efficiency. Or, the purchase of an autoscrubber may have tied up funds that could better have been utilized in another way, because of inability to utilize the autoscrubber in congested areas or winding corridors.

2. Development of equipment specifications. It must be remembered that equipment and materials *combined* amount to only a small percentage of the housekeeping dollar. Thus, it is poor economy to provide other than the best equipment available. Of course, *best* is not synonymous with *most expensive.* If the use of a 20-inch floor machine is called for, it would be unsound to settle for a 15-inch machine on the basis of "economy." Some of the features which should be considered when determining specifications are:

Appearance
Size
Weight
Horsepower
Electrical characteristics
Attachments
Maneuverability
Flexibility
Finish
Safety features
Noise level
General ruggedness and quality
Warranty

3. Selection of the supplier. Proper regard should be given for the reputation of the supplier and his ability to furnish service and parts. The equipment furnished should be of a type that will permit ease and rapidity of maintenance and will be manufactured over a long enough period of time to permit general standardization.

4. Purchasing. Equipment should be obtained by the purchasing department under the arrangement that best suits the organization. Equipment may be purchased outright, leased, rented, or perhaps obtained on a product tie-in.

5. Training. The manufacturer or his representative should demonstrate the equipment and train selected custodians in its use initially. The supplier should be encouraged to provide additional retraining from time to time (this should unquestionably be provided where standardization on one brand of equipment has taken place). Where equipment is purchased on the basis of price alone, the supplier likely will be unable to provide service of any kind. Supplier training should only be a supplement to regular internal training, which will involve:

Selection of the proper equipment for the job
Selection and use of attachments
Storage of equipment when not in use
Care of equipment
Safety procedures

6. Equipment repair. Preventive maintenance will keep the equipment in condition to provide a long life of efficient service. Although basic repairs are performed by maintenance personnel, certain duties should be performed within the

housekeeping department. It is desirable to assign these duties to an individual having a mechanical aptitude. This might be the central stock clerk, or the group leader of a special projects team. Within the housekeeping department, the equipment should be properly lubricated, painted, cleaned, tightened, and adjusted (unless other departments desire to perform these functions). If the plant employs a lubrication engineer, or if suppliers provide suitable services, help should be obtained in setting up lubrication records and procedures, including a record tag for each machine.

7. Inspection. Equipment should be inspected regularly to be certain that it is in safe and correct operating condition. Equipment should be checked at least weekly for cleanliness and general condition.

8. Parts stock. Once efficient equipment has been purchased, downtime must be held to a minimum. In addition to selecting suppliers who can provide parts on short notice, it is also desirable to keep spare parts on hand in the housekeeping stock area. New parts which are subject to breaking or malfunction, such as attachment plates, switches, v-belts, pinions, and electric plugs, should be purchased. Also, worn-out equipment should be dismantled, and parts in good repair cleaned, properly marked, and put into storage.

9. Storage facilities. Proper storage and service facilities should be provided for all equipment to prevent damage, soiling, and loss, as well as to permit ease of inspection.

10. Attachments. Suitable attachments should be provided in order to give as much flexibility as possible. Attachments should be carefully controlled by permanently issuing them to specific custodial closets. Color-coding is also helpful.

11. Resurvey. A periodic reanalysis of equipment needs and conditions is very worthwhile. This should be done every one to three years, depending on the size of the operation. The survey should cover these points:

Extent of use. Idle equipment on hand should be investigated, and the problem corrected.

Equipment condition. Inefficient or unsafe equipment should be replaced.

New developments. New equipment is coming on the market regularly and should be investigated and evaluated.

Additional requirements. The housekeeping department may be maintaining additional areas or performing additional or chanted functions which could benefit from additional mechanization.

## Acquiring Capital Equipment

Assuming that an organization has sufficient working capital, and intends to remain in business, it will purchase capital equipment which will return its own cost within a specified period of time. Probably no responsible firm will permit a pay-out of longer than five years; most companies restrict the pay-out to two or three years, but vary the time depending on the type of equipment.

Capital purchases are normally a management decision because of their fixed nature. They do not vary with the production rate as wages or materials can be varied and thus a decision to purchase equipment must be "lived with" until the equipment is completely amortized. In order to properly control fixed expenses, many organizations require the approval of a special committee or board for the funds purchase of capital equipment. These boards normally have formulas as a guide to decisions on capital purchases, involving such factors as wage rates, interest charges, amortization periods, pay-out period, salvage value, etc. The group or person responsible for passing on capital equipment purchases deserves, and should be provided, a factual presentation of the benefits of the equipment. As many companies operate with a limited capital expenditure budget for any given year, the probable return from the purchase of housekeeping equipment will be compared with requests or recommendations from other departments.

Special consideration may be given to the purchase of capital housekeeping equipment if these points are made:

1. The housekeeping effort does not suffer obsolescence—it must go on despite any changes in the nature of the manufactured product or the physical facility.
2. New housekeeping equipment is a great morale builder.
3. Housekeeping wage rates keep going up.
4. Although the economic analysis of the equipment will be based on the speed of the job, it must be pointed out that housekeeping equipment often also provides a better or more thorough job, sometimes difficult to analyze.
5. Housekeeping equipment is normally quite long-lived if it is of quality manufacture and properly cared for. Many floor machines and vacuums, for example, have been in use well over ten years.

Consideration should be given to leasing housekeeping equipment when it cannot be purchased because of restrictions on capital expenditures. Although this does raise the total cost of the equipment, it has the advantage that moderate payments can be made on a periodic basis.

It is not necessary to save half a man's time in order to justify the purchase of a piece of equipment. The National Association of Bank Auditors and Controllers points out that, on the basis of an annual salary of 5,000 dollars, a daily time saving of only ten minutes justifies the purchase of equipment costing 1,000 dollars!

## Floor Machines

Single-disc floor machines account for roughly half the dollar value of all powered housekeeping equipment. The basic purpose of the floor machine is the cleaning of many types of floor surfaces through a scrubbing action, but the same machines, with different brushes or attachments, are also widely used for buffing, scarifying, rug shampooing, and sanding. Their wide range of uses, sizes, and attachments make them the most versatile piece of equipment available to the custodian.

Double-disc machines are also available. In this type of machine, the self-propelling feature of the single-disc machine is lost, although it requires less skill to maneuver. Because it has more working parts, the dual-brush machine is more likely to give trouble. Finally, the size of the double-brush machine is misleading. Where the horsepower and brush pressure of a machine are equal, the efficiency will be proportional to the area of the brush in contact with the floor (a function of the square of the diameter). Thus, a double-brush 16-inch machine has two 8-inch brushes covering a total area of 75.38 inches which takes into account the 4-inch hole in the center of the brush where the attachment plate is located. On the other hand, a single 16-inch brush, also taking into account the 4-inch center hole, covers 188.49 square inches. This is a ratio of almost exactly 2½ to 1, yet both pieces of equipment are sold as 16-inch floor machines!

A third type of floor machine utilizes a cylindrical brush, which has the advantage of concentrating the pressure over a small floor area. This machine does not have the flexibility and maneuverability of the single-disc machine and is comparatively rarely used in the smaller sizes. In the larger sizes, it becomes an autoscrubber, a variety of which is the common street-cleaning machine.

#### SELECTION

The most important factor in the selection of a floor machine is its size, in brush diameter, and this leads to the most common and costly mistake which can be made in selecting a floor machine—the choice of a machine which is too small. The price of a floor machine, typically, increases roughly twenty dollars per inch of diameter. Thus, a 17-inch machine will cost about forty dollars more than a 15-inch machine, yet will be approximately 30 percent more efficient. The small annual cost difference can be repaid in the performance of a single floor-cleaning project. Over the life of the machine, the additional cost becomes ridiculously low.

Too many floor machines are purchased on the basis of the smallest machine which it is possible to get by with. A much better economy is to purchase the largest possible machine which can be used within the limitations of the size of the area and its congestion. The typical floor machine in industry is 15 or 16 inches in diameter—it should be 19 or 20 inches in diameter.

Special considerations must be taken into account when determining machine size:

1. The total area to be maintained.
2. Congestion within the area.
3. The weight and portability of the machine. This is particularly important when the machine has to be transported up and down stairs or between buildings.
4. Larger floor machines are less likely to wobble and "buck" than the smaller machines.
5. For uneven floor areas, it will not be possible to use the larger machines, as they will span over depressions.
6. The weight per square inch generally decreases above the 16-inch size. Therefore, for heavy-duty operations such as sanding, grinding, or scarifying, a 15- or 16-inch machine should be used.
7. For carpet shampooing, the machine should also be in the middle-size range.

The ideal floor machine would be completely flexible, having variable speed, pressure, and brush diameter, as well as built-in conversion systems to change from scrubbing to polishing. Undoubtedly, future machine designs will contemplate some or all of these refinements.

In addition to the size and weight of the machine, a number of other factors are to be considered:

1. *Horsepower:* No floor machine should be purchased for industrial use of less than one-half horsepower, and this would be suitable only below 16 inches. Three-quarters horsepower is adequate through the 18-inch size; one horsepower through the 21-inch size; and one and one-half horsepower through the 24-inch size. Proper horsepower is required for an effective weight-rotational speed ratio.
2. *Motor type:* Most floor machines use either the *capacitor*-type or *induction-repulsion*-type motor. Either is acceptable although the capacitor type has the advantage of not requiring a separate starter, which may give trouble.
3. *Electrical characteristics:* Most floor machines are manufactured for use with "house current" (110 volts, 60 cycle a-c). Special motors are available for higher a-c voltages, or direct-current use.
4. *Arrangement:* In order to scrub or polish under furniture, the type of machine known as the "divided-weight" type is used, where the motor is placed back from the brush over the wheels, the wheels carrying a portion of the motor weight. Thus, the brush housing projects and is able to get under low-hanging furniture. This is opposed to the "concentrated-weight" machine, the much more usual variety, where the weight of the motor is entirely supported over the center of the brush.
5. *Wheel retraction:* In the "concentrated-weight" type machine, the wheels

must be off the floor to permit the full weight of the machine to rest on the brushes. Some wheels are spring-loaded and can be retracted by the operator while in a standing position by the application of foot pressure; other wheels must be retracted manually. Some designs keep the wheels in a fixed position.

6. *Wheel size:* Larger wheels provide easier maneuverability when the machine is being transported, and are less likely to become damaged.

7. *Switch:* Most machines are now equipped with a heavy-duty deadman type switch. Other types of switches are dangerous.

8. *Adjustable handle:* Those machines which have an adjustable handle provide the greatest comfort to the operator, regardless of his height.

9. *Ruggedness:* This term is hard to define and includes such things as the quality of bearings, gears, and belting, as well as the type and design of the housing and other parts. Also important is the quality of the motor and its permissible temperature rise, as well as its electric cable.

10. *Noise level:* This is particularly important when the machine is to be used in administrative areas during normal business hours.

### CARE AND USE

Proper care and use of floor machines will extend their life and efficiency. It will also provide a safer operation. Here are some pointers:

1. Floor machines should be plugged into a suitable wall outlet and properly grounded. Do not plug into a light socket as this may overload the circuit or lower the efficiency of the motor. If an extension cord is used, it should be sized at least as large as the cord on the motor, and preferably a size larger.

2. Water or other liquids should not be splashed onto the machine. If acids or caustic solutions come into contact with the machine, they should be washed off immediately.

3. The machine should be kept clean. A dirty piece of equipment is not likely to receive good care. The electric cord should be wiped clean each time it is wet to keep from marking floors or furniture.

4. If a solution tank is used, both the tank and the feed lines should be flushed at the end of each day of use.

5. The brushes should be attached to the machine manually. If a brush is attached by placing the machine on the brush and then starting the machine, injury or damage will be likely to occur.

6. When the machine is not in use, it should be rested on the wheels, rather than on the brush. If the machine is permitted to rest on the brush, the bristles of the brush will flatten out at the point of greatest pressure and cause the machine to wobble in use. This reduces the machine efficiency and puts a greater strain on the operator.

7. Where a floor machine is used in a large medical area, it should be properly color coded and marked for that area and not used elsewhere. The machine should also be washed at least once per week in a disinfectant solution. These procedures will greatly lessen the possibility of cross-infection from this source.

8. The cord should trail behind the operator when the machine is in use. It should never be wound around the handle, thus making the deadman switch inoperative.

9. A machine should be transported and operated in such a way so as not to strike walls, doors, or furniture.

10. In transporting the floor machine, it should be handled carefully. It should never be forced over a rough surface, allowed to bang about in a truck, or bounced down stairs.

11. Care should be taken when operating the machine to keep it off the cord. If the cord winds on the brush, it may damage both itself and the machine, as well as shock or otherwise injure the operator.

12. Machines should be stored in a clean, dry room that is properly lighted. Cords should be wound on the hooks provided on the handle, and brushes properly cleaned and hung up to dry. Machines should be lubricated as required. Overlubrication should be avoided.

13. To prevent shock, the use of rubber gloves or rubber shoes is recommended when working with solutions.

### FLOOR MACHINE ACCESSORIES

*Attachments.* The most useful floor-machine attachment is the solution tank. Usually mounted on the machine handle, this tank dispenses a cleaning solution onto the floor through a special brush. The tank, control valve, and feeder tubing should all be corrosion-resistant. Carpet scrubbing is the perfect application for the solution tank, but opinion is divided on its use for general floor scrubbing. Although theoretically a very sound idea, it is practical only under certain conditions. The solution tank does have the advantages of avoiding the necessity of changing from mop and bucket to floor machine, and the placing of the solution at the point of use requires less stepping in the detergent solution. On the other hand, the mop in the hands of an experienced worker permits more accurate placing of the solution so that it does not run and allows the solution to remain on the floor before scrubbing begins, so as to gain the advantage of the chemical cleaning action of the detergent. Finally, where large areas are being cleaned, it is quicker to prepare a mopping solution in a pail than to prepare and pour one into a solution tank. The latter operation often involves spillages or foaming over. When not in use, the machine with solution tank cannot be rested on its wheels but must remain with the weight on the brush.

Some floor machines are fitted with a small vacuum to remove dust created by the buffing operation. This is sometimes assembled as a homemade job. Up to this point, it has seen rather limited use.

*Grinding and sanding attachments.* These are available in the middle-size range. They are not applicable to the smaller machines, which do not have sufficient weight and power, or for the larger machines, which do not have sufficient pressure. These attachments utilize specially cut carborundum stones or sandpaper. Floor machines used for light-duty grinding and sanding will be damaged if so used except for a short period of time.

*Brushes.* Floor-machine brushes are available in a considerable variety of types for each of the various machine sizes. Brushes are measured by the bristle diameter, as spread under the weight of the machine. It is this diameter which determines the size of the machine, rather than the diameter of the wooden or metal backing of the brush. Floor machines can utilize smaller brushes than the size of the machine, but not larger brushes. Smaller brushes are actually desirable when the floor machine is utilized for scarifying, sanding, grinding, or heavy-duty scrubbing of problem areas over a considerable period of time, to prevent undue strain on the motor and increase the brush pressure.

The brush is manufactured by setting bristles, wire, etc., into the drilled holes in the brush block. Most bristles are set vertically to the plane of the block. As the brush rotates, the bristles bend. Some bristles are now being set at an angle to the block so that in rotation the bristles bend to the vertical position, thus improving their action.

A wide variety of brush-filling materials are in use:

1. Bassine fiber. A stiff fiber for heavy-duty scrubbing, dark red to black in color.
2. Nylon. A very durable synthetic fiber which can be varied from very soft to very stiff. It is used for floor scrubbing and carpet shampooing, and is usually colored black.
3. Palmetto fiber. A moderately stiff fiber used for medium to heavy scrubbing and also for rug shampooing. Light red or reddish brown in color.
4. Palmyra fiber. A medium stiff fiber, somewhat stiffer than palmetto, generally used for scrubbing. It is mottled brown in color.
5. Steel wire. Made from various grades of carbon steel wire, this brush is used for very heavy-duty scrubbing. A brush made from "butcher's wire" can be used for scarifying. Steel wire may be blended with fibers such as bassine. Steel wire brushes must never be used on any smooth finish floor susceptible to staining or scraping. These brushes must be dried carefully after each use to prevent rusting.
6. Tampico fiber. A medium soft fiber for buffing, often blended with other fibers to provide more stiffness in the larger sizes. Tampico is generally white or tan in color.
7. Union mixture. A fiber mixture, usually of palmyra and tampico, and therefore generally mixed brown and tan in color. This combination is medium stiff and is normally used for polishing.

When a solution tank is used with the floor machine, a specially drilled and channeled brush must be used to permit the flow of the solution through the brush and onto the floor. These channeled brushes may be made of wood (usually in the smaller sizes) or metal (usually in the larger sizes). Generally speaking, wood-backed brushes should be considered expendable, whereas metal-backed brushes should be refilled.

A little care will pay brush dividends:

1. Brushes should be combed frequently to remove foreign matter.
2. Brushes should always be placed on the machine by hand.
3. When not in use, brushes should be hung on a nail or rested on the block.
4. Brushes should be air dried. Drying by heat, such as on a radiator, may damage the block.
5. Brushes should be washed occasionally, but care should be taken so as not to overwet the wooden block.
6. Brushes designed for dry cleaning should never be used in liquids.

#### FLOOR MACHINE PADS

*Lambswool.* Circular lambswool is used on a special holder or under a brush where buffing to an unusually high degree is desired. Lambswool is not recommended for normal buffing operations.

*Nylon pads.* These pads are made of nylon impregnated or treated with silicon carbide and are made in various grades for scrubbing or polishing. Some types of pads require a special holder, whereas others may be used under a regular floor-machine brush. The pads have the advantages of being rustproof, safe to handle, not "loading" quickly, not accumulating static electricity, and not spinning off water. Those who feel that there has been nothing new in the housekeeping field should take a close look at these pads!

*Steel wool pads.* Woven steel wool is available in eight grades, #0000 to 4. For floor work, #0 is normally used for buffing, #1 for light scrubbing, and #2 for medium scrubbing. Grade #00 is sometimes used for fine polishing, while grade #3 may be used for heavy-duty scrubbing or for "shaving" cork floors. Grade #3 should not be used on composition floors. When in doubt as to the proper grade of steel wool to use, it is safest to begin with the finer grades and work up to the grade which works most efficiently, yet still safely.

### Vacuum Cleaners

Vacuum cleaners are available in a large array of styles, types, and sizes. The equally wide range of attachments makes possible a combination for almost any conceivable industrial need. Yet, the industrial housekeeping department which operates without even a single wet vacuum, for example, is seen regularly. Space permits only a limited description of this versatile tool.

### VACUUM SELECTION

A number of factors affecting vacuum selection are identical with those for equipment in general. In addition, these items refer specifically to vacuums:

1. The vacuum should have a high suction rating. This is normally expressed in inches of water lift.
2. A motor should be of adequate horsepower to permit continuous operation without damage. The motor should be the by-pass type.
3. If the vacuum is to be used for wet pick-up, it must be provided with safeguards so that neither water nor suds are ever drawn into the motor.
4. The vacuum tank should be large enough so that frequent dumpings may be avoided.
5. A dump valve is desirable, eliminating the need to remove the vacuum head and turn the entire unit over when handling liquids.
6. The unit should be easy to transport and maneuver. Casters should be large and ruggedly mounted.
7. The vacuum should be adaptable to a wide range of attachments.
8. When in operation, the vacuum should be reasonably quiet.
9. The unit should be provided with white rubber guards to prevent marking or damage to walls and furniture.
10. Style, finish, and trim should be taken into consideration.
11. Stainless tanks should be selected for wet use.

### VACUUM CARE AND USE

Vacuums will provide better service when properly used:

1. Vacuums should be emptied at the end of each day, no matter whether the unit was used for wet or dry service.
2. On wet vacuums, the automatic cut-off mechanism should be checked frequently to make sure that water cannot enter the motor.
3. Tanks on wet vacuums should be washed at least monthly with a disinfectant solution to prevent the growth of bacteria and algae.
4. The vacuum should be stored in an open position to allow entry of air.
5. Strings, hair, or soil should be removed from vacuum attachments at the end of each day.
6. Reversing the hose every few months will prolong its life because of the change in the wear pattern.

TYPES OF VACUUMS

*Drum adapter.* A standard 55-gallon open head steel drum becomes a vacuum container when a regular industrial vacuum head is placed on it with a special adapter cover. Where this combination is to be used for wet use, the drum should be equipped with a dump valve near the bottom. Maneuverability is achieved by fixing casters onto the drum, or by placing the drum on a dolly. This arrangement is particularly useful in industrial operations where large quantities of dry or wet waste are to be collected, such as in cleaning sumps, removing sawdust from woodworking equipment, vacuuming overhead with extension tools, etc.

*Dry tank type.* Some vacuums are designed for dry use only. They must be carefully marked for this purpose, as they are easily damaged by water. For carpet or general floor dry cleaning, this type of vacuum may be mounted or carried on a utility cart. A special class of this vacuum, of unusually rugged construction and high power, is used for collecting metal and plastic chips. Unless a dry vacuum can be restricted to use by a custodian who will never need a vacuum for wet cleaning, it is desirable to always provide wet-dry vacuums for greater flexibility and safety.

*Hospital vacuum.* Any of a wide variety of standard vacuums may be converted for hospital or similar critical use through the provision of a special filter system. These filter systems may remove dust particles and microorganisms as small as 0.5 micron. In this range, the hospital-type vacuum also becomes useful in the cleaning of a small "gray room," where the room is not of sufficient size, or the cleaning problem sufficiently acute, to warrant a special stationary system. The suitability of the hospital-type vacuum for the cleaning of areas requiring special attention can be determined through an analysis of the exhaust air for dust count and particle size.

*Pack type.* This is a high-power dry vacuum worn on the back like a knapsack, or slung over the shoulder with a strap. The types having an internal filter bag seem to have the greatest acceptance and use. These units are most suitable where the worker will be cleaning in confined spaces or moving a great deal from place to place. It avoids the necessity and hampering effect of having to move a vacuum cleaner on casters to a location very near where the employee is working. The biggest drawback with the pack-type vacuum is the relatively small tank capacity.

*Scoop type.* This is a wet-dry vacuum having a built-on squeegee mounted either in front of or behind the tank. As the operator walks forward, pushing the tank, the squeegee scoops up the water. This arrangement eliminates the wand and hose and permits a very rapid cleaning operation. It may also be used for dry floor cleaning in a similar fashion. This type vacuum is limited in its maneuverability, and its use should be restricted to long wide aisles and open spaces. It also requires even, smooth floors for efficient operation.

*Stationary system.* Built-in systems have a central vacuum machine, normally located on the lowest floor, with pipes running to vacuum stations, permitting cleaning of any area of the structure. Stationary systems are particularly well suited for multistory buildings where a good deal of vacuum work will be needed, such as in precision assembly work. Stationary units are applicable to a large range of industrial uses and may actually connect a number of buildings. Open vacuum units may be stationed above inspection stations so as to keep these areas free of dust. The stationary system eliminates the need for a number of different vacuum units, but, on the other hand, if it becomes inoperative, no vacuuming can be performed. Care must be exercised in using long hoses so that they are not dragged or bent, and so that they do not mark thresholds, furniture, door jambs, etc. The vacuum stations provide a specialized advantage in that they may be used for the removal of dust and lint from dust mops at periodic intervals. The stationary system is almost a requirement for large superclean rooms.

*Suction sweeper.* This is a specialized style of scoop-type vacuum for removal of dry waste and litter. This unit is normally driven by a gasoline or electric motor, the former being preferred where conditions permit. It has a very large collector bag and is particularly suitable for outside policing of trash and leaves, and internal collection of waste paper scraps and similar light materials.

*Wet-dry vacuum.* This is the most common and versatile of the vacuum types. Constructed with a round or square tank ranging from 5 to 20 gallons in size, the wet-dry vacuum has a wide variety of uses when equipped with the proper attachments. Some of these vacuums have built-on carrying trays for these attachments. The use of this vacuum for wet cleaning, for example, produces a much cleaner and drier surface than can be obtained with a mop, even for removing soil and water in the joints between composition floor tiles. The use of this vacuum will also reduce the surface and airborne microorganism count, contributing to better industrial hygiene. No housekeeping department should be without at least one vacuum of this type.

*Tilt-type vacuum.* Where floor drains are not available, a wet vacuum with a tilting tank, so as to dump into a commode, is very useful.

The pack vacuum permits effective cleaning in congested areas; an open area of this type should utilize a floor-standing machine (Advance Machine Company).

The scoop type wet vacuum removes large amounts of water quickly and efficiently (Clarke Division, Clarke-Gravely Corporation).

A litter vacuum is useful for the removal of loose, light-weight soil, either in grounds care or inside a non-obstructed facility (Parker Sweeper Company).

VACUUM ATTACHMENTS

A vacuum without attachments is as useful as a milling machine without cutters. Vacuums should be selected so that the attachments will fit all machines, thus improving flexibility and reducing the necessary stock of parts. Attachments should snap on and off easily, rather than having to be screwed on or off. Many of the attachments are available in various sizes and materials.

Basic components, without which the vacuum will not operate under the usual conditions, are:

1. Filter bag. The filter bag is designed to permit the flow of air but to entrap most soil particles. A type of soft cloth known as "moleskin" is suitable, but the newer synthetic materials are preferable because they do not require removal for wet pick-up and may be more easily cleaned. The hospital-type vacuum has a special filter medium.
2. Hoses. Vacuum hoses are available in the 1½-inch and 2-inch diameters, and the various tools must be sized to fit the hose, which in turn must be sized to fit the vacuum. Hoses are normally 10 feet in length, but shorter lengths are available for close work. These are available in several different materials, but the most useful hose is made of neoprene, as this can be used on solvents which would damage other materials.
3. Wands. Wands are made of tubular metal and connect the hose to the various attachments, acting as a handle. Wands may be bent or straight, and should be light in weight. Extension wands permit the custodian to do overhead cleaning without leaving the floor.

Specialized tools are available for the following jobs:

1. Cleaning carpets, wet or dry.
2. Dry cleaning floors. Tools of this type use a sweeping action provided by bristles, a felt strip, or cotton yarn.
3. Removing liquids from floors. This attachment is fitted with squeegee blades.
4. Cleaning chalkboard erasers.
5. Cleaning dust-mop heads.
6. Dusting the tops of books while on library shelves.
7. Cleaning venetian blinds while in place.
8. Dusting walls.
9. Removing loose soil from the tops of overhead piping. These tools are available for various pipe diameter ranges.
10. Crevice cleaning, such as between radiator fins.
11. Upholstery cleaning.

Power scrubbers are available in many sizes, such as this large machine (Tennant Company).

12. Cleaning of bins and drawers. This tool is designed to remove dust but not heavier materials such as hardware fasteners.
13. Boiler tube cleaning.
14. Cleaning irregular surfaces, such as wire partitions.
15. Some vacuums may be fitted with a spraying attachment on the exhaust side of the unit.
16. A blower attachment may also be fitted on the exhaust side of the unit.

## Automatic Floor Machines

The automatic floor machine, often also called the autoscrubber, is essentially a combination of the scoop-type vacuum with a floor machine. The autoscrubber dispenses a detergent solution from a storage tank onto the floor, the floor is scrubbed by single, double, or multiple brushes, and the solution is then vacuumed into a waste reservoir. The machine is also suitable for dry cleaning, in which case no cleaning solution in involved.

Such a machine is self-propelling. A sulky may be attached on which the operator may be seated, but this is only successful for very large operations. The need for a sulky might indicate the desirability of obtaining a rider-type machine, which resembles a street cleaner.

Automatic floor machines are powered by storage battery, electric cable, propane, or gasoline engines. The cable-powered unit is the most trouble-free, although its range of operation is limited by the length of the cable. The battery-operated models provide the greatest flexibility of use, but they require battery recharging and are unusually heavy.

The purchase of an automatic floor machine on principle alone can tie up a good deal of money with no return, if conditions are not suitable for its regular use. This machine can only be used in large exposed areas such as long corridors, expansive lobbies, wide aisleways, etc. An elevator or ramp is required to move the machine from one floor to another. Floors must be smooth and level to permit proper operation of the brushes and squeegee.

In a single pass, the autoscrubber picks up detergent solution almost immediately after it is placed on the floor. Thus, because there is no time for the detergent to provide a chemical cleaning action, a scrubbing on a single pass will perform only a medium cleaning job on a heavily soiled floor. For heavy-duty cleaning, two or more passes will be required, and these steps must be taken into consideration when considering the efficiency of this machine. As with all equipment, manufacturers' claims for cleaning rates are often for an ideal situation over a carefully selected period of time.

Power sweepers are useful for removing dry soil in unobstructed areas. (Tennant Company).

In some cases, greater efficiency can be obtained if a second worker "services" the autoscrubber operator, thus permitting the machine almost uninterrupted use. The support worker removes obstructions, prepares cleaning solution and removes waste solution, mops corners which the machine cannot reach, removes water spots which it could not pick up, etc. In some cases, more than one machine can be serviced by a single custodian.

Under ideal conditions, the automatic floor machine can save a great deal of time. Under other conditions, it will be more trouble than it is worth. Undoubtedly, automatic machines of the future will be designed so as to be more maneuverable, applicable to more situations, and more reliable.

## Power Sweeper

Power sweepers fall into a natural classification of types, on the basis of their small, medium, or large size:

1. Walking type. The operator walks behind the smallest of the power sweepers, which clean a path from about 20 to 32 inches wide. These sweepers may be manual or powered. It is only the mechanical type of sweeper which includes a built-in vacuum, which is desirable for keeping down the dust. A manual sweeper is tiring to operate, creates a good deal of dust, and should only be used for the most restricted situations.
2. Riding type. In the intermediate size of power sweeper, the operator rides either at the front or rear of the machine, which is self-propelling. The most popular sizes of this sweeper clean 36-inch and 48-inch paths.
3. Municipal type. This is the large power sweeper of the type often used in street cleaning, sweeping a 6-foot path. Even larger sweepers are available, such as an 8-foot sweeper used to sweep aprons and runways at airports. The larger machines are generally designed for outside sweeping and may include a water spray system for further controlling the dust.

Selection of a power sweeper will depend on an economic analysis of its dollar saving potential, including such factors as:

1. Type of surface to be cleaned.
2. Degree and type of soiling and litter to be removed.
3. Congestion of the area.
4. Planned frequency of cleaning.
5. Ease of operation, maneuverability, and maintenance.
6. Efficiency of soil removal.

Selection should include consideration of the various methods of powering the sweeper, such as gasoline, propane, and storage battery; by the method of dumping the waste, which may be either manual or mechanical; and by the availability and cost of accessories such as:

1. Engine-hour meter.
2. Fire extinguisher.
3. Oxy-catalyst exhaust filters to cut down on carbon monoxide.
4. Side brushes for sweeping against curb or walls.
5. Attachments so that the vacuum system may be utilized as a large vacuum cleaner.
6. Warning lights and horns.
7. Water-bath mufflers for use in fire hazard areas.
8. Water-spray system for dampening the soil to keep the dust down.

## Special Powered Equipment

A number of types of mechanical equipment has been designed to perform specialized cleaning functions.

*Can washers.* These are automatic devices for the washing of garbage cans and waste receptacles. The can is inverted over the unit and held in position, while a foot pedal actuates the cleaning mechanism. A hot detergent solution is sprayed with considerable force against all inner surfaces of the container. The heat assists in drying the container. This type of equipment is also useful for the cleaning of small mixing kettles, milk cans, stock pots, and the like.

*Grounds care equipment.* The use of grounds equipment such as tillers, seeders, mowers, and trimmers is not normally a function of the housekeeping department, although common supervision is used at times. For a full treatment of this subject, see *Grounds Maintenance Handbook* by H. S. Conover (F. W. Dodge Corporation, New York).

*Sprayers.* Although sprayers are primarily designed to dispense insecticides, they are also useful for spraying herbicides on weeds, deodorizing, treating dust mops and cloths, and applying detergents and disinfectants. The types of power sprayers include:

1. The electric vibrator type for spraying in small quantities.
2. The electric aerosol type which disperses a fine mist. This machine can be turned off automatically with a timing attachment.
3. Fogging guns which utilize compressed gas as the propellent.
4. Fogging machines, either portable or vehicle-mounted, which utilize hot exhaust gases from a pulse-jet-type engine.

Sprayers and foggers, when handling flammable chemicals, should never be used in a closed or poorly ventilated area. When the machine is not in use, all of the insecticide or other material, as well as the propellent in the fogging type, should be removed and properly stored. Sprayers should be cleaned regularly, as the small orifices tend to clog.

*Steam cleaners.* Steam cleaners consist of a steam generator or steam source, a "gun" or nozzle attached to the machine by a flexible steam hose, and a solution tank for detergent. Steam cleaning is a very rapid method of removing heavy soils from certain types of equipment and floors, and it does this by the combined action of heat, turbulence, force, and chemical action.

Not all materials or surfaces may be safely steam cleaned, and care must be taken to prevent damage through the cleaning of fragile or delicate equipment, or the spoiling or damage of adjacent materials, equipment, or surfaces.

Steam-cleaning materials should be carefully selected. Operators should always wear goggles. Since the steam-cleaning action causes sprayback and floods the floor, the wearing of a waterproof cap, apron, gloves, and rubber boots is also desirable.

*Tank cleaners.* These are special devices which are lowered into tank cars, storage tanks, or vats, or similar enclosed vessels. Detergent spray is forced out of a revolving head to completely wet the interior of the vessel.

*Transport cleaners.* Special guns and spray systems are constructed for the cleaning of truck and trailer, aircraft, railway car, and ship exteriors.

*Venetian blind cleaners.* Special equipment is available either for cleaning individual venetian blind slats, or for immersing the entire blind in an agitated detergent solution.

*Wall washers.* Wall-washing machines usually consist of two tanks, one for the cleaning solution and the other for rinse water. Air pressure is used to force the cleaning or rinsing liquids through flexible hoses to a trowel which is covered by a porous cleaning pad.

The wall-washing machine is only efficient for cleaning large expanses of smooth walls on a project basis. Where irregular walls or relatively small areas are to be cleaned, or where the work is subject to interruption, the walls should be cleaned manually using equipment such as sponge mops, dry mops, sponges, and cloths.

*Waste-disposal equipment.* The field of waste-collection equipment is extremely large. To treat it adequately is beyond the scope of this work. In general, standard and specially constructed equipment is available:

1. Balers for waste paper, corrugated cartons, and other waste materials.
2. Slitters and grinders to reduce corrugated boxes and other containers and waste material to shreds.
3. Pulpers to reduce cellulose-containing waste to a pumpable material.

A scarifier may be required to remove heavy buildup on an industrial floor (Tennant Company).

4. Conveyors of all types to move waste material.
5. Incinerators for safely burning waste.

Special cleaning equipment is available or can be constructed for unusual needs. For example, the Broadway Tunnel in San Francisco is cleaned by a fork truck which has been specially adapted with a washer unit.

The National Parks Service, as an example of adapting equipment to needs, has designed and constructed mobile comfort stations containing complete rest-room facilities.

# 35 Manual Equipment

MANY of the common hand tools of the custodian have remained almost unchanged over the years. Improvements and innovations that have been made are not startling and have not contributed markedly to overall custodial efficiency, except in certain cases where specialized types of equipment have been developed.

Despite this, the thoughtful selection and provision of manual equipment can ease the burden of the custodian and permit him to work at a higher level of productivity. The carpenter, for example, may be equipped with a vast array of power equipment which will save him a great deal of time and improve the quality of his work; yet, he selects his hammer, hand saw, and level carefully and treats them well.

An impressive number of manual equipment items are available for housekeeping use. Many of these are old standbys such as mops, brushes, and brooms. Others are more infrequently used, such as applicator pans, plow mops, and proportioners. This chapter can only provide a general survey of the field.

## Brooms

"Straw brooms," or "corn brooms," are so named because of their manufacture from undyed corn-straw stock. Brooms have changed little over the centuries, except that the fibers are now arranged in a fan shape whereas in olden times they were wrapped in a circle around the handle.

For most uses, brooms have been supplanted by push brushes, dust mops, and vacuum sweeping tools. For certain situations, however, the broom still serves best, such as for the cleaning of ribbed rubber matting, removal of certain types of soil from rough concrete floors, for moving bulky litter, and the handling of lint and nondusting soil in congested places. The use of corn brooms should be avoided where dust will be raised.

### BROOM SELECTION

Brooms are classified by their weight per dozen. The 40-pound broom should be selected for use by custodians in industrial areas. Maids should utilize brooms from 24 to 32 pounds in weight. Smaller brooms are available for use in more congested areas or for light-duty purposes.

A special type of broom is the "toy broom," having a handle only 24 inches long, which is used in conjunction with a lobby dust pan for removing cigarette ends and other litter from public areas on a policing basis. Hand whisk brooms, which may be made of palmetto fiber or nylon bristles, are used for brushing fabric where vacuuming is impractical or unsuccessful.

### BROOM CARE

A few pointers:

1. Brooms should be stored, to permit free circulation of air, and should be hung up rather than allowed to rest on the straws.
2. Brooms may be cleaned by washing in a mild synthetic detergent solution, followed by a rinse with clear water. They then should be hung up to dry with the straws down.
3. Wet corn brooms should not be used, and brooms should not be used on wet materials unless unavoidable.

## Brushes

Thousands of years ago, a cave man rolled up an animal hide. He was about to make a painting on the wall of his cave. He used the brushing action of the hide's bristles to wipe the dust off the cave wall. This may have been the first brush, our oldest cleaning tool.

Since that time, the animal skin brush has developed into an enormous variety of brushes of all types, some hardly even recognizable as brushes.

Those who like to toy with mathematics can calculate the possible number of combinations on the basis of these factors:

1. Block material. About ten basic materials are involved, including hardwoods, plastics, aluminum, and various types of rubber.

2. Block shape. About twenty basic shapes are available, including geometric forms such as round, rectangular, triangular, and oval as well as a number of special shapes such as wing tip, short-handle grip, long-handle grip, etc.

3. Filling material. At least ten different varieties of each of five basic types:
   Animal. Bristle, horsehair, etc.
   Metallic. Brass, carbon steel, stainless steel, etc.
   Mixtures. Metallic and vegetable, plastic and vegetable, etc.
   Plastic. Nylon, PVC, saran, styrene, etc.
   Vegetable. Bass, cactus, palm, etc.

4. Length of filling. Five basic possibilities: stub, short, medium, long, and extra long. Actually, length of filling material is measured in inches out from the block face. A single brush may contain more than one length of filling.

5. Setting methods. There are about ten basic methods of setting brush fillings into the block, such as with the use of adhesives, nails, staples, woven wire, etc.

6. Special designs. At least twenty different variations on basic designs are available for special applications: combination brushes, where one side of the block is filled for one purpose and the other side for another purpose; detachable handles; fountain brushes, where a cleaning solution flows through both the handle and the brush; bumpers, to protect the brush from marring or damaging a finished surface; gang brushes, which can be locked together to form a longer brush; angle-set brushes, with the bristles set at various angles to the block; shaped fillings, where the filling materials are of varying lengths, etc.

A special class of brush is the push brush, which is taken up in the next section. Even the broom, discussed in the previous section, is a special class of brush. Floor-machine brushes were discussed in the previous chapter. Outside the field of housekeeping, of course, are all the various types of paint brushes, wheel brushes, stencil brushes, artist's brushes, industrial brushes, and, to complete the picture, hair, clothes, animal, and tooth brushes!

The scope of this work only permits a general listing of the various types of brushes available to the custodian:

1. Bowl brushes. These are also known as toilet-bowl brushes, commode brushes, and sanitary brushes. The common bowl brush is made of stiff bassine fibers fixed with wire to a hardwood handle about 24 inches long. Some of these brushes are equipped with "collars" for cleaning under the rim of bowls. Others

of a special design include a tip of sponge. (Bowls are also cleaned by the use of cotton swabs or sponge applicators built onto bowl-cleaner containers.)

2. Counter brushes. These are made of horsehair, fiber, or nylon filling in a wooden block with a hard grip. This brush is most often used for sweeping soil into dust pans, or for reaching into corners. Its earlier use for dusting surfaces, such as desks and counters, has been replaced by the use of dusting cloths since brushes tend to stir up dust.

3. Deck scrub brush. This brush may be used by hand or with a long handle for scrubbing floors and other badly soiled, coarse surfaces. Scrub brushes may include a built-in squeegee for moving the scrubbing water. They are available in various block lengths. A special type of deck scrub brush is the baseboard brush. Triangular in shape, it is able to clean the baseboard and adjacent floor area.

4. Hand scrub brush. Also known as "scrubs," these brushes are used for the hand cleaning of equipment, fixtures, and small floor areas of moderate coarseness.

5. Fountain brush. This brush is designed to dispense a detergent or rinsing solution through the brush by means of a hollow handle. Fountain brushes are used for washing windows, vehicle bodies, and other surfaces such as badly soiled industrial walls. Long extension handles are now available so that windows may be cleaned as high as three or four stories. Both windows and vehicle bodies may be cleaned with brushes that are not of the fountain type. Vehicle body brushes should always be equipped with a rubber bumper to prevent damage to the finish.

6. Special brushes. The need to clean numerous types of surfaces has required the design of brushes for specialized applications. A special brush has been designed to clean each of the following:

- Bed springs
- Bottles
- Drinking glasses
- Fenders
- Pots
- Pipes
- Radiators
- Venetian blinds
- Walls

When considering brushes, alternate cleaning methods should also be borne in mind. For dust removal, for example, both vacuuming and the use of a treated dust mop or dusting cloth will create less airborne dust than a brush.

Brushes should be cared for by combing or cleaning them occasionally and hanging up or resting them on the block when not in use. Brushes should be washed and dried carefully when soiled. Bowl brushes should be washed daily in a disinfectant solution.

## Push Brushes

Floor brushes are also known as push brushes, push brooms, and "sweeps." The floor brush, if properly used, will clean large areas of floors without the raising of a considerable amount of dust or overly tiring the worker. Floor brushes are available in a range of sizes and types to fit individual requirements. Because of lack of control, some industries arbitrarily purchase the cheapest push brush available in a rather small size and consider this item expendable. The implications in terms of efficiency and cost are obvious.

The proper application for the push brush is in the sweeping of the industrial plant floor in areas which cannot be reached by a powered or push-type sweeping machine. Where dust is raised, a sweeping compound may be used sparingly. Push brushes should not ordinarily be used for the cleaning of resilient floors or any types of floors in administrative, medical, or dietary areas, as too much dust is raised. The proper tool for this job is the treated dust mop or vacuum sweeping tool.

When sweeping with a push brush, the custodian should stand with a straight back, with feet about 12 inches apart. The body weight should be swung on to the right foot with the left side of the body toward the clean floor. The weight will be swung to the left foot as the brush is extended. The stroke of the brush starts close to the body and should extend for about 6 feet. Shorter strokes, which are tiring, should only be used on congested areas, or where the dirt is composed of fine sand and soil. Dust "boiling" can be held down if the brush is not flipped at the end of the stroke.

Push brushes require a little more care than most hand equipment:

1. The brush handle should be rotated in the block frequently so as not to unduly wear one side.
2. The handle should not be leaned on for resting.
3. If the brush is permitted to rest on its fibers when not in use, it will bend them out of shape and make the brush useless.
4. Push brushes should be used only for sweeping. They should never be used as a mop, squeegee, lever, or hammer.
5. Selection of as large a push brush as practicable for a given job will get the job done quicker and cause less wear of the brush.
6. The brush should be combed at least weekly to keep the filling material clean and flexible.
7. Brushes with natural bristles can be attacked by moth larvae if in storage for some time. They should be protected with moth balls, paradichlorobenzene crystals, or insecticides.
8. In storage, brushes should be rested on the block, hung up by the handle, or hung on special attachment rings.
9. Very soiled brushes should be washed in warm neutral detergent solution and rinsed in clear warm water. The brush should be shaken to straighten the fibers, and should not be used until completely dry.
10. Brushes of 24 inches in length and longer should be fitted with a handle brace to avoid the handle breaking at the block.

11. Some brushes are fitted with scraper attachments. These should be used for light-duty scraping only. A special long-handled scraper should be used for heavy work.

## Carts

Carts are used to transport materials and equipment from one area to another. Their use eliminates frequent trips between work and storage areas and helps insure the use of the proper materials and equipment in the various cleaning operations.

Carts can be classified in three basic types: the waste cart, equipment cart, and utility cart.

The waste cart is primarily used for waste collection and often also for waste transportation. In moving from one area to another, the custodian will empty wastebaskets and litter into the cart. When dumping wastebaskets, they should be held down into the cart at the level of the waste and upended. Dumping from above the cart will create dust and possibly strew bits of litter on the floor. The most popular type of cart is made of a collapsible tubular frame from which is hung a fabric or plastic litter container. The frame is on casters for mobility. The litter bag normally holds about six bushels, and an extra bag should be kept on hand to be used while the soiled bag is being laundered, or while a torn bag is being repaired. For policing purposes, such a waste cart may be fitted with a small carrier, resembling a shoe bag in which can be placed tools and materials such as sponges, small brushes, bottles of detergent, scrapers, etc. Waste carts may also be constructed of the rigid type such as a hamper, with a frame of metal or wood, with sides of cloth, plastic, fiber, wood, sheet metal, etc. An openhead fiber or metal drum may be used as a waste cart when placed on a dolly with casters. A drum or pasteboard box should never be rolled or slid on the floor for this purpose.

Equipment carts are used for the transportation of mopping buckets, floor machines, vacuums and accessories, and quantities of cleaning materials from the storage area to the work place. Special equipment carts are generally restricted to use by project workers, as they are not required by custodians performing daily cleaning. Equipment carts may be fitted with a small litter bag for waste that is collected during the performance of the project.

Utility carts are designed to perform a combination of functions. The ideal utility cart contains a litter bag, has a tray for housekeeping chemicals, a storage shelf for items such as replacement paper towels or tissue, and also is able to transport manual equipment such as a mopping bucket, wringer,

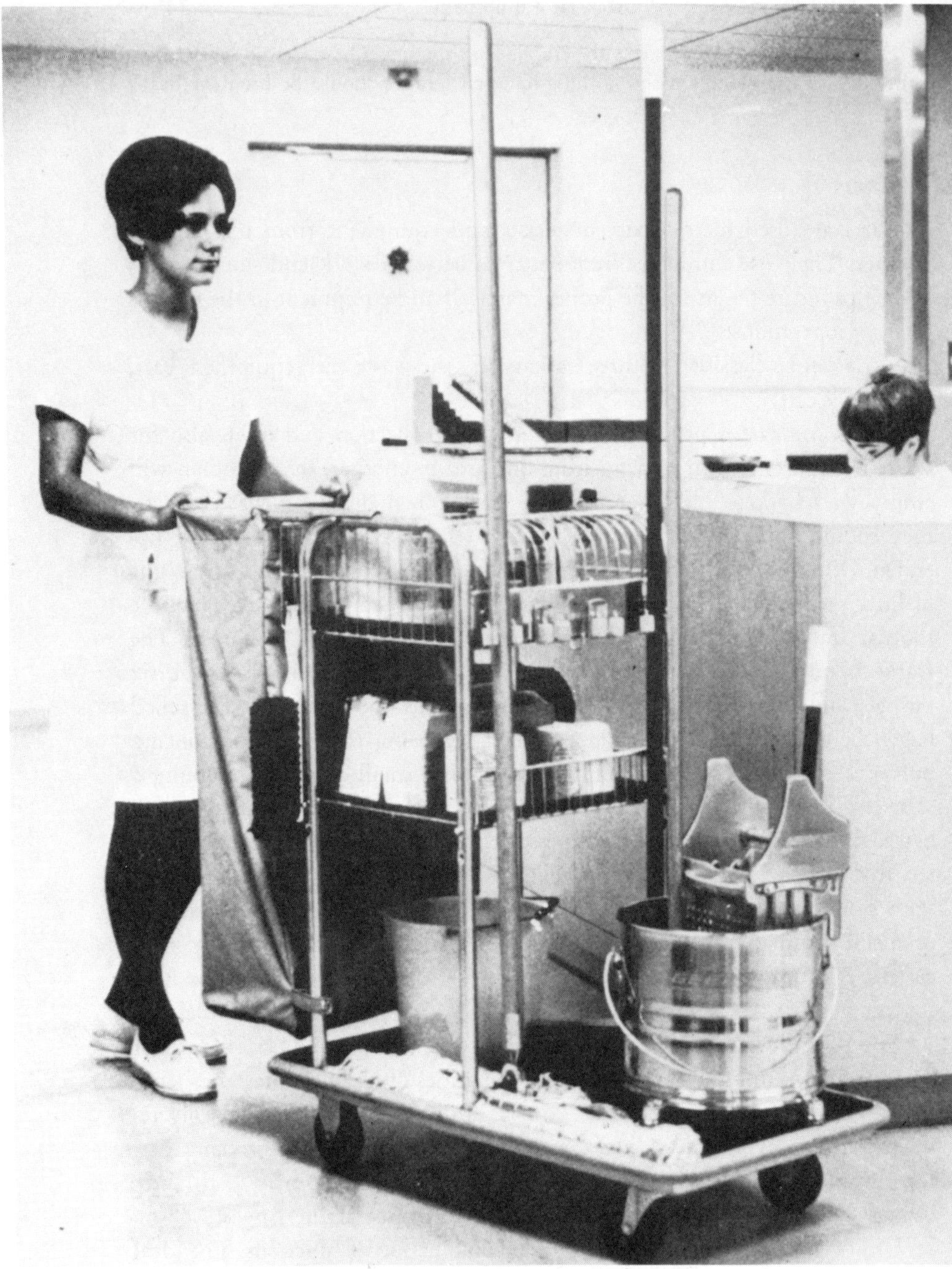

Custodial carts eliminate a good deal of walking from the work area to a custodial closet, reducing non-productive time (Geerpres Company).

mops, and broom. Some utility carts for use by maids, or for special use by custodians, omit the storage shelf and may also omit the carriage of the equipment.

Utility carts should be cleaned completely each week, with the wheels being checked for strings and lubrication. If the cart is used in a dietary or medical area, it should be washed with a disinfectant solution twice weekly.

A well-selected and properly equipped and maintained cart can pay for itself in just a few months.

## Dispensers

There are four basic types of soap dispensers:

1. Liquid soap dispensers, for dispensing liquid soap in liquid form.
2. Lather soap dispensers, for dispensing liquid soap in lather form.
3. Powdered soap dispensers, for dispensing powdered soap.
4. Detergent dispensers, for dispensing detergents, hand lotions, or soap with a detergent content.

For specialized uses, there are also available paste-type cleaners and dispensers, impregnated cloths and papers, numerous types of bar soaps, protective creams and lotions, etc.

Generally speaking, soap dispensers are available for mounting as follows:

1. Wall mounted: For mounting on the wall with screws or adhesive pads. Available for all four basic types.
2. Basin mounted or for countertops: For mounting in the basin or countertop. This type of installation is generally not for use with powdered soap.
3. Tank-type gravity feed systems: A battery of liquid or lather soap valves with a central soap supply.
4. Recessed equipment: For flush wall mounting. Available for dispensing liquid, lather, or powdered soap. Recessed units may include other necessary washroom accessories, such as shelf, mirror, light, towel dispenser, etc.

Selection of the proper type of soap dispensing equipment should be made after a careful study of the needs and requirements of each particular installation, based on these considerations:

1. Economy of operation: The following information is based on a usage test report, conducted by The Institute of Sanitation Management:

   a. Powdered toilet soap provides approximately 260 hand washes per pound.
   b. Cake soap ground off provides approximately 390 hand washes per pound.
   c. Liquid soap dispensed in liquid form provides approximately 595 hand washes per gallon.
   d. Liquid soap dispensed in lather form provides approximately 1,950 hand washes per gallon.

Mobile hampers assist in waste collection; note the support to the bag which can also carry small tools (Walton-March, Inc.).

2. Economy of maintenance. Since labor costs account for most of the maintenance dollar, this becomes a basic consideration.

   a. The dispenser should be easy to fill and should have an unbreakable window to indicate to the custodian when it is time to refill.
   b. Depending upon the number of prospective users, the unit should have a large enough capacity to assure refilling only periodically, thereby allowing the custodian more time for other duties. In heavy traffic installations, the larger capacity dispensers have proven themselves to be far more economical to maintain than the customary 12-ounce size.
   c. The valve mechanism should be as trouble-free as possible. For instance, a high quality liquid or lather soap dispenser should avoid the use of packings or washers which require replacement.

3. Design. The dispenser should employ as many tamperproof features as possible. It should be locked on the wall with a concealed wall plate. The container should be designed so as to be locked with a special key. The unit should be constructed completely of metal (preferably stainless), and when translucent areas are required, they should be made of unbreakable plastics.

4. Powdered soap dispensers should be used when there is a need for abrasive cleaning action. These features should be considered:

   a. The unit should be designed to dispense any free-flowing soap powder and should not be designed to dispense one particular brand or type only.
   b. The unit may incorporate an "adjustable-output" mechanism, thereby allowing proper adjustment, depending upon requirements.

5. Paste-type dispensers should be used by mechanics, machine operators, and others whose hands become oily or greasy.

6. Detergents (or soap containing a detergent content) cannot be used in regular soap dispensers, unless the dispenser is specifically designed for their use.

7. In medical or dietary areas, the "hospital-type" dispenser should be used, which is operated by a foot pedal.

8. Aerosol cans, for use with hand cleaners and lotions, may also be considered a type of dispenser.

The following conditions may periodically occur:

1. Liquid or lather soap dispensers:

   a. An inoperative valve is usually caused by soap coagulation due to the use of a liquid soap with too high a soap solids content, the soap valve having been unused for a long period of time, or having corroded from the use of a detergent. Rinsing coagulated residue from internal parts is easily accomplished by filling the soap container with hot water, allowing it to stand for about ten minutes, then releasing it through the valve mechanism. Repeat if necessary.
   b. A dripping valve is usually caused by packings which are worn with age and merely require replacing. It may also be caused by uneven coagulation in the valve.

2. Lather soap dispensers: Poor lathering action in a lather soap dispenser is usually caused by the use of a soap solution having too low a soap solids content

for the hardness of the water. In the case of a new lather soap dispenser, it is due to a special preservative compound containing oil, which kills the lather. In a new dispenser, work the soap through the valve several times, after which proper lather consistency will appear.

3. Powdered soap dispensers: Some types of powdered soap dispensers have a strong tendency to clog, due to water splashing into the discharge opening, thereby causing a build-up or caking of the powdered soap. The dispenser mechanism should be checked for this periodically.

## Dust Mops

The ideal tool for the sweeping of smooth, lightly soiled, dry floors is the dust mop. When properly treated and cared for, the dust mop will remove loose soil from the floor quickly and efficiently, without leaving a residue.

Several types of dusting mops are available:

1. The rectangular cotton dust mop. This is the most popular and useful of these mops. It consists of a mop head made of cotton yarn tied onto a wooden block or metal frame with handle. These mops are available in widths up to 60 inches.
2. Dust mops are made in materials other than cotton, such as saran. The saran mop is normally factory treated, but once it is used and washed, it loses its dust-gathering efficiency and cannot be easily re-treated.
3. U-shaped mop frames and mop heads are available which, while removing dust from the floor, collect litter in the enclosed area.
4. For working in congested areas, small dusting mops of a triangular shape are available. These mops are sometimes used for dusting or even washing walls.
5. Also for congested areas, the swivel-type mop permits the custodian to change the position of the mop head on the floor by twisting the handle.
6. For sweeping open areas such as wide corridors, gymnasiums, and lobbies, a large dust mop in the shape of a "V" is available. Some times these mops are equipped with double handles and are called "plow mops." A similar effect can be achieved by holding two rectangular mops together in a "V" position.
7. Treated mopping towels and papers are available to be placed over the dust mop or over a push brush for dusting purposes. The appearance of these types is better than the dust mop, but they are less efficient since a smaller surface is exposed to collect dust and soil. As the yarn-type dust mop is utilized, new surfaces of the treated yarn are exposed, providing a continuing cleaning efficiency.

A few simple rules will extend the life and improve the efficiency of the dust mop:

1. Soiled dust mops (also dust mitts and wiping cloths) should be collected at the end of each working shift and properly treated.
   a. Shake out the mop carefully into a waste receptacle, dust bin, or in an area where the soil will not spread, or vacuum.

b. Remove splinters or other foreign materials from the mop strands.
c. Spray on 1 ounce of treatment for each 4 inches to 6 inches of dust mop length. Be careful not to overtreat the mop head.
d. Roll up the mop head snugly and place in a closed metal container reserved and marked for this purpose.
e. Allow at least twelve hours for the mop treatment to permeate the strands before reusing.

2. Launder the mop when it becomes soiled:
a. Place in a netted bag, or bind the strands in two places to prevent tangling while the mop is being washed.
b. Wash in a solution of warm water and detergent.
c. Dry the mop head carefully.
d. Comb the strands.
e. Treat the mop head (treating may be done during the rinsing cycle when washing, in some cases).

3. Mops should be stored in a clean, dry area. Hang the mop so that it does not touch the walls and equipment and will not be brushed against by personnel. Mops should be hung with the mop heads down.

4. Keep the tie cords in proper repair and in place.

5. Do not allow the strands to become knotted or matted. Keep the strands free from splinters, metal particles, and keep them combed regularly.

6. When shaking out a mop, do not strike the handle or any part of the mop against a hard surface. Such action might damage the frame and weaken the handle.

7. Do not use a dust mop on wet or oily floors.

8. While mopping, keep the back as straight as possible and bend at the hips. Do not overreach. Following these rules of work posture will reduce fatigue and prevent strain.

9. Do not lift the mop from the floor unnecessarily; some of the gathered soil may drop from it.

10. Do not bear down on the mop; pressure is not necessary to proper mopping action.

11. If a vacuum is available, use it to remove loose soil from the mop frequently. If no vacuum is available, the mop head should be frequently shaken into a large waste receptacle or a dust box to remove and collect loose surface soil.

12. When both the large and small dust mop is to be used by a custodian, the smaller dust mop can be carried on top of the larger one in a piggy-back fashion. Thus, in the cleaning in an administrative area, a large dust mop is used for sweeping the corridors, while the smaller dust mop is carried on top to be used in the individual offices as they are passed.

For dust removal functions on surfaces other than floors, dusting cloths and dusting mitts are used. These should be treated and laundered in the same way as dust mops.

## Matting

From a housekeeping standpoint, the purpose of floor matting is to entrain soil at entrance areas, or where the soil is produced, so that it is not tracked throughout the facility. The use of matting is based on the principle that it is easier to keep dirt out than to remove it once it is inside.

The most common type of mat is the door mat. Door mats are available in the following types:

1. Rubber link. These mats are very often made from old tire carcasses, but they may also be manufactured from special molded rubber links. This type of matting, because of its many openings, will collect a good deal of soil. It has the disadvantages that it is difficult to clean and may cause accidents by catching women's heels.
2. Molded rubber with perforations. The openings in this type of matting perform the same function and present the same hazards as the link mat, but to a lesser degree.
3. Molded rubber with squeegee-type ribbing. This is the safest type of matting as it presents no tripping hazard. The design tends to wipe water and soil from the bottom of the feet and collect it in the grooves.
4. Carpet surface. This type of matting is suitable only for use at inside doors, such as the entrance to a private office.
5. Coco-mats. Coco-mats are safe if recessed, and particularly efficient in collecting water. Their chief drawback is they are very difficult to keep clean.
6. Woven grass. This type of mat is chiefly decorative. It quickly becomes unsightly and is difficult to keep clean.
7. Gelatin mat. This is a very specialized mat for extremely efficient removal of fine soil particles, such as in entrances to dust-free areas.

Runners are matting material manufactured in length, for use in corridors, aisles, behind counters, etc. They are generally made of the same material as the first four categories of door mats. In addition to these, there are special types of runners:

1. Grease-resistant materials are available.
2. A decorative-type of vinyl runner may be had in a variety of colors and patterns.
3. Steel link runners are useful for the removal of metal chips from the soles of the shoes.

To complete the matting picture, we have these special types:

1. Decorated entrance mats, containing a company trade-mark, various kinds of wording, or colorful designs.

2. Matting may be cut to special sizes and shapes or fitted to special situations, such as around columns or fountains.
3. Molded rubber stair-tread coverings may be considered as a type of matting.
4. Shelf mats prevent the damage of fragile materials such as laboratory wear. Their design also allows the draining of washed parts.
5. Fatigue mats are placed around machinery and other places where personnel must stand for greater comfort.

## Mopping Equipment

The arsenal of mopping equipment can be categorized as: (1) buckets, (2) mopping outfits, (3) mops, (4) squeegees, (5) wringers.

### BUCKETS

Mopping buckets are usually constructed of galvanized steel, but they are also available in aluminum and stainless steel. Buckets may be either round or oval in shape. The former type has the advantage of being stronger and easier to clean, whereas the latter type takes less space and offers a wider opening to receive the mop.

Most mop buckets are equipped with "gliders," which are small beads of metal which permit the bucket to slide on the floor. Buckets are available with attached casters of various sizes, which protect the floor from such sliding marks. The 3-inch caster size is generally suitable. The smaller buckets, such as pails, are not so equipped.

Buckets should be selected on the basis of the metal gauge and the quality of the construction. Certainly, both the inside and the outside of the bucket should present a smooth surface that will not cut or scrape the hands of the user.

Buckets, as well as all types of metal mopping equipment, should be washed, rinsed, and dried after use. If the bucket develops a leak, it should be soldered or patched right away, and the patch spots filed smooth.

Liquids should not be permitted to remain in buckets overnight. Some liquids will cause odors, or, such as alkaline solutions, damage the bucket.

### MOPPING OUTFITS

For everything but the smallest mopping jobs, one of the two basic types of mopping outfits should be used.

The larger of the two types is the mopping tank, which is a compartmented tank on large casters, with a built-on wringer. These units have two or three compartments and usually range from a total of 30 to 60 gallons in size.

The smallest operation will require two buckets, one for the detergent solution and the other for rinsing. These two buckets can be mounted on a special carrier or truck and fitted with casters and a pulling handle. A wringer is placed in one of the buckets to complete the outfit. Systems consisting of three buckets are also available, and are very effective in some cases.

### MOPS

The mop is the basic hand tool for wet cleaning.

Mops may be either permanently attached to a wooden handle or may be attached and removed from a special frame having a wooden or metal handle. The detachable mop head has the considerable advantages of permitting laundering in a machine, placing the entire strand length in contact with the floor, ease of cleaning and drying, and reusability of the handle. The old-fashioned stick mop should not be considered for industrial use.

The mop head itself may be constructed of several different materials:

1. The cotton mop is most extensively used. It is inexpensive and easy to launder. It is also well suited for the application of wax by trained workers. The cotton mops, as well as other mops, are available in various sizes and are usually classified by weight. A 16- or 20-ounce mop should be utilized by a maid or handicapped custodian, while a 24- or 32-ounce mop should be used by the usual housekeeping worker.
2. Mops made of rayon are lighter in weight than cotton and more absorbent. They are not as suitable as cotton for applying wax, and where mops are to be used for both wet cleaning and wax application, standardization should be made of the cotton mops.
3. Cellulose sponge mop heads, although considerably more expensive than cotton or rayon, are much more absorbent and durable when used on smooth floors.
4. A special type of mop is the sponge mop with a built-on levered wringer. This mop is useful for policing operations and stair tread cleaning, as well as for certain types of wall washing.

It is interesting that most mops do not wear out through use, but rather are rendered useless through misuse. Although mop damage can sometimes be contributed to mopping equipment, such as a poor choice of wringer, it is most often caused by abuse to the mop itself. Here are some pointers on mop use and care:

1. The basic purpose of the wet mop is to transfer liquid to and from floors. The use of mops for other purposes should be discouraged.

2. Soak a new mop in warm water for at least twenty minutes before use, in order to remove excess oils and expel entrapped air, thus providing better absorbency.

3. Change mopping water frequently while mopping to prevent the water from becoming overly dirty. Rinse the mop each time the water is changed. Dirty water will cause a redeposit of soil during mopping, leaving soil streaks.

4. While mopping, the mop should stay on the floor and should not be flung about.

5. The mop should be turned from one side to the other frequently while in use in order to expose more clean, moist strands.

6. Mops should be squeezed or pressed carefully to remove water. Twisting the mop will tear or weaken the strands.

7. Mops can be used for scrubbing small areas and spots by folding the mop over on itself so that the tips of the strands are in direct contact with the floor, under pressure. Never scrub with the part of the mop closest to the holder because this will tear the strands.

8. Avoid the use of mops with lye, caustics, or strong undiluted cleaning solutions because these materials attack the mop strands.

9. If a rough surface is to be mopped (such as splintered wood), mop very carefully and try to mop in a direction which will avoid snagging. It is best to reserve a special mop for use on such a surface.

10. It is not economical to continue to use worn-out mops—they should be discarded.

11. Always rinse a mop carefully after use and squeeze it dry.

12. Mops should be stored in a warm dry area where air circulates freely.

13. A mop in storage should be hung with the yarn away from the wall, with the strands down.

14. Do not let wet mops touch each other or come in contact with walls or equipment.

15. If mop strands become loosened, they should be removed or cut off with scissors in order to prevent snagging and splattering.

16. In addition to rinsing the mop whenever the mopping water is changed, the mop should be carefully washed periodically (daily if possible). The frequency of washing will depend upon the use to which the mop is put, but mops in regular use should be washed carefully at least weekly, preferably before the weekend. Where a number of mops are in regular use, a mop-washing machine may be used which exposes the mop to a high-pressure clear water bath.

Suitable training in mopping techniques will permit a custodian to obtain rapid coverage with a large mop without fatigue. Unsound mopping procedures will give backaches, even with the use of smaller mops in restricted areas.

### SQUEEGEES

Squeegees, through the wiping action of their rubber blades, are used to move large quantities of water on a floor. Thus, the squeegee can collect water in a confined area for pickup with a vacuum, or can move the water to a floor drain.

Squeegees are straight or curved. The curved squeegee has the advantage of moving more water without spillage around the sides, but it is a little more difficult to handle and cannot be pulled back as the straight squeegee can.

Water may also be picked up from a floor with a squeegee and a special squeegee pan, or pick-up pan, which traps the water for later disposal.

Pick-up pans have a liquid capacity of 3 to 5 gallons and should be used in conjunction with the straight type of squeegee.

Concerning squeegee use:

1. After each use, the rubber blades should be wiped clean and dry.
2. When the squeegee is not in use, it should be stored in a cool place away from liquids.
3. In storage, the squeegee should not rest on the rubber blades, which are easily damaged. It is best to hang up a squeegee by its handle.
4. The rubber blades are damaged by sunlight, so the squeegee should never be hung up to dry outside.
5. It is good practice to keep a spare set of squeegee blades on hand. Squeegee blades should be replaced when efficiency is impaired.

### WRINGERS

The name "wringer" should be a misnomer, since the wringing of a mop to remove water loosens and damages the strands. The proper drying action for a mop is squeezing, which is the design of most of today's "wringers."

This squeezing is accomplished either by moving the mop between rollers, like an old-fashioned clothes wringer, or squeezing the mop between the perforated metal plates of a box-type wringer. The squeezer-type action of the latter is more efficient and less destructive to the mop. Squeezer-type wringers are either of the horizontal or downward pressure types. Here are the factors involved in their choice:

1. The horizontal type can take more abuse without failure.
2. The downward-pressure wringer is more efficient.

3. A mop wringer needs to be rather closely sized to the mop to be used.

4. The mop wringer of plastic construction has proven to be quiet, nonrusting, light weight, and durable.

From a construction standpoint, the roller type seems to be more suited for the compartmented mopping tank because of the height of the tank.

For the proper care of wringers:

1. Keep working parts properly oiled.
2. Keep screws and bolts tightened.
3. Clean the wringer after each day's use.
4. Too much force on a wringer can break it; do not lengthen handle or press too hard on the foot lever.
5. Make sure the wringer is large enough for the mop that is being used.
6. Remove loose mop strands and other articles caught in the wringer.

## Waste Receptacles

In addition to the waste and utility carts previously mentioned, a wide range of waste receptacles is in use:

1. The "torpedo" receptacle is named for its shape, as it is cylindrical with a round top. Refuse is placed in the container through a single swinging door. The waste is collected in a separate inner container which may be made of fiber, galvanized iron, plastic, etc. The torpedo receptacle is an attractive and durable container for small to medium quantities of waste, such as in administrative areas.

2. Rectangular containers. These metal receptacles, in various colors, are usually fitted with a swinging top, through which waste can be deposited from either side. The inner container may be a burlap or duck bag, metal, or plastic. The smallest of this type of container is classed as a sanitary napkin disposal unit, while the largest has a volume of about 60 gallons.

3. Special types of rectangular containers, either with open or self-closing tops, are fitted with brackets to support paper towel dispensers, mirrors, and the like.

4. Another variation of the rectangular container includes a tamping mechanism which compresses the waste to a fraction of its original crumpled size. This permits a reduced frequency of waste disposal.

5. Wall-hung receptacles, although limited in the amount of waste they can accommodate, are very handsome in appearance and do not have to be removed for floor maintenance.

6. Flush-type receptacles are built in during construction or renovation and can be had in combination with soap dispensers, mirrors, and towel dispensers. They are usually fabricated of stainless steel.

7. The common galvanized garbage can is still in widespread industrial use, particularly for the handling of wet waste and food waste. The sizes in most common use are of approximately 20- and 32- gallon capacity.

8. Fiber waste cans are also available, but they create odors and deteriorate rapidly if misused for wet waste.

9. Plastic garbage cans have the advantages of not denting, rusting, absorbing liquids, or being noisy. The newer plastic materials can be steam cleaned successfully. Some plastics have the disadvantage of supporting fires.

10. Drum covers are available which will convert an openhead drum to a waste receptacle. The cover has a torpedolike head with a self-closing door.

11. Waste receptacles for classified material are constructed of heavy-gauge metal and cannot be opened for disposal without a key. Waste is inserted through a small opening so that the materials cannot be easily pulled out.

12. Containers for oily waste are also made of heavy-gauge metal. They have a special lid and are properly vented underneath to avoid spontaneous combustion.

13. Woven-wire receptacles for paper litter are inexpensive, but rather unsightly. Some people will use these receptacles, however, while paying no attention to the closed units.

14. Wastebaskets are generally used in office areas, although they may be used in the plant as well where small amounts of dry waste are created. Empty buckets and pails which may have contained raw materials or chemicals are usually substituted for this purpose in the plant. If this is done, they should be painted. Wastebaskets are available in metal, fiber, and plastic.

15. Special metal cans are made for edible waste. These cans are not affected by steam cleaning, and they have no edges or lips which might interfere with cleaning.

16. Cigarette urns are a special class of waste receptacle (so is the ash tray, for that matter). Cigarette urns may rest on the floor, or may be mounted on the wall. The wall-mounted type is certainly preferable in administrative areas, as floors may be cleaned without their removal. It is also preferable in plant areas where malicious mischief is not a problem. If cigarette urns are not located wherever cigarette smoking is permitted, the natural result is that cigarette ends will be thrown to the floor.

## Other Manual Equipment

There are a number of specialized hand tools available for the custodian. Here are some of them:

1. Applicators, made of lambskin, fitted to a wooden block and handle, are used for application of wax and seals. They should be used with an applicator pan which places the proper amount of liquid onto the applicator. Unless lambswool applicators are carefully cleaned after each use, they harden and become worthless. The same material, often called chamois, is available for hand cleaning and drying, as well as polishing.

2. Deodorant cabinets are mounted in rest rooms and other areas. These cabinets are vented and disperse deodorants from a solid or liquid state.

3. Detergent guns dispense cleaning solutions through water or air pressure, or a combination of both

Waste receptacles can be colorful and attractive (Clean City Squares, Inc.).

Wall washing is an often-overlooked aspect of an effective custodial program (Geerpres Company).

4. Drum accessories include faucets, pumps, level indicators, racks, and wrenches.

5. Dust boxes can be used for the shaking out of dust mops and push brushes, although other methods are generally preferable.

6. Dust pans are fitted with a hand grip for the removal of chips and granular waste, or fitted with a short handle for lobby policing.

7. Feather dusters are still being used, although the appearance of other tools has made them obsolete.

8. Foot-bath trays are used to prevent athlete's foot in shower areas, through the immersion of the feet in a suitable disinfectant. Mechanical foot baths are also available which spray the feet.

9. Funnels and measuring cups are used for the pouring and measuring of small quantities of liquids.

10. Gloves, aprons, hard hats, goggles, and face shields should be issued permanently to custodians who must work with hazardous materials.

11. Ladders, scaffolds, and hoists are required for the cleaning of walls and overhead areas. Their use should be restricted to specified individuals. Any work requiring this type of equipment should be inspected frequently.

12. Proportioners mix a measured quantity of detergent or other chemical water by an aspirating action.

13. Safety signs provide warning when floors are wet and slippery.

14. Scrapers are used to remove chewing gum and other sticky substances from floors. The common putty knife should be carried by a custodian in office areas, whereas the long-handled floor scraper may be used in plant areas.

15. Sponges are indispensable for general cleaning. The natural sponge has been generally displaced by the synthetic sponge, which appears in a number of sizes and styles. A sponge cloth is now available which combines the advantages of the cloth and the sponge.

16. Sprayers are used to dispense chemicals in a fine fog, mist, or spray. Sprayers are used for lacquers, insecticides, deodorants, dust-mop treatments, and detergents. Some larger sprayers are carried by a strap over the shoulder, whereas most small sprayers are carried by hand.

17. Steel wool is best used by hand for metal cleaning. Where rust is to be avoided, stainless steel wool is preferred.

18. Tool holders permit the hanging and the storage of brooms, dust mops, squeegees, and wet mops by their handles, rather than resting them on the floor.

19. Window squeegees are used for windows and other glass, either by hand or with a wooden handle.

# 36

# Housekeeping Chemicals

IN any given building service department, housekeeping chemicals can be a morale builder or a morale destroyer, an asset or a liability, a legitimate expense or a financial burden, a collection of time-wasters or time-savers, a complex and confusing or simple subject, a minor or time-absorbing part of the manager's work and other plus or minus characteristics depending on the skill and understanding with which these materials are handled.

The typical custodial department devotes from 3%-10% of its budget for cleaning chemicals. The figure will be higher or lower depending on the quantity of chemicals purchased, and thus price reductions based on that quantity—on custodial wages, since as they go higher they tend to reduce the percentage cost for chemicals—on consumption control devices that prevent over-use—on the prevalence or absence of pilferage and on the actual amount of cleaning being performed. In the last case, it is important to recognize that an improved cleaning operation will result in a more frequent performance of certain operations, which will utilize more chemicals. Conversely, a strict minimization of chemical use may limit the amount of cleaning work that can actually be done.

The building services manager, like anyone else in management, is well advised to devote his time to various other responsibilities proportionately, as these responsibilities represent shares of his budget. Since payroll amounts to

90% or more of the budget, then 90% or more of management time should be devoted to "people problems". Yet, many managers will devote 30% or 40%—perhaps even half of their time—to interviewing sanitary supply salesmen! Undoubtedly, one reason for this is that these sales representatives are affable, friendly people, easy to talk with and easy to be with. Some of them can be helpful in terms of providing information and other services, but spending time with them can certainly be overdone.

But if sufficient time is not spent with the subject of chemicals, an improper selection may result, with results all out of proportion to the small percentage of the budget which chemicals represent. For example, if an inadequate floor finish is selected, that does not level well or powders, that shows scuff marks or water marks easily, then the workers will be demoralized because they will understand that the difficult and time-consuming job of finish stripping and reapplication will not yield good results. They will understand that certain activities will be required more often because of the poor chemical quality than would have been necessary had a better material been selected.

When an organization undergoes an austerity phase, a finger is often pointed at custodial chemicals as a logical place to make savings. One should be very careful in this regard, because of the problems that can arise in work performance when lesser-quality chemicals are introduced. A change in chemicals can create a change in procedure and thus create a different time allowance for cleaning. This, on top of changes in frequency can have a marked effect on total man-power requirements, and the paradoxical results might be that where one is attempting to save, he is actually creating more difficult work. In the custodial field, cost reductions are best made in custodial staffing, with frequency changes to compensate (thus unavoidably resulting in quality changes, if optimum efficiency is assumed in both cases). It would not be unthinkable, as the reader can visualize, in a cost reduction program to actually increase the unit purchase price of certain chemicals, so as to make it possible to expend less labor in the performance of given tasks.

### Chemicals Selection

The sanitary supply industry, which long ago topped the billion dollar annual sales level, consists literally of thousands of firms, most of which are quite small but many are multi-million dollar operations. The yellow pages of the telephone directory in any large city will turn up scores if not hundreds of manufacturers, distributors, and jobbers (those who sell products manufactured by others, but often under their own name). The competition is fierce, one reason for this being the comparatively large percentage of salesman's commission, typically

15% and often more (compared with typically 5% in many other types of industries).

As in all types of businesses, there are both reputable firms and reputable salesmen prepared to serve their customers, as well as disreputable firms and salesmen out to make a "quick buck" by any means. The typical supply firm literally has hundreds of products, perhaps several brand names of each type: for example, a given company may market a half-dozen detergents, all generally for the same purpose, but at different concentrations, viscosity, and ingredients. The claims made for chemical products are often exaggerated, and it sometimes appears to the buyer that every salesman has "magic products" to offer.

No wonder the buyer is dazzled and confused!

The problem is further compounded by television commercials, which tend to perpetuate obsolete products and refute statements of reputable supply salesmen. For example, television commercials still extol the virtues of pine oil as a disinfectant, yet for a number of years this product has been eclipsed by two or three other types of disinfectants which perform much better, and without the drawbacks of pine oil, as will be discussed below.

Perhaps all these things makes it easier for one to understand why so many people buy chemical supplies on such an emotional basis, making decisions on factors that have little or nothing to do with performance. For example, some buyers are only interested in soaps that are colored green, possibly because their grandmothers used green soap when they were children–others like only the odor of lemon–some buyers are impressed with viscosity or thickness, thinking that the more syrupy a liquid the more concentrated it is and many are impressed with how much foam a detergent will create. Yet, it is possible to formulate a chemical solution that is green, has a lemon smell, is so viscous that it can hardly be poured, and fills the room with foam, yet is such a poor detergent that it cleans practically nothing! Obviously the color and odor contribute nothing to cleaning–the other factors are a little more elusive and many buyers will rely on them even after agreeing that they may be of no consequence whatever. For example, it is possible to thicken water through the addition of cellulose derivatives, which have no active ingredient at all from a detergency standpoint. And there are many cases where soap suds would be a detriment rather than an advantage, such as in an automatic scrubbing machine, where the foam would choke out the vacuum system (some of the best detergents have no foam at all).

Ideally, purchase should be made of tested products from reputable suppliers, on an annual commitment basis in order to obtain a quantity price, with the products delivered in convenient and safe containers.

Suppliers can be evaluated on these factors:

1. Reputation in the industry. Consider such items as how the firm is regarded by its competitors, and by its customers, and whether any adverse publicity has arisen concerning shady or illegal dealings.
2. Whether or not the firm is engaged in interstate commerce—this is especially important when considering germicides, insecticides, and other Federally-controlled products.
3. Delivery—does the supplier have a local outlet, and can he provide emergency deliveries when the need arises? (Remember that emulsions, such as waxes, can be ruined when a long delivery causes the material to be exposed to repeated freezings.) The manufacturer typically has better delivery than the jobber, since it doesn't have to go through a second pair of hands.
4. The performance of the supplier, in terms of fullfilling commitments.
5. The qualifications of the salesmen that represent the company: their experience, training, and understanding of their own product line (many of the salesmen in the sanitary supply industry concentrate on selling a very small percentage of their own company's products, and know very little, if anything, about other than their own "pets").
6. Price (this must not become the overriding factor to the exclusion of all the other items).
7. Laboratory control, to assure uniformity of product and its performance (again, the manufacturer has the edge over the jobber).
8. Whether or not the manufacturer or seller is willing to provide demonstrations.
9. The type of packaging offered.

As with the purchase of any type of material, the purchase of sanitary chemicals should follow three basic criteria: cost, performance, and delivery—but the other factors mentioned above can be important.

As with equipment purchasing, standardization has an important part to play. If products are continually changed in order to effect a small cost reduction, the savings can be lost several times over in terms of retraining personnel, errors that may be commited, deterioration of quality, and the like. A change in products should only be made if a significant cost reduction is indicated.

**Selection Methods**

1. Reliance on a reputable supplier. As already suggested, a reputable firm represented by a knowledgeable salesman can save a great deal of time in product selection. For the smaller housekeeping operation, this may well be the best approach, where time is not available for the more lengthy and complex procedures described below. For other than the smallest operation, it might be desirable to purchase from two or three suppliers, so as to retain the advantages of competition.

2. Product testing. For the larger operation, where the purchase of chemicals represents a sizable investment, it is economically feasible to devote enough time to this subject to be more selective, and to test products under various conditions supplied by manufacturers (it is suggested that all samples of one-gallon size or larger be purchased, even for testing, so as to eliminate any obligation to the supplier). Practically speaking, the housekeeping supervisor or manager desires to purchase products that will do a given job–his interest lies little in the field of chemical specifications except in terms of any hazard that specifications might avoid. Most supervisors have neither the chemical background nor the means of testing samples to be sure that specifications are fullfilled. To them, the product test tells all. In chemicals testing, as with equipment evaluation, it is very desirable to involve the actual workers who will use these materials–if they are part of the decision-making process, they will be much more apt to support the use of the chemicals selected. A number of suppliers will offer to apply the materials themselves, such as a floor finish, to give the customer a test–this should be very carefully controlled. It would be better if the supplier would provide instructions in the use of the material, and then one's own housekeeping personnel actually perform the test application since, after all, it will be they who must perform it in the future.

3. Purchase by specification. Where chemicals are purchased on the basis of chemicals specification, the custodian typically complains of the regular deterioration in quality from year to year, to the point where the material being used is very poor in performance, and means are sought to circumvent the specifications. The General Services Administration, which is responsible for maintaining public buildings for the U.S. Government, has a continual struggle with this problem, for example. The composition of the product, such as its solids content, pH, content of certain chemicals, etc. may have a limited effect on a given product's performance–and only performance specifications are really meaningful. Yet, performance tests are difficult to establish–such tests for cleaners, waxes, and floor sealers have not been fully standardized or accepted. Laboratory tests to simulate such performance are sometimes useful but often erratic. Where bid purchasing is required by law, such as for governmental bodies, it is understandable that specifications are necessary, but the specifications should include both chemical and performance specifications, and surely a system of rejection should be possible through performance failure, even where the chemical specifications are in full accord. Further, where specification buying is necessary, at least two varieties of certain materials should be available for the user to select from, especially for such key chemicals as floor finishes, hand cleaners, and disinfectants. For those organizations which are not required to purchase on a bid basis, specification buying of chemicals (although perfectly

applicable for other types of materials in other departments) should be avoided.

4. Purchase of certified or approved products. Various organizations will develop approved purchasing lists for use by their member organizations. There are some serious problems with such an arrangement–for one, the standards are often minimal, and the products approved may not represent the best investment in terms of utilization of labor although they may represent the more economical purchase from a "dollars per gallon" standpoint. Then, such approvals are only as good as the criteria for their establishment, and the laboratories utilized for their policing (non-existant in a number of cases). No system may be in effect to cross-check and upgrade such lists, yet most sanitary supply manufacturers will change their formulas every couple of years or so and no provision may be made for the introduction of new and improved products from the same or other suppliers. From the viewpoint of the chemical supplier, it is a costly and unnecessary expense, and many reputable manufacturers refuse to submit products for such listings, while other manufacturers find this a means of having marginal products listed on par with top-quality performers. As will be described later, a product that performs well in one locale, may not perform at all well in another. The certified products list, then, has too many drawbacks to make it a valuable purchasing tool unless a great deal of time is spent on a continual basis in its establishment and updating (perhaps a better service of various organizations would be to list suppliers who generally produce inferior merchandise, do not operate in a reputable fashion, or who have no laboratory control over the quality of their products).

5. Independent testing agencies. At this writing, only one single independent testing agency specializes in cleaning chemicals and equipment: Buyer's Laboratories, located in Hackensack, New Jersey. For the larger purchaser, the results of this apparently objective organization can save a good deal of time in product evaluation, but in the final analysis, it should be supplemented with one's own product evaluation.

### Special Considerations

In the selection of products, the buyer should consider these facts which have a special place in decision making:

1. Location and size of the central storage area (and satellite central storage areas) may control the quantity, type, and packaging of materials under consideration.

2. Similarly, the size of custodial closets and cabinets (often designed below the barest minimum requirements) may impose some purchasing limitations. For example, the type of shelving may eliminate the possibility of purchasing a certain product.

3. The size or weight of a container may represent a safety hazard which must be avoided.

4. The material from which the container is constructed may have its effect on decision making. For example, there may be a scrap or re-use value in steel containers–glass bottles represent a safety hazard that should be avoided–plastic containers may not be digestable in the waste grinding system etc.

5. Shelf-life can be a strict limitation on some products, especially emulsions which may tend to break after a period of time (don't forget to rotate products on a "first in, first out" basis).

6. The weather may dictate that certain products be purchased only in a given season due to freezing which may occur in harsh climates.

7. If the product is flammable–especially if it is a "red label" product–only certain quantities in certain containers may be stored at any one time, under controlled conditions.

8. Likewise, products representing safety hazards, such as strong acids and caustics, must be carefully controlled in terms of quantity and packaging, and storage conditions.

9. Many organizations find a considerable benefit in arranging an annual purchase commitment for products, at least those which represent a sizable percentage of the total chemicals cost. Under this arrangement, investigation of new products may only occur once a year for a period of a couple of months, before the next annual purchase commitment is established. Ideally, this arrangement obtains a considerable price reduction on the basis of the quantity purchased, but it is an estimated quantity, and delivery is made according to the customer's needs and storage capacity.

10. As mentioned, services available from suppliers may be the deciding factor where other things are merely equal–and they very often are in the sanitary supply industry.

11. In order to increase discounts due to quantity purchase, some organizations arrange a joint purchase where two or more buyers pool their orders (this can be done just as successfully for equipment as it can for chemicals).

12. The supplier who is willing to provide demonstrations of the use of his product would naturally receive preference over that supplier who avoids this.

13. One seeks to purchase from suppliers who have sample and smaller demonstration quantities available for purchase, as well as pertinent literature available.

14. It is not unusual to seek chemicals marketed with "tie-ins", such as where a free electric sprayer is given with each drum of deodorant. Some buyers appreciate this, since they are limited in purchase of capital equipment but not in the purchase of expendable supplies! Obviously, nothing of this sort is ever free, and the product is priced to cover the cost of the "tie-in". When certain suppliers devote much of their sales effort to these tie-ins, one is led to suspect that the emphasis of such a company is not in the production and marketing of quality chemicals. Where tie-ins are of a personal nature, such as a "free" radio, we may move from the unethical to the illegal.

15. Sometimes purchases are made in order to beat a price deadline, due to increased freight rates, higher raw materials costs, or the like.

16. The system of stock distribution and control, tied in with packaging mentioned above, may also be a factor in the product selection.

## The Top Five

The typical housekeeping department will have a storeroom containing a couple of dozen varieties–in many cases still more–of cleaning chemicals. Most of these, however, are used either infrequently, or in small quantity–or both. But in terms of dollar value and importance to the custodial effort, just five products completely dominate the picture:

Floor finish
Finish stripper
General detergent
Disinfectant
Hand cleaner

Let us discuss each in turn.

## Floor Finish

There are a few types of flooring under special conditions that can be utilized without finish, such as special pedestal floors for data processing areas, and pure vinyl floors under lightly traffic and soiled conditions–but in all other than these few cases, something must take the wear, and the choice simply lies between the flooring itself and a protective coating, or finish, or "wax" depending upon the terminology being used.

Organizations whose floors are primarily vinyl asbestos, asphalt tile, or terrazzo find that the floor finish is the most important single product when the factors of unit cost, total cost, impact on custodians, and the effect on the public is considered. A selection of an acceptable floor finish can be so tedious and expensive, that once a suitable product is found, a change should not be made unless a very significant cost improvement is effected. Certainly do not consider changing floor finishes if only a 10c or 20c per gallon saving will be enjoyed over an existing good performer (to assist the reader in floor finish testing, Floor Finish Testing Record and Floor Finish Testing Instructions are illustrated).

These are the varieties of floor finishes now available, and their applications:

1. Solvent or spirit waxes. These are yellowish-colored solutions of natural wax, such as paraffin and beeswax in solvents, such as mineral spirits. Their principal application is for wood, which would otherwise be damaged by the

application of water and water systems. Solvent waxes are also usable on cork, which is, after all, a form of wood. Since wooden floors can also be protected through the use of penetrating and surface sealers, thus eliminating the need for waxes, the use of the solvent waxes, either in liquid or paste form, is minor.

2. Water emulsion waxes. When composition tiles, such as asphalt tile, were first developed, it was necessary to find some protective surface other than the solvent waxes which would soften and damage the tile. Fortunately, it was found that carnauba wax, a natural wax scraped from the leaves of a plant found in South America, had formerly been used in solvents and could also be utilized in an emulsion form in water, much as milk is an emulsion of fatty materials in water. When applied to composition tile or other surfaces, the water evaporated and the wax particles more or less coalesced into a thin surface. Various additives were added to the carnauba or other natural waxes to control their leveling, slip-resistance, softness, sheen, and other such factors. Thus, all such finishes—and most chemical products, for that matter—are compromises which stress one or more factors at the expense of others. These water emulsion waxes serve their purpose fairly well, but drawbacks were noted, such as darkening, slipperiness and the requirement for dry buffing to produce a good gloss. These materials have been superseded in their use by the polymers mentioned below, except in cases where gritty or sandy soil will scratch the harder polymers, and the softer characteristics of the water emulsion waxes provide an advantage in healing the worn spots through their mobility while buffing.

3. Polymer finishes. By far the most successful floor finishes in use are the synthetic polymers which, like the water emulsion waxes, are suspensions of particles in a water system, but rather than being natural waxes, they are the products of chemical invention. Various chemicals go into making these products, the basis item being acrylic plastic, but other plastics are utilized as well, such as styrene and ethylene (a polymer is simply a long chain of these molecules strung together—a co-polymer involves chains of more than one chemical). The advantages of the polymer finishes (and many people still call them "waxes" out of habit) over the water emulsion waxes are:

a. They form into a single film on application, rather like a sheet of cellophane, and thus dry to an initial gloss without buffing.
b. Polymers are more durable and tougher than the waxes.
c. They discolor much more slowly.
d. Of great importance is the fact that polymers can be spray-buffed most successfully (as elsewhere described), a technique that is most efficient for floor care.

4. Metal-link polymers. A special form of polymer floor finishes is the metal-link material, a chemical invention that makes polymers much less susceptible to removal from the floor by detergents. This has been accomplished by making the polymer molecule much heavier through the interlocking of metallic ions, typically zinc or zirconium. Fortunately, these products generally have all the advantages of the polymer finishes, including spray-buffability, so represent an excellent investment.

5. Solution finishes. The latest development has almost turned the wheel full circle, since it is now possible to produce a polymer finish in a water solution (involving alkali soluble resins and coalescing agents) rather than as a water emulsion. The pros and cons of polymer finish and solution are not nearly as definitive as was the split between solvent waxes and water emulsion waxes, and the solutions are still in the state of development. At this writing, wear characteristics in heavy traffic areas are not too good. A compromise exists with translucent finishes that incorporate polyethylene for better wear.

6. Water emulsion pastes. Both waxes and polymers can be prepared in a pasty form, which represents not so much a change in formulation, but a change in consistency and in the method of application, since these products can be smeared around a floor under a floor machine pad, the pad also acting as an absorbent to remove some of the older material and soil, somewhat like spray-buffing. A moderately good job can be done with this product and technique, although there are often complaints of slipperiness in certain formulations. One would expect to be most successful in conditions of light traffic and soil, as a floor so prepared tends to become more and more dingy in appearance.

7. "Lock and key" finishes. Another special development of the polymer finishes was one which required an acidic stripper, thus being resistant to detergents which are primarily alkaline in their action and composition. Now that the metal interlocks are available, this type of product has become obsolete, especially since the use of acidic materials can become injurious to people and surfaces alike.

### Floor Finish Testing Instructions

1. Select area, such as a corridor, where several test sections can be given nearly equal traffic and exposure.

2. Strip entire test area completely of any old finish and rinse thoroughly (at least twice).

3. Set up test "bands" the width of the corridor and preferably ten feet or more in the long dimension of the corridor. By some convenient method—such as plastic tape or masking tape attached to the baseboards—mark each test strip.

4. Each finish to be tested should be delivered to the person responsible for testing in a clean container identified only by a letter or a number so that opportunity for prejudice will be eliminated.

5. Uniformity in application of all products tested is important. After the floor has dried, apply two light coats successively, allowing the first coat to dry thoroughly before application of the second, and noting any special drying characteristics. (First coat should be fully dry in 45 minutes unless inside hunidity is high, or temperature excessively low.) Enter on a Floor Finish Testing Record the product code and the exact area where test product is applied, along with date of application.

6. Using the rating scale indicated on the Floor Finish Testing Record, rate each product individually for the first six quality characteristics shown in the table.

## FLOOR FINISH TESTING RECORD

Product Code ____________

Area Identity ____________

Date Applied ____________

| Time of Evaluation After Original Application | Qualities Rated (1) | | | | | | | | | | | Effects of Maintenance | | | | Total Score at Each Evaluation |
|---|---|---|---|---|---|---|---|---|---|---|---|---|---|---|---|---|
| | | | | | | | | | | | | Buffing | | Spray-Buff | | |
| | Good Levelling | Dries 30-60 Minutes | No Cloudiness | No Discoloration | Gloss | No Slipperiness | No Powdering | Little Scratching | Little Scuffing | Little Black-Marking | Strips Readily (2) | Resultant Gloss | Removal of Scuffs & Marks | Resultant Gloss | Removal of Scuff & Black Marks | |
| As Soon as Dry | | | | | | | x | x | x | x | x | x | x | x | x | |
| 2nd Day | x | x | x | x | | | | | | | x | | | | | |
| 1 Week | x | x | x | x | | | | | | | x | | | | | |
| | x | x | x | x | | | | | | | | | | | | |
| | x | x | x | x | | | | | | | | | | | | |
| | x | x | x | x | | | | | | | | | | | | |
| | x | x | x | x | | | | | | | | | | | | |
| | x | x | x | x | | | | | | | | | | | | |
| | x | x | x | x | | | | | | | | | | | | |
| | x | x | x | x | | | | | | | | | | | | |
| | x | x | x | x | | | | | | | | | | | | |
| TOTAL RATING (3) | | | | | | | | | | | | | | | | |

(See separate Floor Finish Testing Instructions)

NOTES AND COMMENTS:

(1) Rating values: 2 = Good; 1 = Acceptable; 0 Poor

(2) Strippability: See Instruction 9

(3) Highest Total Rating usually will indicate best performance; but, product may be disqualified for unsatisfactory performance in one important category, such as "No Slipperiness" or "No Powdering"

The floor finish is the most important of the impressive array of housekeeping materials; a testing system may be required to select the proper brand.

7. Do not buff the product before the second day after application. At that time, rate all characteristics except strippability and spray-buffing. (Spray-buffing should not be needed so soon after application, but if it is, because a very heavy traffic area has been selected, those characteristics should be rated on all products beginning at the same time).

8. At the end of one week, rate all characteristics except the first four and strippability.

9. Continue ratings about once a week for whatever period deemed necessary to reveal a meaningful difference among the products tested. With careful spray-buffing, stripping should not become necessary within any reasonable time. However, at the end of the period arbitrarily chosen, it will be desirable to strip all products and rate them as to strippability.

10. As noted under Footnote 2, the highest total rating will usually indicate the best performance—however, products may and should be disqualified for unsatisfactory performance in any one important category, such as slipperiness, powdering, or inadaptability to spray-buffing.

11. If bad weather and heavy soil necessitate an unusual amount of wet cleaning, this should not affect the validity of the test if all test surfaces are subjected to comparable procedures.

### Finish Stripper

Periodically, floor finishes must be removed, although the time between removal and refinishing can be greatly lengthened by trapping soil before it reaches the floor (such as with the use of gratings, mattings and runners), and through techniques such as using spray-buffing. Some floors which receive great traffic and soiling and have a poor finish may be stripped every month or so; others may be stripped only after a number of years if properly protected and cared for - and minimization is certainly the rule, since the stripping activity is injurious to floors, as it involves a harsh, alkaline chemical and a great deal of water.

The stripper works by attacking the surface of the finish, much as paint is attacked by paint removers.

The formulation of a stripper is a chemist's job, and should not be done by using a higher concentration of a general detergent, or simply by adding boosters to other types of chemicals; simply adding ammonia to a general detergent may strip the finish, but it may damage the floor as well.

Ideally, the floor finish stripper should be provided by the manufacturer of the wax or finish being used, as the two would be formulated as companion products. Each finish formulation should require a specific stripper formulation; for example, the metal interlocks require aminos or other chemicals to break the bond, whereas other finishes may not require this.

The stripper is used in far less quantity and with far less frequency than the floor finish, but as can be seen is an important custodial chemical.

### General Detergent

The original and fundamental cleaning material is a detergent: water. Since a detergent is a material that provides a cleaning function, water fits this bill admirably, not so much because it is abundant, but because it dissolves more substances than other chemical materials.

The job of the chemist, then, is to make water a still better detergent through the addition of "boosters" of various types (for example, television commercials may claim that a given product "makes water wetter"—and it may). These additional builder chemicals are only required in a very small quantity to do an effective job. In a use-solution of a detergent diluted to two ounces per gallon, the active ingredients present may be no more than one fourth of one percent.

Detergents can be formulated to provide various functions, depending upon the nature of the soil, the cleaning process, and the characteristic of the water. From a simple listing of these functions, we can see that it is the role of the chemist to properly balance them, rather than the custodial personnel:

| | |
|---|---|
| Dispersion | Rinsing |
| Emulsification | Saponification |
| Foaming | Sequestration |
| Mineral control | Suspension |
| Peptizing | Wetting |

It can be seen, then, that there is no such thing as an "all-purpose detergent"—detergents are formulated to perform specific jobs. A floor finish stripper and a hand soap are both detergents—no one would think of reversing their roles or trying to have one product for both jobs.

The housekeeping department does need a general detergent which can be used for such jobs as mopping floors, washing fixtures, and other general cleaning functions. Just as we have progressed from natural spirit waxes to polymer floor finishes, so have we progressed from natural soaps made from animal and vegetable fatty material, to synthetic detergents produced in the laboratory. Such detergents are not generally difficult to select, one must simply be aware of their performance and the effect on the user.

Since we are basically talking about water in any discussion of detergents, the formulation of the product must reflect the condition of the water being used. The primary factors are water hardness, which is controlled at the place of use, and the pH of the formulated detergent.

Water is hard or soft depending upon the quantity of metallic salts in solution, principally formed from the elements calcium and magnesium. The measurement is usually taken as parts per million (ppm) when calculated as calcium carbonate. Soft water is considered to have less than 200ppm, moderately hard water from 200-400ppm, and hard water over 400ppm. Occasionally one sees water hardnesses of 2,000ppm, but fortunately this is a rare condition. Water hardness may also be calculated in grains, with one grain equal to approximately 17ppm of calcium carbonate. Where hard water is used, detergents must be formulated to sequester this hardness, or to precipitate it if the precipitate is not objectionable.

The acidity or alkalinity of a solution is identified as its pH, which is shorthand used by chemists to indicate the presence of hydrogen ions. An arbitrary pH scale has been developed from 0-14, ranging from the strongest acid at 0 to the strongest caustic at 14, with 7 being neutral (see illustration). The scale is complicated by the fact that it is logarithmic—that is, the pH of 5 is ten times as acidic as a pH of 6. It is up to the chemist, not the custodian, to determine pH requirements for various uses. While a bowl descaler would be acidic and thus of a lower pH, a finish stripper would be alkaline and of a higher pH, while a hand soap should be nearly neutral. One custodial supervisor once described to the author his preparation of a cleaning solution by mixing ammonia with vinegar—he did not realize that the alkaline ammonia counteracted the acidic vinegar!

Once the detergent has been properly formulated, other factors affecting or creating cleaning are:

1. Time. Increased contact time provides greater chemical action.
2. Temperature. The higher the temperature, the quicker the chemical reaction.
3. Mechanical action. Some cleaning is done by rubbing, scratching, and scraping.

The emotional approaches to product selection mentioned above are most prevalent in the selection of a general detergent—sometimes it is found necessary to re-color a detergent because color preferences are so strongly engrained in some custodial personnel.

Further, be sure that custodians are not putting additives into an otherwise acceptable product. Sometimes workers throw in handfuls of scouring powder, which can scratch a floor or other surface due to its abrasive particles—or put in some TSP (trisodium phosphate, which can leave a white film on a floor due to the precipitation of metallic salts), etc. One custodian was seen to stir in broken-up urinal discs, which are completely insoluble in water, and pH OF SOME CLEANING CHEMICALS

| | pH | |
|---|---|---|
| ↑ | 14 | |
| | 13 | SODIUM METASILICATE |
| | 12 | TRI-SODIUM PHOSPHATE<br>SODIUM CARBONATE |
| | 11 | AMMONIA |
| INCREASING ALKALINITY | 10 | POLYPHOSPHATES |
| | 9 | SOAP<br>BORAX |
| | 8 | SODIUM BICARBONATE<br>SODIUM TETRA PHOSPHATE |
| NEUTRALITY | 7 | DISTILLED WATER |
| | | SODIUM HEXAMETA PHOSPHATE |
| | 6 | BORIC ACID |
| | 5 | |
| | 4 | |
| INCREASING ACIDITY | 3 | ORGANIC ("WEAK") ACIDS |
| | 2 | |
| | 1 | MINERAL ("STRONG") ACIDS |
| ↓ | 0 | |

The pH scale describes alkalinity or acidity of a solution. Being a logarithmic scale, each number is ten times the value of the preceding number, with 7 being neutral.

have no detergent action whatsoever.

Where a building is primarily carpeted, the importance of the general detergent would be supplanted by the carpet shampoo material, discussed below.

When selecting a general detergent, consider the following:

1. The pH should normally range from 7.0 to 9.5.
2. Excessive foaming is undesirable, as it requires a secondary rinsing job, or leaves a film that attracts soil and looks unsightly.
3. Avoid natural soap content, and stick to synthetic detergents, which do not leave a precipitate (such as the "ring around the bathtub").
4. For general purposes, use-dilutions should be no higher than two ounces per gallon of water.
5. The formulation should avoid ammonia, free alkali, solvents, abrasives and free oils or fats.

### Disinfectant

It is natural to consider that germ killing is one of the most important aspects of housekeeping in a hospital—yet it is also one of the most important activities in all types of buildings, since the control of disease-causing bacteria is necessary in rest rooms, locker rooms, food service and vending areas, and even on telephone mouthpieces.

In the past, disinfection was performed by first cleaning with a soap (and later a neutral synthetic detergent), and then a second step of disinfection was performed. Now it is perfectly feasible to utilize products that perform both cleaning and disinfection steps in one—these products are typically called germicidal detergents or cleaner-disinfectants. In order to gain the advantages of standardization, and to assure that germ killing will always occur during cleaning, a number of health care organizations have eliminated the general detergent and have decided to use a combined product for all cleaning.

To discuss what disinfection is, we shall also discuss what it is not. In the killing of micro-organisms, various degrees of control are possible:

1. Antisepsis is the killing of germs on living tissue, such as with mercurochrome, or tincture of iodine.
2. Sanitization involves reduction of bacteria levels to numbers which are generally considered safe—a 50% reduction in bacteria count qualifies one to use this term, which is often applied to food service surfaces, such as counters, and glassware.
3. Disinfection is the term that housekeeping personnel are most concerned with, as it relates to killing bacterial organisms that are potentially harmful. The strict definition of the term varies but is now generally taken to include the killing of most (but not necessarily all) disease-causing bacteria and fungi, but does not necessarily include the killing of spores (germs which encase themselves

in a hard shell) or viruses (the simplest living things, resembling a complex molecule). Disinfectants as well as other materials that kill are controlled by the United States Department of Agriculture—a chemical can only be classified as a disinfectant if it kills at least 99,999 out of 100,000 organisms under controlled conditions. Thus, complete kill is not necessary (or even possible under most use conditions). Bacteria are usually found in such enormous numbers that to leave only a single organism alive out of 100,000 might still leave millions alive on a given surface.

4. Sterilization signifies the complete killing of all organisms which may be present, and is generally achieved in an autoclave through the use of steam or lethal chemicals or perhaps by irradiation.

Thus, it is disinfection that is the primary concern of the housekeeping department, in whatever type of building. It might be well to define a few other terms generally associated with this type of activity in order to eliminate confusion:

1. Bactericides kill all bacteria, those which are beneficial as well as those which are harmful. One should not forget that the body cannot survive without the beneficial bacteria which lives in and on it.
2. Fungicides kill fungus growths, such as that which causes athlete's foot.
3. Germicides kill those bacteria—or germs—which cause disease.
4. Sporicides kill spores, which is the protective form of certain bacteria which build a shell around themselves, difficult to penetrate, and these are the hardest of the bacteria to kill.
5. Virucides kill viral organisms, such as those which cause viral pneumonia and influenza. Some forms of viruses are very difficult to kill.

Since the control of disease-causing micro-organisms is so important to man, it is not surprising that the marketing of disinfectants, a term generally synonomous with germicides, would be rigidly controlled by the Federal Government. Since the Federal Government cannot exercise this control except over products shipped in interstate commerce, it is most desirable to purchase such products from companies doing business in more than one state, so one can be assured of controlled manufacture and testing. The Department of Agriculture requires all pertinent information to be shown on the label in such cases.

One cannot have control without testing procedures, and the original test for determining the kill characteristics of disinfectants was the Phenol Coefficient—this was a measurement that compared the killing ability of a chemical at a standard use dilution with that of pure carbolic acid, or phenol, which was the original disinfectant first used in England by Sir Joseph Lister. This was a satisfactory system of measurement until the advent of the quaternary ammonium compounds mentioned below which yielded in laboratory tests coefficients of 200 or more. These are simply not attainable in practice. Therefore, another form of test was necessary, and we now utilize the Use Dilution Confirmation Test, as developed by the Association of Official Agricultural Chemists, and thus called the AOAC Test. This is a laboratory test that simulates field conditions and measures the kill of the given chemical at control dilution rates over specific periods of time for individual test organisms.

This last item is of considerable importance—the AOAC Test does not qualify a chemical to kill all bacteria, but just those against which it was tested. In selecting disinfectants from one of the types which will be mentioned below, it is suggested, to the reader, to require the following:

1. At normal use-dilution (this is usually two ounces or less per gallon) the product should kill Staphylococcus aureus and Pseudomonas aeruginosa, as well as Salmonella typhosa and usual test organisms.
2. The product should kill fungi, such as those causing athlete's foot, and should inactivate many of the viruses that are pathogenic to man.
3. In a general hospital environment, the product should also kill the tuberculosis organism.
4. Bearing other requirements in mind, the product should be selected for minimum skin irritation.
5. The germicidal activity should not be affected by the available water supply.

One has the choice of several types of chemicals that can be used for germicidal purposes, and despite the advantages of standardization, the housekeeping department may be required to use more than one variety. Generally, however, one should basically rely on either a phenolic type disinfectant or a quaternary ammonium type for most usages, with the use of an iodine disinfectant for very special cases under hospital conditions.

1. The earliest disinfectants were strong, odorous products such as simple phenol, cresol, and pine oil. These products led to the feeling still prevalent, that for a germicide to be effective it must necessarily smell strong! All of these materials are obsolete, even though great amounts of money are spent promoting pine oil to the housewife, who still believes it is an effective germicide. It simply does not kill, for example, Staphylococcus aureus, typically the greatest culprit in cross-infection.
2. Synthetic, or modified phenols, have been developed to overcome the disadvantages of the earlier materials, and are quite effective in doing this. They are essentially odorless, and kill a wide range or spectrum of organisms, significantly including tuberculosis germs. They have a fairly low degree of skin irritation, although some people develop a rash with continued use—just as some people can be irritated by pure water. The synthetic phenols—now simply called phenols—are the most popular disinfectants.
3. Quaternary ammonium chemicals. In order to overcome the limitations of phenols, the chemists have developed the "quats", but at the same time have

introduced limitations that the phenols do not have. The "quats" have the advantage of good odor control, very low skin sensitivity and a high dilution ratio—but they do have the disadvantages of not being able to kill the tuberculosis organism, and being inactivated by a smaller quantity of soil than the phenolics. The "quats" are non-staining, odorless and non-toxic. Their characteristics make them ideal for use where food is processed or served, and are very useful in hospitals where tuberculosis is not considered a problem—yet, hospital users, in the author's opinion, should stick with phenolics because of the legal implication of using a material that will not control TB.

4. Sodium hypochlorite chemicals, being corrosive, should be restricted to specialized use in certain processing operations, and should not be generally used by a housekeeping department.

5. Iodine is a powerful disinfectant, and can be prepared for general use in the form of "Iodophors", a solution based on a non-ionic syndet (synthetic detergent). When in use, these materials release active iodine which will kill a wider range of pathogens than the phenols or "quats"—but their acidic nature limits them to specialized use, such as for "dirty cases" in hospitals where one would fear that the phenol or "quat" would not be sufficient. One would use the Iodophor, for example, for a gas gangrene or spinal meningitis case. General use of this type of material is not recommended and where biological laboratories exist, their use should be carefully avoided, since vapors from the chemical can interfere with such tests as protein-bound Iodine measurement.

At this point, a word about electric charge might be helpful. Both synthetic detergents and soaps, when they are dissolved in water, break up into particles which are charged positively or negatively. Charged particles are called ions, and in a solution the positive charges counterbalance the negative charges. In a solution of soap or synthetic detergent, however, there are large and small particles or ions, the large ones being the active constituents, and these materials are classified according to the charge on this larger ion—thus, ions are negatively charged, and cations are positively charged. Where there is no charge at all, these are called nonions. Ionic classification is important in the manufacture of chemicals, because a cationic material can neutralize an anionic material when they are mixed together. Thus, the housekeeping department is once again urged not to do its own formulation, but to leave this to the skilled chemists.

This is even more true when one considers the numerous factors involved in germicidal activity:

1. The presence of soil and organic matter.
2. Water hardness.
3. The presence of other chemicals.
4. The pH of the solution.
5. Contact time (which should be a minimum of two minutes for adequate kill).

6. The temperature of the solution and the surface being cleaned.

## Hand Cleaners

Although hand cleaners are not a material used by the housekeeping department in its service of environmental control, it is in the category of paper towels or toilet tissue: a dispensed product for use by other employees and the public in rest rooms. Hand soaps are such a personal product that many people judge the function of the entire housekeeping department on the basis of the type of hand cleaners being offered. Because of the many varieties of hand cleaners available and because of personal preferences, it is not unusual to see a housekeeping supply room with as many as a half dozen or more varieties of hand cleaners available. In general, however, two products from the following list should be sufficient: a hand cleaner for lighter soils, such as is used by administrative or clerical people, the liquid hand soap dispensed in lather form being a good example—this to be supplemented by a heavier-duty material for greasy or oily soils, the lotion-type cleaner with solvent content being a good choice.

Here are some of the varieties of hand cleaning materials available:

1. Bar soap. This is the oldest of the hand cleaners, made from natural fatty materials. The bar soaps are inexpensive, but have the considerable disadvantage of being a "community" product which many people wish to avoid. It is possible that the use of bar soaps can lead to cross-infection, and even if this is unlikely, people still give it thought. The use of bar soap also complicates the cleaning of the lavatory surface.
2. A dispenser is available which will grind up bar soap into a powdered or chipped form, thus removing the community aspect of the material. Most people do not like the slippery, sticky feel of the resulting product, and the dispenser tends to become clogged and soiled.
3. Synthetic detergent bars are also available, having the same advantages over soap bars as neutral synthetic detergents have over liquid soaps. For example, when used in a shower room, a natural soap film or curd is not formed on the walls. But these bars have the disadvantages of defatting the skin and thus causing irritation, and also they decompose rapidly in the presence of water.
4. Soap can also be dispensed in leaf form, from either disposable or permanent dispensing devices. Again, the user complains of a sticky, slippery film in the hands which is difficult to rinse away, and when such leaves fall to the floor they cause cleaning problems.
5. Soap towelettes are also available: soap impregnated into rectangles of paper which are rubbed onto the skin when wet and then rinsed off. The disadvantages include a feel which is unpleasant to some people, and the litter which the discarded towelette may cause, as well as the time involvement in its use.

6. Powdered soaps, in numerous formulations, can be dispensed from a device by lifting up on a plunger, turning a crank, or otherwise. The shortcomings include clogging of the dispenser, dispenser rusting if it is made of steel, powder falling on the lavatory or floor requiring further cleaning, and a tendency toward overuse.

7. Similarly, granular products are available in dispenser form for heavy duty cleaning, and these products may contain such abrasive materials as wood flour (a finely ground sawdust), cornmeal, pumice, etc. Their rough nature can be injurious to the hands, not to mention the face.

8. Pasty hand cleaners may simply be scooped out of a large open-mouthed container, or dispensed by extrusion, much as is toothpaste from a tube. These materials can be useful for cleaning grease and oils from the skin, as they may contain a solvent component, although this tends to de-fat the skin and lead to irritation or chapping.

9. Lotion-type cleaners generally have the heavy-duty cleaning advantages of the paste-type materials, but are milder, and thus more generally accepted. Some of these contain emollients to protect the skin.

10. Liquid soaps are most frequently used in industrial and institutional situations, and by the public in airline terminals, office buildings, and the like. Glass dispensing globes should be scrupulously avoided from the safety hazard standpoint, and plastic used instead—or a central tank provided for remote dispensing. Care must also be taken that a concentrated soap is not introduced into a tank system or the individual dispensers, as they will clog up. It is best to have such materials diluted immediately upon arrival in the storeroom, and then marked "ready for use". From the standpoint of public relations, a higher quality liquid hand soap is certainly desirable.

11. Lather dispensers are a cut above liquid dispensers, as they take the same liquid soap, but a special valve mixes air and produces the material in a foam or lather. Not only is this form of dispensing more acceptable to the public, but it also uses less than half as much material as the liquid soap dispenser. Some liquid soap dispensers can be converted to the lather type by the purchase of a special valve. The lather dispenser also eliminates the problems of selection by color, as the color of the original material is not visible—but the odor problem remains!

### Specialized Products

An imposing, if not confusing, array of specialized products are available for use by the housekeeping department. Whether they are of major, minor, or no consequence depends upon the surface to be cleaned and the activities which are the responsibility of custodial workers.

*Anti-Static Materials*—In climates where low humidity causes static problems with carpet, an uncomfortable situation for tenants and visitors, it is possible to spray the carpet with a material which reduces this static. Such chemicals may be based on quaternary ammonium bromides, in the same chemical family as the quats which are used for disinfectants. Where carpets are purchased with conductive fibers built in them, there is no need for such chemical agents. Sometimes the anti-static agents tend to cause faster re-soiling—for this reason they may be combined with anti-soiling agents.

*Carpet Cleaners*—A number of chemicals are available in this category alone, and the reader will find their selection and use covered in Chapter C.11. Carpet care products include:

Spot cleaners (often available in kit form)
Powdered or granular cleaners
Surface brighteners
Foam-producing cleaners

*Chalkboard Cleaner*—In general, products of this type do more harm than good, as they tend to fill the "tooth" or pores of the surface with an oily or clay-like material, that eventually leaves the board too smooth and too slick for good writing and reading. For the most part, chalkboard care should be dry, without even washing, except on some of the newest synthetic boards, and then only as recommended by the manufacturer.

*Chelating Agent*—These are special materials formulated to remove scum and other metallic deposits, through the ability of the chemical to encapsulate these materials. They are often marketed under the name of "terrazzo scum remover", etc. When available in a jellied form, they may be used on vertical surfaces.

*Cleanser*—The cleanser, by convention, is a scouring material—that is, one which cleans primarily by abrasion, but also containing detergent components to remove the loosened soil. The ingredient that performs the scratching may be silica dust, volcanic pumice, or granular minerals. Cleansers are available in three physical forms:

1. Cleansing powders: These are most familiar in terms of products advertised on television and used by the housewife. But they also contain the largest particle size, and therefore are the most injurious to surfaces. Cleansing powder regularly used on plated restroom fixtures will completely remove the metallic plating, and will scratch the finish on the fixture itself so that it is much more readily soiled and becomes progressively more difficult to clean—a vicious circle indeed.

2. Paste Cleansers: This type material has a lesser particle size than powdered cleansers, and also avoids the problem of overuse that the powdered materials have, since they are often shaken into a sink and they clog the drain. The material should be put onto a damp sponge or cloth if used at all.

3. Lotion-type Cleanser: These have the smallest particle size of all and should be first choice for general use. Graduation to paste and then to powdered cleansers should only be on a basis of unavoidability because of the nature of the surface or the soil.

*Concrete Curer*–Products of this type are often marketed by janitor supply companies and sometimes applied by housekeeping departments, but basically it is a material for use in concrete construction, to harden and cure concrete floors. The material forms a latex-like structure that permits the migration of water molecules to the point where the seal will not lift, but retains enough moisture over a long enough time to allow good curing. Actually, such material should be applied by a construction contractor.

*Deodorant*–The housekeeping department is regularly called upon to control odors, and this can be done through several approaches, all of which should be used simultaneously where possible:

1. Providing impermeable surfaces through design or through sealing that will not permit the harboring and growth of bacteria.
2. Ventilation to remove transient odors.
3. Killing the bacteria from which most odors derive by disinfection and sanitation before the use of odor-controlling chemicals.

The last item provides us a number of possibilities.

1. Deodorant blocks. These are made of paradichlorobenzene, and provide a masking odor–that is, one which is perhaps more pungent than the odor that is objectionable. With rare exception, there is no worthwhile application of this product, and its use often indicates a failure to clean and disinfect.
2. Drip fluid. This is another masking chemical, made by adding a perfume (typically oil of eucalyptus) to a paraffin mineral oil. The installation of unsightly cabinets, the damage to ceramic walls and the overpowering odor created when fixtures have not been used for some time, are some of the reasons that this material should never be used.
3. Jellied deodorants are available, with the active component generally being chlorophyll, which can actually combine with the particles in the atmosphere which cause odors for their direct elimination. When available in containers where the amount exposed can be varied, this product performs well.
4. Wick deodorants are dispensed in liquid form through a cotton wick, the amount dispensed depending upon the length the wick is pulled from the bottle (and of course the amount of air movement). These materials may also be based on chlorophyll, and some have a formaldehyde content which literally anesthetizes the nasal membranes, so that the nose cannot smell! They also may contain a perfume. These products are generally good performers.
5. Aerosol deodorants. Although these may contain some of the same ingredients as the jelly or wick deodorants, all too often they rely on perfume, and thus may not actually remove odors. Further, any chemical dispensed in aerosol form, whether manually or on a timer, will be expensive.

*Descalers*—Acid products used for removing mineral deposits, such as are found in toilet bowls and urinals—they attack minerals directly, so that they form new compounds with the acid which was washed away. It is desirable that descalers contain less than 10% equivalent hydrochloric acid, so that they do not need to carry the poison label, and can be used with less hazard to personnel—some of these products are marketed with the acid content as high as 20% or more. Also, their formulation with organic acids tends to make them less hazardous than when they use mineral acids, such as muriatic (which is a commercial grade of hydrochloric acid). Descalers should be used only periodically and not on a regular basis for cleaning, and their use should be by designated personnel only, wearing proper protective clothing and face shields.

*Drain Opener*—Although many housekeeping departments are issued products for unstopping drains, this is really a plumber's function. A number of accidents have been caused by "blow back", that is the formation of gasses in the piping which create enough pressure to blow back in the face of the user—some of these accidents have resulted in blindness. Again, drain opening by chemical means should be performed by a qualified plumber, rather than a custodian.

*Dust Mop Treatment*—Materials are available for impregnating the yarn of the dust mops so that they will pick up and hold dry soils more easily. These treatments are of two types: oil-base, and wax-base. The hazard with oily materials is that over-treatment (especially if the mop is used before the material has penetrated the fibers—24 hours should be allowed for this purpose) may leave a film on the floor which can cause falls or deterioration of a resilient floor through actually dissolving the material. For best results, the housekeeping department should either rent treated dust mops which are carefully controlled by the supplier (but not all suppliers provide this careful control), or have the treatment done, perhaps in one's own laundry machine, by a designated person for all dust mops (and perhaps dust cloths as well). Individual custodial workers should not perform their own dust mop treatment.

*Foam Depressant*—Sometimes it is necessary to put an additive into a detergent so that too much foam will not be formed—this would especially be important where the detergent will be removed by a wet vacuum, such as in an automatic scrubbing machine—too much foam can damage the equipment. Foam depressants are typically based on silicone type chemicals. A properly formulated detergent for use in automatic scrubbing machines, however, will eliminate the need for this special product.

*Glass Cleaner*—Where glass is to be washed in small quantities, aerosol cleaners or spray-bottle cleaners often containing quantities of isopropanol to improve evaporation can be useful. For larger surfaces, a neutral synthetic detergent can be prepared or purchased for the job, generally in much more diluted form than would be used for, say, floor mopping. Window cleaning detergents also may contain additives for attacking greasy soils or metallic soils. Be certain not to use glass cleaners that contain any abrasive materials, such as pumice.

*Insecticides*—Although the control of insects is not strictly a cleaning activity, it is still the responsibility of many building service departments, especially on a limited basis. Where insect control becomes an important aspect of environmental control, such as in food processing plants, large institutions, and the like, this is very often handled on a contracted basis with specialists in this field. In the housekeeping department, insect control may range from having an aerosol container on each custodial cart, simply in order to kill any insects which may come into view, to a regular fogging or misting of fairly large areas. Insecticides typically consist of a small quantity of active ingredient in a solvent carrier. The buyer is therefore concerned not only with the properties of the active insecticide ingredients, but with the properties of the carrier as well. In general, it is safest to use insecticides, especially in the hands of workers with limited specialized training in this field, that are non-toxic, and these are primarily the Pyrethrins and their derivatives. Especially where insecticides are to be used around food stuffs, the Pyrethrins should be used. Another class of insecticides are the chlorinated hydrocarbons, with such names as chlordane, DDT, dieldrin, and lindane—these have the advantage of residual activity, but the disadvantages of toxicity. For example, a great deal of publicity was developed concerning the problem that DDT caused in the contamination of growing foods and injury to animal life. The United States Department of Agriculture controls insecticides by classing them as "economic poisons", just as it controls germicides, weed killers, and the like. Thus, in selecting insecticides, it is best to deal with an organization in interstate commerce whose products fall within the purview of the Department of Agriculture.

It is also best to deal with specialists in this field when one does not have a qualified expert on the staff.

*Metal Polish*—A number of metal polishes are available for the cleaning and shining of such metals as copper, brass, and bronze. They contain both abrasives and chemical cleaning components. Most polished metals are being "designed out" of new construction, as the cost of maintaining them is excessive. Attempts to avoid metal polishing by the use of clear lacquer usually fail except in very lightly trafficked areas.

*Oil Absorbent*—These materials are made of porous clays or minerals of various types—they are able to absorb almost twice their weight in oil or water, and thus are useful around machine shops and processing equipment to prevent hazardous, slippery floors. In purchasing oil absorbents, consider both the mesh of the clay—too fine a mesh can create dust—and its efficiency of absorption.

*Seal Stripper*—Water emulsion sealers are removed with the same finish strippers used on polymer floor finishes, but the solvent sealers must be removed with special chemicals typically composed of methanol and methylene chloride—these are the basic constitutents of paint removers, and the operation is quite similar. The chemicals attack the surface, blister it, and thereby make it possible for it to be lifted. This removal is facilitated in practice through the introduction of some granular material to which it becomes attached, such as sawdust or cornmeal. The surface must be carefully rinsed with solvent as well. The whole procedure, then, is tricky, messy and to a certain extent, hazardous.

Again, we should call on experts for help, either in the recommendation of chemicals and method, or in the actual performance of the work.

*Sealers*—A seal is a material used to protect a floor surface (as defined for a building service department) over a considerably longer period of time than a floor finish. These materials are called penetrating seals when they are thin enough to sink into the pores of the structure, and surface seals when viscous enough to remain primarily on top of the surface. Seals are of two basic types:

1. Water emulsion sealers are very much like water emulsion polymer finishes, except that the solids content is higher and other provisions made for a more permanent use. Such sealers are useful on terrazzo floors in order to provide a base upon which a surface can be developed. They are also useful on resilient floors, such as asphalt tile and vinyl-asbestos, which have become roughened, dried out, or damaged so that their pores need filling—otherwise, a regular finish would sink into these floors and leave no surface for wear or shine. A water emulsion seal is not normally needed on a resilient floor in sound condition.
2. Solvent sealers, as the name implies, are solutions in solvent and have much longer and better wear characteristics then the water emulsion seals—but they are much more difficult to remove, as well (see seal strippers above). The most popular solvent sealers are the phenolics, epoxies and polyurethanes, each of which has two sub-varieties. These seals may variously be used on concrete, wood, terrazzo, stone and other surfaces. The only one which can be used in the presence of moisture, with any degree of success, is the moisture cured polyurethane, which was developed for the sealing of wooden floors in textile areas where the atmosphere is normally humidified to prevent the breaking of yarn fibers.

The application of seals of various types is a complex subject, the variables involved including the nature of the surface, soiling, moisture, removability, hardness, slipperiness, ease of application, toxicity, etc. The adjoining chart indicates some characteristics, but the proper sealer for a given job should be the recommendation of someone skilled in their use.

*Stainless Steel Polish*—Stainless steel is a metal that belies its name—the multitude of microscopic parallel scratches introduced in the metal in its forming hold oily soils that cause unsightly stains. In many cases, the chief culprit is finger marks. The proper solution to this problem is to first clean the stainless steel, then dry it carefully, then fill these scratches and grooves with a product made of silicone material or some oily material, so that the metal will no longer become stained.

*Sweeping Compound*—Sweeping compounds are mixtures of oily or waxy materials to attract and hold particles of dust and soil, along with granular materials such as sawdust and sand to provide a suitable carrier and surface for the oil or wax. In the past, sweeping compounds were utilized on almost all types of floors since no better cleaning method was available, but with the advent of vacuum sweeping, treated dust mops and cloths, automatic scrubbing machines and better floor finishes, sweeping compounds are now used primarily on unfinished floors, such as concrete and stone—their use on finished floors would be where a great deal of soil has been introduced and no other means of removal is available.

Sweeping compounds can be quite injurious to resilient floors if the wrong type is selected. Sweeping compounds are primarily of the waxy type on a sawdust base, designed for use on floors which are easily damaged, and the oily type on a sand base for rougher floors. If the latter is used on a finished floor, the sand will cause scratching, while the oil will soften the resilient material.

Further, sweeping compounds are typically overused and thus may create soils of their own through the raising of a dust cloud. The material should not be spread liberally over an entire area to be swept, but rather formed into a line of compound which is then moved forward with a pushbroom.

*Solvent Cleaner*—Solvent cleaners are needed when the soils consist of oily or greasy materials, or tars, which may not yield to cleaning by soaps or synthetic detergents. The purpose of the solvent is to dissolve the soil, then to suspend it so that synthetic detergents can then emulsify the soil and flush it away. This dissolving action may be provided by kerosene, pine oil and certain alcohols, for example. The products may or may not be formulated in association with water.

One problem associated with solvent cleaners is the effect on the skin, since the solvent will tend to de-fat the skin and cause irritation—where such irritation is present or suspected, protective skin cream or gloves should be provided.

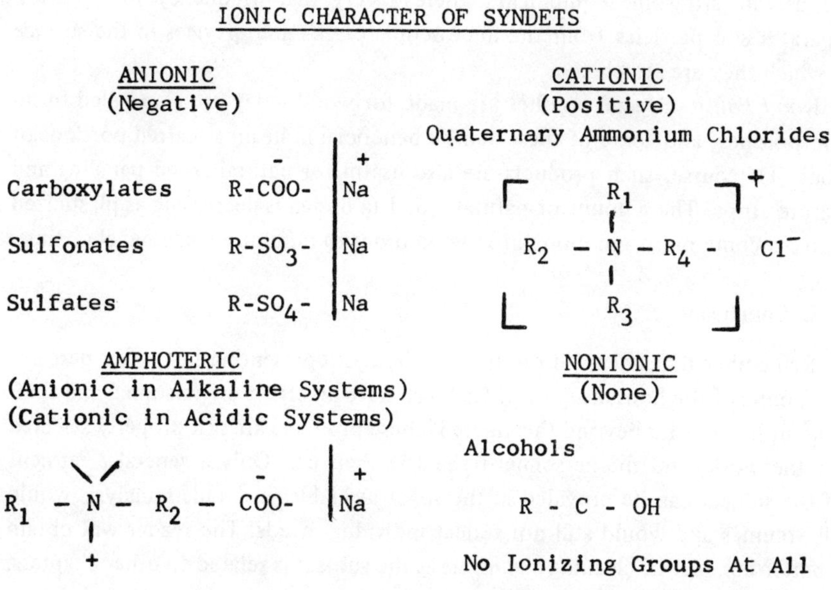


The complexities of chemical formulation indicate a job for the chemist, not the custodian.

Another problem deals with flammability, but fortunately safety solvents have been developed which reduce the flammability hazard to a reasonable minimum and also overcome noxious vapors associated with some solvents. Highly toxic materials such as carbon tetrachloride should never be made available in a housekeeping department.

Solvent cleaners should be carefully used around resilient flooring, as their action can soften and damage the tile. Special equipment has been designed to utilize solvent cleaners and their vapors—these cleaners are especially adaptable for use in ultrasonic equipment, where a very high frequency of vibration separates soil particles from the microscopic cracks and crevices in the surface to which they are attached.

*Wood Polish*—Special polishes are made for wood—they are also called furniture polishes—and some of these can be beneficial in healing scarred portions of wood. Of course, such products are also useful for natural wood paneling and feature strips. The amount of natural wood in design is decreasing as plasticized surfaces come more and more into use, in order to minimize cleaning labor time.

### In Conclusion

Remember that chemical products, although representing but a few percentage points of the housekeeping budget, can have an effect on personnel and their time utilization far beyond this figure if these products are not properly selected for the work, and the personnel trained in their use. Only a general treatment of the subject can be provided in the space available—a detailed analysis would fill volumes and would still not reflect individual needs. The reader will obtain a better concept of cleaning chemicals as the subject is related to other chapters, which also discuss their selection and application.

# 37

# Floor Care

THE care of floors represents the average housekeeping department's greatest single expense. The International Sanitary Supply Association indicates that floors consume 40 percent of the housekeeping dollar. The cleaning of the next largest category, equipment, requires only 24 percent. A survey conducted by one maintenance industry magazine showed that floor-care items brought sanitary chemicals suppliers 70 percent of their business and 80 percent of their profits.

Thus, the care of floors is of great importance to both the housekeeping department and to its suppliers. It is important to others as well because the condition of the floors is the first thing noticed when a person enters a building. Often, the level of maintenance of an entire facility can be judged by the condition of the floors.

## Floor Hardness

It is sometimes convenient to categorize floors as "soft" or "hard." Soft floors include composition materials such as the various resilient tiles and cork. Hard floors include concrete, wood, terrazzo, and stone.

The relative degree of hardness is not a direct indication of comfort, however, as has been pointed out by Dr. Arne J. S. Lundervold of Oslo University. Dr. Lundervold set up tests to determine the muscular activity required for walking on various types of floor surfaces. He used the techniques of electromyography and the recording of electric muscle-action potentials. His findings indicated that walking on very hard floors caused the greatest discomfort. But, the greatest comfort was not achieved in walking on the softest coverings, such as rubber and carpeting, but rather on floors of medium hardness such as cork, vinyl, and spruce wood.

A fellow countryman of Dr. Lundervold, Mr. R. Schjodt, Research Engineer at the Norwegian Building Research Institute, collaborated in these studies. He pointed out that an individual's opinion of his comfort relates to the situation with which he is familiar or used to. In *ASTM Bulletin,* Number 247, July 1960, Mr. Schjodt proposes the following hardness scale for floor surfaces:

| *Relative Hardness* | *Material* |
|---|---|
| 10 | Concrete, terrazzo, stone |
| 7 | Vinyl, asbestos tiles |
| 6 | Linoleum, soft rubber |
| 5 | Cork tiles, soft wood, cork linoleum, high-grade vinyl |
| 4 | Sheet vinyl on felt base |

## Installation

The installation and repair of all floor types is a job for professionals.

The skill with which a floor is first installed has a considerable bearing on the appearance of the finished floor, as well as on the effort necessary to keep it properly maintained. This is not only true of the final floor surface itself, but also is true for subfloors. An irregular concrete or wooden subfloor will show up very quickly in any composition covering, which immediately assumes the contours of its support. This is particularly true of materials such as vinyl. Attempts to properly maintain a resilient floor with an irregular subfloor are quite frustrating. The depressions fill with wax and soil and take on a darker tone than the balance of the floor. Floor-machine brushes span these depressions so that they cannot be scrubbed or buffed properly. The peaks of the floor receive almost the entire wear so that they thin out quickly, exposing color and texture variations. If such a floor is not properly protected by wax, it will wear through to the subfloor very quickly. On the other hand, proper wax protection under such conditions calls for inordinately frequent waxing.

The provision of a good subfloor and quality floor covering material, both installed by skilled workers, represents a worthwhile investment.

## Floor Types

The following listing indicates the variety of major floor types now in use. The consideration or 'selection of flooring types is somewhat simplified by the fact that all floors can be put into one of the following categories:

Cemented (such as concrete, ceramic, terrazzo)
Carpet
Composition (asphalt or vinyl tile, rubber, cork, linoleum)
Metal
Stone
Wood

#### ASPHALT TILE

*Description:* A resilient floor tile is made of asphaltum, asbestos fiber, lime rock, colored pigments, and sometimes wood flour. It is the asphaltum which binds the materials together. Asphalt tile is usually cemented directly to a smooth-troweled (preferably machine-troweled) concrete floor, or to a felt underlayment over a wood subfloor. Where asphalt tile is to be laid over wooden planking, an intermediate layore of plywood, $\frac{3}{8}$ inch or thicker, should be used to obtain a smooth surface.

*Grades:* Normally supplied in 9-inch square tiles (occasionally 12-inch square tiles) in $\frac{1}{8}$- or $\frac{3}{16}$-inch thickness. Colors are graded from A to D, with A being the lowest and cheapest color grading. The colors are generally mottled or striated. A special coloration gives asphalt tile the appearance of cork. A special grade of greaseproof asphalt tile is available for use in kitchens or in precision assembly areas where greases or lubricants may be spilled.

*Cost:* Installed cost for regular 1/8-inch asphalt tile will range from twenty to fifty cents per square foot, depending on such factors as wage rates and color grade. The greaseproof tile runs from 80 cents to a dollar per square foot. A well-maintained asphalt floor will last for many years. Removal costs about the same as installation.

*Characteristics:* Solvents and oils attack asphalt tiles, causing them to soften and break down. Strong alkalis are also injurious, and the Asphalt

Tile Institute recommends that cleaners be limited to a pH of 10, although general resistance to alkalis is considered good. Resistance to most acids is also good. They are deformed easily by heavy loads (limit 25 pounds per square foot), and are one of the noisiest of floors. They have fair light reflection.

They are: inexpensive, available in a good range of colors, can be used on or below grade; easily and quickly installed; and long-wearing when properly maintained.

*Maintenance:* Most natural soaps contain materials that form a film over asphalt tile. Therefore, neutral synthetic detergents should be used for wet cleaning this material. For dry cleaning and dust removal, care should be used in treating dust mops. A wax-based treatment is the safest, although most users still prefer the oil-based treatment. Since oil is injurious to asphalt tile, it must be carefully applied. Asphalt tile should always be protected with natural water-emulsion waxes or polymeric floor coatings.

*Cautions:* Avoid solvents and oils. Remove factory coating before applying wax. Do not use any liquids within two weeks of installation. Always use water sparingly. Avoid harsh alkalis. Use rubber or plastic cups under heavy furniture to avoid indentation. Temperature extremes are injurious. Sweeping compounds containing sand will scratch. Mopping will dull the finish, requiring rebuffing or new wax application.

### CARPETS AND RUGS

*Description:* Carpets and rugs are fibrous floor coverings made by various weaving and back-covering processes. Wool and cotton were, and in many cases still are, the most common type of carpet fiber. However, the advent of synthetics has introduced nylon, acetate rayon, and viscose rayon as new fibers for the carpet industry. These synthetic fibers are as great a boon to durability, appearance, ease of maintenance, and cost in carpeting as they have been in their other fields of use.

Some of the more popular types of weaves employed by carpet makers are:

1. Wilton. This weave contains buried yarn beneath the pile surface. This gives the carpet a sturdier and longer life.
2. Velvet. All the pile yarns in this weave are on the surface. It is a less expensive version of the Wilton weave.
3. Axminster. Most of the pile yarn in this weave is on the surface. This is the most versatile weave for color, pattern, and design; it is also inexpensive, but not too durable.

4. Chenille. The most expensive weave, which involves two separate operations in its preparation. It is the most durable of weaves.

Wilton and velvet weaves have loop or uncut pile counterparts. Wilton loop and velvet loop are very similar to them in characteristics, except the loop-type is more durable and harder to clean. Other types of carpeting are the coated-back carpets which sometimes are made of a normal weave with a rubber of plastic coating on the back, or simply with tufts set in a rubber or plastic cement solution.

*Grades:* The various grades of carpet are determined by these interrelated factors:

1. The number of rows of pile per inch lengthwise.
2. The number of rows of pile widthwise.
3. Pile weight.
4. Pile height.
5. The type of mat construction. Generally, the firmer the backing, the more durable the carpet.
6. The nature of the fiber, whether natural or synthetic.

*Cost:* Carpet and rug costs can vary considerably. A complicated weave with an expensive fiber, laid permanently in the traditional wall-to-wall fashion, cannot be realistically compared to a coated-back scatter rug. The use for which a carpet is desired should be the determining factor for cost. Associations representing the carpet industry can provide the most varied range of carpet costs and uses.

*Characteristics:* In sound absorption, carpeting is in a class with acoustical tile and is far superior to any other floor covering in this respect. Carpeting is very slip-resistant but it tends to fray and unravel, and this can create a tripping hazard, especially on the edges. The prestige and dignity carpeting lends to the decor of an office, lobby, reception room, conference room, etc., are generally recognized. Carpets may also be permanently laid or merely placed without any physical attachment to the building. Carpets are available in a practically unlimited range of colors and designs. Daily maintenance is inexpensive.

*Maintenance:* Carpets should be vacuumed at least once a day. Spot cleaning should be performed as necessary. Dry cleaning, which can be accomplished fairly easily and quickly, is recommended at intervals of approximately four to eight weeks, depending upon soiling conditions. Wet cleaning, or shampooing, is a more difficult method of cleaning and should be performed once or twice a year by competent personnel trained in this technique.

*Cautions:* Avoid spilling liquids of any kind. Wet cleaning or shampooing should only be done by trained personnel. Frayed and unraveled edges are a tripping hazard. Watch for undue wearing and soiling in heavily trafficked and exposed areas.

### CERAMIC TILE

*Description:* A hard floor tile is made of clay which is kiln-baked to the hardness of stone. The two types of ceramic tile are glazed and unglazed. The grout, or binding cement, is available in acid-resistant, stain-resistant, and standard types. Ceramic tile is generally found in areas that are subject to many spillages and where rapid and easy cleaning is desired, such as rest rooms, kitchens, cafeterias, laboratories, and workshops.

*Cost:* Ceramic tile has a high installation cost but is very inexpensive to maintain.

*Characteristics:* Various coloring processes make a great many colors available. Glazed tile has an impervious, glassy finish much the same as glazed china. The unglazed tile has no surface finish and is the same throughout. This tile is durable, long lasting, and very easy to maintain.

*Maintenance:* Daily cleaning should be accomplished with a wax dressing for dust mopping, dusting, and sweeping. Oily dressings are not recommended. Clean with a nonalkaline synthetic detergent. Soap tends to build up films which are dust catching and slippery. A penetrating floor seal is definitely recommended. Waxing with a water emulsion finish is desirable in some cases. Waxes with a colloidal silica, nonslip component often make hard, smooth surfaces less slippery than before the application of the wax.

*Cautions:* Alkaline cleaners cause disintegration of the grout. Grout is adversely affected by acids (however, acid-resistant grouts are available). Grit, abrasive powdered cleaners, and other cleaners sometimes causes scratching.

### CONCRETE

*Description:* Concrete consists of a chemically bonded mixture of cement, sand, and gravel. It can be of ordinary hardness or compacted for extra heavy-duty use. Concrete is usually gray to white in color, but it may be pigmented to obtain various colors.

*Grades:* Concrete grades are determined by the installation. This entails such factors as thickness, percentage of cement to sand and gravel, type of finishing process (machine trowel, broom finish, float finish, etc.), and location.

*Cost:* Concrete costs are mainly dependent upon the thickness of the slab, type of finish, and percentage of cement. This involves only the slab cost and does not include labor, ground or surface preparation, forms, or reenforcement. Other variables affecting cost include distance to a mixing plant, the quantity of concrete being poured, forming requirements, the type of joints (such as expansion) and labor rates. Pigments, hardeners, curers, and seals will also increase the cost.

*Characteristics:* Concrete is the hardest and most durable floor surface, if properly cared for. It is also inexpensive to install. One of the inherent disadvantages of concrete floors, perhaps with the exception of good portland cement concrete properly installed, is its dusting or blooming.

*Maintenance:* New concrete should be properly cured. There are several ways to promote the proper curing of newly laid concrete. In the past, it was often done by covering the concrete floor with about one inch of water or by covering it with strips of wet burlap bags or kraft paper.

In recent years, a better method of curing has been developed through the use of a polymeric concrete curing aid that is sprayed or mopped directly on a new concrete surface an hour after troweling.

One or two coats of this curing aid is sufficient to promote the complete cure of the floor. The polymer forms a thin membrane over the floor and promotes more efficient chemical reaction in the setting of the concrete, thus effecting a more complete cure. Simultaneously, the curing aid helps to seal and dustproof the floor. If a sprayer is not available, the combination curing aid and seal may be applied with a mop or lambswool applicator. This can be done about four hours following the final troweling or as soon as the surface is hard enough to permit use of an applicator.

The action of the curing aid is such that curing goes on thereafter for many weeks and produces the maximum structural strength. This seal should be nonskid and must easily accept other surface finishes and filling materials over itself after the cure is complete. The seal will prevent dusting or blooming. It also prevents the penetration of greases into concrete, or the adhesion of paint, soil, or plaster that may fall from walls or be tracked in during construction.

The cost of maintenance of a floor that has been covered with a polymeric curing aid an hour after troweling is tremendously reduced during the decades that follow.

About sixty days after the curing aid has been applied, the floor should be cleaned and one coat of a permanent concrete seal applied. On old or properly cured concrete floors, it is important that the floor be carefully cleaned before the seal is applied. You should wet the floor with hose or mop, then apply a synthetic powdered concrete cleaner or a liquid detergent to the surface. Let it stand, then scrub the floor with a floor machine or deck scrub brush. Pick up the dirty solution with a wet vacuum. Rinse, then vacuum again.

It is important that the floor be thoroughly dry before the new seal is applied. If no wet vacuum is available, use a squeegee to remove the water from the floor. If the floor has not been evenly laid, or if it was not designed to drain properly, extra care has to be taken to see that the squeegee removes all water. Let the floor dry overnight with no traffic before applying the new coat of seal.

Seals should not be applied on a rainy day or a day with very high humidity. Humidity directly affects the drying time of the seal. A sealing operation is also aided if the area being sealed is well heated. Spread the seal evenly with a lambswool applicator or mop, being careful to "feather" overlapping strokes to avoid lap marks or an uneven appearance.

If for some reason a new concrete floor is to be sealed, but no combination curing aid and seal used, the floor should be neutralized or "etched" with 10 percent muriatic acid solution before sealing. Mop on the solution and allow it to remain for five to fifteen minutes, pick up, and rinse *thoroughly* with clear water.

After the neutralizing or "etching" is completed, a penetrating type of seal should be used as a primer coat at this stage. After this primer coat has dried, a permanent seal should be applied. Application procedure is the same as described above.

In the sealing of old concrete floors, the first step is to remove the soil. Apply detergent that will chemically, as well as mechanically, clean the floor surface. After the detergent has been applied to the floor, scrub the floor with a floor machine, using a wire brush. If you do not have a floor machine —and any building with a large area of flooring, whether concrete or any other type, should have one—you may use a heavy deck scrub brush for this cleaning.

If you are cleaning an old concrete floor where there is a heavy crust of grease, as in a machine shop, use a strong detergent. In severe cases it may be necessary to scrape the floor first to get the top crust off. Then proceed with the cleaning operation just described. *Be sure to remove all grease deposits.* In many cases it is wise to clean the floor, then wait several days to see if oil comes to the surface of the floor by capillary action. If it does, do the cleaning operation a second time. After cleaning, the seal is applied as above.

Neutralizing is not usually necessary with old floors, although it certainly can do no harm. Sometimes it is considered a good safety precaution. If the floor is smooth troweled, it *must* be neutralized to open the pores and provide a good bind for the seal.

A routine maintenance program should be utilized for concrete floors, sealed or unsealed. If the concrete floor is subject to heavy soiling or heavy traffic, it will have to be cleaned more frequently, with a heavy-duty cleaner. Set up a schedule for doing this cleaning job at regular intervals. Some plants clean their concrete floors once a month to remove grease and provide maximum cleanliness. Others clean much less often, depending on the kind of use the floor gets and the level of maintenance desired. On the other hand, in most food-processing plants, schools, hospitals, and in some factories, cleaning is done daily.

To clean the floor, move as many items of furniture out of the way as is practicable. Dust mop the floor with a treated dust mop. This prevents frequent changing of cleaning solution and rinse water. Then apply the cleaning solution to the floor with a mop. Let it stand for four to five minutes. Scrub to remove soil. Pick up the solution with a wet vacuum, a squeegee, or a mop. Rinse the floor with a hose or wet mop. If you use a mop, change the rinse water frequently.

Where high gloss and added protection is desired, a sealed concrete floor may be waxed regularly. (Unsealed floors should normally not be waxed, but should be sealed if more gloss and protection is desired.) This is especially recommended in personnel areas such as lunch rooms, locker rooms, etc. Contrary to popular opinion, waxing does not create a slip problem on sealed concrete floors if a slip-resistant wax is used.

The waxing procedure for sealed concrete floors is the same as for composition floors. If certain spots or lanes are subject to very heavy traffic, you may want to clean and wax them more often, taking great care to feather out the wax so that it joins the old coat smoothly. For this purpose, it is permissible to dilute a polymeric floor wax somewhat with water and spread a thin coat over the trafficked area every few days.

Concrete is alkaline, and when new the lime salts are very active. Usually for a period of six months the presence of these salts will prevent proper drying and adhesion of the conventional type of floor paints. There are many pros and cons about the painting of concrete. Generally, a clear seal is recommended instead of a colored seal or paint. However, many colored seals are available if you especially desire to make a floor blend into a decorative color scheme. Colored seals may be applied over cured concrete, using two coats of the colored seal. Or, if the floor has not been cured and is very smooth or glazed, you should clean and etch the floor and apply two or three coats of colored seal. The first coat may be thinned with turpentine for better adhesion. Paints and colored seals are surface finishes and should *always* be protected with wax if heavily trafficked.

Severe crazing and cracking of an old concrete floor may require grinding off the top surface. This is then replaced with a new topping. However, this is seldom necessary if good patching is done. Use a regular surface mixture of concrete as recommended by your concrete dealer. Clean out all holes and cracks. Square off the bottom edges of the hole so that new cement poured into it is actually a plug set down into a chiseled-out hole, rather than just a surface layer. The hole and the surrounding area should be saturated with water for several hours before patching.

If the concrete floor is sealed, patching should be done at a time just before you plan to strip off the old seal and reseal the floors.

*Cautions:* Avoid strong acid and alkalis. Seal to prevent dusting.

### CONDUCTIVE FLOORS

*Description:* Conductive floors are employed in areas where explosive vapors are present. Sparks resulting from the accumulation of static electricity constitute a very real hazard in these areas. Generally, these floors are found in the operating and delivery rooms of hospitals where explosive anesthetics are used, but industries dealing with explosive vapors or substances also utilize conductive floors. At present, there are quite a few specially compounded proprietary flooring materials which meet the requirements for conductive floors.

*Grades:* The resistance of a conductive floor should be less than 1 million ohms and more than 25,000 ohms. The National Fire Protective Association recommends a standard special tester for measuring the conductivity of floors.

*Cost:* Conductive floors are available in ceramic tile, terrazzo, and oxychlor in hard floors, and rubber and various types of vinyl tile in soft floors; the price of a conductive floor will be only slightly higher than the cost of the selected floor type.

*Characteristics:* Conductive floors have the same characteristics as floor surfaces in which no conductive properties are incorporated, plus the desired quality of conductivity.

*Maintenance:* Conductive floors should be mopped each day or after any spillage or use of the floor. Soap is *not* recommended for such cleaning as it leaves a film on the floor. A synthetic detergent is the correct material for cleaning conductive floors. *Conductive floors should not be sealed or waxed.* Although conductive waxes and seals are available, the risk involved in the floor retaining its conductive properties by incorrect application or the erroneous substitution of a nonconductive seal or wax is neither worth the savings in floor wear nor the improved appearance.

*Cautions:* Metal floors, although completely conductive, present another hazard, as they can cause a severe shock from faulty electrical equipment or wiring. Conductive floors must be cleaned regularly. Avoid soap and other film-forming products.

### CORK

*Description:* Ground cork bark is molded, compressed, and heat-treated, and made into sheets or tiles. Tiles are usually $\frac{1}{8}$- to $\frac{5}{16}$-inch thick, with either plain or tongued and grooved edges. Different shades, from tan to brown, are obtained by changing the temperature levels of the heat-treating process. Normally available in 9-inch square tiles.

*Cost:* Installed cost for 9 x 9 tiles, 3/16-inch thick, would be about twice that for asphalt tile, depending upon wage rates and the type of surface on which installation is made. Factory-finished cork will increase prices.

*Characteristics:* Cork is an excellent acoustical insulator. Its thermal conductivity is very low. Cork is the "springiest" of resilient floors but does not have quick and permanent recovery from static loads. However, its maximum static load without permanent indentation is three times that of asphalt tile.

*Maintenance:* A neutral cleaner should be used to clean cork floors, preferably a synthetic detergent, which cleans well in soft and hard water and leaves no scum. Solvent cleaners may also be used to clean cork tile floors. Factory finishes are available, or cork may be sealed with a penetrating sealer. Use solvent or water-emulsion waxes. To remove stains or discolorations, buff with an abrasive pad or dry #0 steel wool pad. Sanding may even be necessary for badly discolored areas. Sanding is best accomplished with #00 grade sandpaper under a regular floor machine. Do not attempt to sand a cork floor with a drum sander unless you are an expert with this type of equipment. Sealing is definitely required after such abrasive treatment of cork floors.

*Cautions:* Avoid greases and excessive water. Avoid alkaline cleaners. Do not wax unless the cork is factory finished or well sealed. Do not drag heavy or sharp objects over cork. Cork naturally tends to fade and discolor.

### GRANOLITHIC

Granolithic floors are similar to terrazzo, except that granite chips are used instead of marble chips, which are used in terrazzo floors. These type of floors use no metal dividers, although they have expansion spaces which run end to end. For maintenance, see the section on terrazzo floors.

### LINOLEUM

*Description:* Linoleum's primary ingredients are linseed oil and a filler. Originally, the filler was cork, but now manufacturers have largely replaced cork with wood flour, whiting, and other inert materials. This mixture is compressed onto a backing, and the linoleum is then heat-treated to form the finished product.

Linoleum is produced in sheets and tiles. The sheets are up to 6 feet in width, the tile 9 inches square, and both are made in thicknesses from $\frac{1}{16}$ to $\frac{1}{8}$ of an inch. An excellent range of colors and designs is available.

*Cost:* Installation costs would be about the same as cork and would vary depending upon the thickness, grade, and subfloor surface, and labor costs. The thinner tiles cost about the same as the thin sheet linoleum. The thicker tile costs about five to ten cents more per square foot than the same thicknesses in sheet linoleum.

*Characteristics:* Static load capacity of both tile and sheet is good (75 pounds per square inch). There is a hard linoleum tile available with excellent static load characteristics (200 pounds per square inch). Linoleum's grease resistance is excellent. Its resilience, quietness, and thermal conductivity is good, and it is available in a wide variety of colors and patterns.

*Maintenance:* Mild soaps or mild detergents that are not alkaline, or only very mildly alkaline, should be used for cleaning. Avoid ammonia-type cleaners as they may cause discoloration. Linoleum should be rinsed thoroughly after cleaning. Do not flood the floor as linoleum will deteriorate after repeated use of excess water. Linoleum floors should always be waxed after cleaning with a good water-emulsion wax, polymeric floor coating, or solvent wax. Keep cleaning frequencies to a minimum, as overcleaning reduces the life of linoleum. Abrasive-type cleaners may be used for spot cleaning stains, but are not recommended for general use.

*Cautions:* Avoid alkaline cleaners. Avoid excessive use of water.

### MAGNESITE

*Description:* A hard floor very similar to terrazzo. The binder is magnesium chloride with fillers such as cork, asbestos, wood flour, sawdust, and sand. Grades and cost of magnesite vary according to the filler used.

*Characteristics:* The great variety of fillers that may be used to manufacture magnesite offers a wide range of appearance, durability, and resilience. The harder the filler, the more durable and less resilient the floor.

*Maintenance:* The floor should be dusted, swept, and mopped regularly. Because of the nature of most of the fillers, magnesite floors are sensitive to too much water. Magnesite floors should be sealed to prevent blooming or dusting. Regular waxing is also recommended.

*Cautions:* Avoid excessive water. Avoid acids and alkalies.

### MARBLE

*Description:* Marble is a naturally occurring rock composed of carbonate of lime. The rock is cut, polished, and laid to form a smooth surface. Most of the finer marbles are used for walls, window sills, and other decorative purposes. The well-known travertine marble and honed marble are used for floors. The impurities in marble cause the variety of colors. Marble without these impurities is white. The various decorative schemes create the different grades and costs of marble.

*Characteristics:* Marble is found in solid colors such as black, red, and green. There are variegated colors with unusual swirls and wisps that make very beautiful patterns and designs. Marble in all its color forms is one of the most beautiful floor surfaces. The travertine and honed marbles are more rough and porous on the surface and, hence, are safe for foot traffic. They also have less tendency to scratch and wear than other types of marble.

*Maintenance:* Marble is very expensive and very easily damaged by improper care. Therefore, the maintenance of marble floors must be excellent. Regular dust mopping with a wax-treated dust mop should be done at least daily. For removing surface dirt, mopping with a synthetic detergent is recommended. Floors should be sealed and waxed with a slip-resistant wax. For stubborn soil, a mild abrasive may be used.

*Cautions:* Avoid acids of all types, as the calcium carbonate, of which marble is composed, will react with an acid. Even fruit acids will ruin the finish on marble floors. Avoid staining with oils, fats, and greases. Avoid harsh abrasives, especially those containing ammonia and dyes.

### METAL

*Description:* Metal floors are fairly unusual and are generally considered as special-purpose floors. They are almost always ferrous in nature as most other metals would be too expensive or lacking in the properties that are so valuable in steel and iron. The main types are sheet, plate, and grating. Some of the special purposes for this type of floor would be for exceptionally heavy machinery or to cover old floors that have been worn beyond the state of acceptable use. Grating is used where there are excessive spillages. You may also find metal used for temporary floors or where readjustment or redesign of the basic floor plan is desired.

*Maintenance:* Plate or sheet-type metal floors can be swept or scrubbed. Steam cleaning is probably the best method for cleaning metal floors.

*Cautions:* Metal floors are especially dangerous where there is a possibility of electrical shortages. Metal floors can be very slippery, especially where spillages of oil and greases occur.

### RUBBER

*Description:* Natural, synthetic, and reclaimed rubber is made into tiles and sheets. Inert fillers and color pigments are added, and the mixture is heated and rolled out under pressure. Most rubber flooring of today is synthetic rubber, which was developed and perfected during World War II. Rubber is one of the most resilient flooring materials, and one of the safest of the resilient-type floors.

*Cost:* The installation cost of sheet and tile is approximately the same. The cost could be about 50% greater than asphalt tile and would vary primarily on the thickness of the tile or sheet and the surface or subfloor on which it is installed. The cost varies from about 85c to over $1. per square foot depending on these different compositions.

*Characteristics:* The high resilience of rubber makes it one of the most comfortable floors for foot traffic. It is a very quiet floor and one of the most durable of the resilient-type floors. There are many colors available in either a solid, variegated, or mottled design.

*Maintenance:* Rubber floors should be mopped daily with a synthetic detergent. Soaps are not recommended for rubber floors. It is definitely recommended that rubber floors be waxed with a polymer water-emulsion type finish. Occasionally, the color pigment will come loose on the surface of the floor and will appear to "bleed." This should be scrubbed up with a good wax stripper and rewaxed or sealed with a water-emulsion type of seal.

*Cautions:* Avoid oils, grease, turpentine, petroleum solvents, and carbon tetrachloride. Do not use oily dust mops or sweeping compounds. Rubber floors should not be laid in bright, direct sunlight.

### STONE

Stone floors are generally of granite, slate, or flagstone (crab orchard stone) and are used in lobbies, patios, courtyards, stairs, and other outside areas. These types of stone are very durable and also very hard. Stone surfaces, which are not directly exposed to the elements, should be sealed with a penetrating type seal to prevent "blooming" or dusting. On inside areas, stone may be sealed with a penetrating type seal as a base coat and then covered with a surface-type of seal.

### TERRAZZO

*Description:* Terrazzo is a mixture of marble chips and portland cement. It is often separated by metal strips. The mixture is poured and troweled out and allowed to dry for five or six days. It is then ground and polished with a stone grinder. The appearance of terrazzo varies with the color of the marble used. A new and thinner type of terrazzo is the "Epoxy Terrazzo." The epoxy matrix is much denser and less porous than the portland cement.

*Characteristics:* Terrazzo is a very beautiful floor with marble, in its variegated colors, as the major portion of the mixture. It is very easy to maintain and, if installed and maintained properly, it is very long lasting. Terrazzo is a relatively inexpensive hard floor, and is very durable.

*Maintenance:* When new, terrazzo should be mopped or scrubbed on a daily basis. After a few months, the cement or grout will have completed its curing process, and the mineral salts will no longer dull the finish. Sealing will prevent the dulling of this salt film and will actually aid the curing process. Use neutral soaps or detergents. Acidic or alkaline cleaners will attack the floor and cause chipping and excessive wearing. On old or cured terrazzo, a treated (nonoily) dust mop should be used daily. The floor should be mopped at least weekly. Waxing should be done with a slip-resistant wax. Frequent buffing of waxed floors removes traffic marks and restores the sheen and luster.

*Cautions:* Avoid alkalies and acids. Avoid inorganic and crystallizing salts. Avoid oily sweeping compounds and treatments. Do not use steel wool for scrubbing or buffing as small pieces of steel left on the floor will rust and discolor the terrazzo.

### VINYL ASBESTOS TILE

*Description:* Vinyl asbestos tile is somewhat similar to asphalt tile. The binder is a vinyl-type resin. Other ingredients are asbestos, color pigments, and inert fillers. The tile is generally 9 inches square, sometimes 12 inches square.

*Cost:* The installed cost of vinyl asbestos tile same as rubber (see above) depending primarily upon the thickness of the tile and the surface or subfloor on which it is installed.

*Characteristics:* Vinyl asbestos tile is inexpensive but its resistance to permanent indentations is low (25 pounds per square inch). These tiles are available in many colors and designs. They are very resistant to acids, alkalies, greases, oils, soil, and solvents.

*Maintenance:* The floor should be dust mopped or swept daily. It can be mopped as often as necessary with almost any soap or synthetic detergent. It should be waxed regularly with a natural wax or polymer-type emulsion finish. Older floors should be sealed with a water-emulsion seal to fill the pores of the tile and to provide a new, smooth surface to facilitate cleaning. Waxed floors should be stripped at least after every six waxings or semiannually, then sealed (if necessary), and rewaxed.

*Cautions:* Avoid abrasive cleaners on unwaxed floors. Use floor cups on thin table and chair legs to avoid permanent indentations.

### VINYL TILE—HOMOGENEOUS OR PLASTIC

*Description:* This vinyl, sometimes referred to as "pure" or "100 percent vinyl," is somewhat similar to rubber tile. However, it does not have as much inert filler and pigment, and the binder is a vinyl-type resin. The tile is generally sold in 9-inch squares, sometimes 12-inch squares, and it is also available in sheet form.

*Cost:* The installation cost of homogeneous vinyl tile may be twice as much as vinyl depending primarily upon the thickness of the tile and the surface or subfloor on which it is installed.

*Characteristics:* This homogeneous, or plastic vinyl tile, is very expensive. Its resistance to permanent indentation is very high (200 pounds per square inch). This tile, like vinyl asbestos tile, is available in many colors and designs and is very resistant to acids, alkalies, oils, greases, soil, and solvents.

*Maintenance:* Homogeneous, or plastic vinyl tile, should be maintained in the same manner as vinyl asbestos tile, above.

*Cautions:* Avoid abrasive cleaners on unwaxed floors. Dirt and grit will cause scratching and it quickly shows subfloor irregularities.

### WOOD

*Description:* (1) Oak—The most extensively used wood floor. There are two main groups of oak, white oak (nine species) and red oak (eleven species). There is little difference in quality between these two groups. Oak floors owe their beauty to the open cellular construction, which gives the grain its unique character. (2) Maple—The next in popularity of wood floors is maple. There are two species used for flooring, sugar maple (also called sugar tree) and black maple (also called rock maple). The wood is close-grained with small and evenly distributed pores. It is very resistant to abrasion and very hard. For this reason, it is an exceptionally good floor for heavily trafficked and heavy-wear areas. (3) Birch and Beech—These two woods are very similar. They are very hard and abrasive-resistant. Like maple, these two are close-grained with very small and evenly distributed pores. (4) Pecan—The hardest of all wood flooring is the pecan wood. A very durable wood, it is also close-grained. (5) Other species—Walnut, cherry, ash, hickory, teak, and cypress. These woods are not too popular because of their scarcity and high cost. The use of soft woods for the sake of initial economy should be discouraged.

*Maintenance:* The correct maintenance of wood floors is of paramount importance, if their beauty and life is to be preserved. For old wood that is in poor or fair condition, clean with a good solvent-type stripper, using a floor machine with a #3 steel wool pad or abrading-type pad. For oak or open-grain wood, apply a filler, then a penetrating sealer. Wax old wood floors with a solvent-based wax, or, on properly sealed floors, with a water-emulsion wax. New floors should be sealed and waxed as above. All wood floors should be dust mopped, swept, or dry vacuumed daily. Sealed and waxed floors should be damp mopped to remove excess soil.

*Cautions:* Avoid excessive water as the wood will tend to warp and cup after saturation. Avoid oils and greases as they will stain and streak the wood.

### WOOD BLOCK

*Description:* Wood block floors are wood floors with the grain end or cross section on the surface. As we might expect, this cross section would absorb a great deal of water and moisture, so wood block floors are creosoted (or similarly treated) to prevent warping and buckling. Wood block is almost exclusively an industrial floor.

*Maintenance:* Wood block floors should be swept daily. Sweeping compound is recommended for dust control in manual sweeping. Scrubbing should be done when the floor becomes heavily soiled. The scrubbing solution should be vacuumed up immediately after scrubbing to prevent the saturation of the wood block floor. Mopping is recommended for spillages. This floor, because of its block construction, is very easy to patch by removing the old or broken block and replacing it with a new block.

*Cautions:* Avoid excessive water and liquids. Near composition floors, wood block floors should be kept very clean. Entrances to composition floor areas should be covered by mats and runners to prevent trackage of the creosote or other similar treatment.

# 38 Floor Care Methods

## Finished Floors

THE most important types of finished flooring are asphalt, linoleum, rubber, terrazzo, vinyl and vinyl asbestos. It is generally desirable to protect these floors with some type of finish; this includes vinyl flooring, despite the claims of its manufacturers when this type of material first became available. The most important reasons for using floor finishes include protection from damage by traffic and environmental elements, improvement of sanitation, maintenance of more readily cleaned surfaces and improved appearance.

It should be noted at the outset that the use of protective finishes is *not enough.* For adequate and economical protection, walk-off matting should be installed at all entrances where any soil is likely to be carried in. This includes not only entrances from out-of-doors but also entrances from manufacturing, lab, or shop areas. Further, it is very important to consider the need for supplementing normally-used walk-off matting at times when new construction or building renovation may be in progress. Floors should be protected from the mechanical damage which may result from the dragging of desks and heavy furniture from place to place by occupying personnel. Special equipment is available which makes a simple and easy job of the movement of desks, file cabinets and other heavy furniture. A device by means of which one person can safely move a desk from place to place without damage to furniture or floor is available from several manufacturers.

Another consideration in the protection of floors from unnecessary damage is the prevention of the rusting of metal surfaces in direct contact with the floor. Often this type of damage results when water is allowed to run under the bottom of a file cabinet and the sheet metal rusts, quickly producing permanent stains in the pores of the floor. Naturally one preventive measure is the proper training of custodians, but this should be supplemented by the use of plastic corners or angle strips installed on or under the cabinet so that the sheet metal does not touch the floor.

## Dry Floor Cleaning

Beyond the protection of floors from unnecessary soiling and damage, the major factors in floor care are the removal of heavy soil by some form of sweeping or vacuuming, wet cleaning of the surface at the proper intervals and with necessary accompanying mechanical processes, and the application and maintenance of floor finish.

The most commonly used and generally most suitable tool for dry floor care is the treated dust mop. There are many variations to the basic tool, which usually consists of strands of yarn some five inches in length assembled into a mop which may vary from a small wedge shape up to lengths of six or more feet. For dust mops of three feet or greater length, it is generally desirable that the handle be rigidly fixed. For narrower dust mops which are to be used in congested areas, it is desirable that the mounting be designed to allow the mop to swivel through 360°.

The dust mop should normally be treated with a compound which will cause dust to adhere to the strands of the mop and thus minimize the tendency for dust to be distributed into the air during the mopping operation. Mop-treatment compounds generally consist of formulations of light petroleum fractions. It is important that the mop treatment selected not be too oily or too solvent in nature, for this can damage finishes and surfaces or can produce slippery floors. Mop-treatment formulations sometimes contain quaternary ammonium chlorides or other bacteriostatic chemicals.

Variations on the classic dust mop include treated cloths of a patented form which are applied to blocks and handles so that they are used in the same manner as dust mops. The contention of the manufacturers of this type of equipment is that better sanitation and comparable costs are obtained through folding these cloths in different ways to present clean surfaces as those already used become loaded with dust. Another substitute for the dust mop is a device which uses a disposable so-called non-woven cloth for the dust-collecting surface. Both of the devices appear to do a better job of collecting very fine dust than will the string dust mop; however their use becomes less advantageous when trash, paper scraps and larger sized particulate matter are present.

For dense coarse particulate soil, such as sand encountered in large quantities, the use of a push broom and sweeping compound may be advisable in place of dust mopping. Proper selection and proper use of sweeping compound are both important if serious damage to resilient tile surfaces is to be avoided. The dust control component in the compound should be predominantly waxy, rather than oily. Oils can seriously soften and damage tile, especially that composed of asphalt or rubber. Custodians are frequently allowed to make the mistake of scattering sweeping compound over the entire area which is to be swept, a practice which is wasteful of both time and compound and which is likely to result in staining of the floor if the compound is walked on. Actually, just a few handfuls (the total amount being proportionate to the amount of dust or soil) should be put down along a line the width of the area to be swept; then this line of compound is moved forward with each stroke of the broom in the sweeping operation. Working in this fashion, the custodian does not accumulate a large volume of sweeping compound to be picked up with the dust pan and discarded. If he finds that the compound is becoming overloaded with dust before he reaches the end of the area which he is sweeping, a small amount of new compound can be added, or the material already accumulated picked up and all new compound put down. Naturally, it is important that the type of broom fiber used be selected in accordance with the type of soil likely to be encountered. For instance, for a mixture of coarse and fine dirt, the best broom would probably be one with a Tampico center and a casing of horsehair; or a mixture of horsehair and bristles may be used.

In an industrial clean-room or critical areas of a hospital, it may be necessary to avoid dry sweeping and use a special vacuum cleaner. This is a much slower procedure than sweeping, and that fact must be recognized in the establishment of time standards. Two suitable types of vacuum cleaners are available: those which have a dust separation and collection device located remotely from the area being cleaned (central systems) and those which use special filters capable of removing better than 99% of all particles larger than about 0.5 of a micron. Each system has its own advantages and disadvantages, and these should be weighed carefully in deciding which to select. Central systems generally have the disadvantage of requiring the custodian to work with rather long sections of hose; also careful design is required to insure a sufficient number of risers and connections and a system of piping which does not result in too much loss of air flow efficiency.

## Wet Floor Cleaning

Daily wet cleaning (possibly with a germicidal detergent) is normally the rule in hospitals or others areas where rigid microbiological control is required. Frequent wet cleaning may be required during periods of unusual soiling. Except for the foregoing conditions, the frequency of wet cleaning is normally kept to a minimum. Well-designed battery-operated auto scrubbers are now available to apply clean solution to the floor, scrub with a mechanically driven brush, and pick up the soiled solution, all in one operation. Before the purchase of such equipment, careful evaluation of its practicability in the situation is important. If large, open-floor areas exist without obstacles or congestion, if frequent wet cleaning appears to be necessary or desirable, and if the inevitable removal of a significant amount of the floor finish layer with each use of the equipment can be tolerated—if all these conditions exist or there is an important combination of at least two of these conditions, then the purchase and use of an auto scrubber is probably desirable. If one of these mechanical devices is used, then the largest equipment readily maneuverable in the open space should be selected, even though there must be some supplemental manual wet cleaning in certain congested areas.

Where manual wet cleaning methods are used, they should be varied according to the amount of soil encountered. If the soil is light, damp mopping or mopping with a single bucket with a fairly light solution of neutral synthetic detergent may be entirely satisfactory. With heavier soil, it may be desirable to put down more solution and pick it up, wringing it into an empty bucket to avoid soiling the clean solution. Under conditions of very heavy soil load, rinsing will be essential; then it will be desirable to use a three-bucket system with detergent solution in one bucket, rinse water in another bucket and a third bucket into which both the mop for applying detergent and the mop for applying rinse water may be wrung.

If only cleaning of the floor is necessary, without accompanying disinfection, a neutral synthetic cleaner will normally be used. Such a product formulation is usually based primarily on wetting agents with moderate emulsifying powers which do not produce scums with hard water, as do soaps. The pH of a neutral cleaner should generally be less than ten.

The two classes of cleaner-disinfectants most commonly used are modified phenolics and the quaternary ammonium chlorides. The selection of the product should be based on the types of organisms which must be controlled and the cleaning requirements.

The remaining type of cleaning compound most frequently used in floor work is the stripper for removal of wax or finish. Generally strippers are formulations with high pH (up to about 12.0) provided by either ammonia or organic amines; sometimes organic solvents are used, but these must be carefully selected so that they do not result in deterioration of resilient tile. The so-called "metallized" floor finishes normally require the presence of ammonia in the stripper. Some companies have developed floor finishes formulated with acid systems, so that an acid-type stripper must be used for their removal; however these have not proven widely popular.

Formulations of cleaners used in auto-scrubbers usually vary from those used in manual operations in that the wetting agents employed do not produce as much foam. High foamers will fill the receiving tank of the machine too rapidly, causing trouble with the vacuum system as a result of foam and moisture getting into the bearings of the fan shaft or motor bearings.

## Waxes, Finishes and Sealers

While waxes in organic solvent systems have been used for many years to coat floor surfaces, it was not until about 1940 that water emulsion systems containing finely divided wax particles came into popular use. Various natural and synthetic waxes have been used in such formulations, but the greatest (and probably a disproportionate) emphasis has generally been placed on the content of carnauba wax, a natural derivative from the leaves of a palm plant. One of the misapprehensions widely existing about water emulsion waxes relates to the solids content: usually the total solvents in a good product range between twelve and eighteen percent, and only a portion of these may be made up of carnauba or other natural wax. Because of the tendency for waxes to produce slippery surfaces, finely divided silica may be introduced into these products, under the trade name "Ludox." Practically all of the natural waxes used on this type of floor finish tend to darken with exposure to light and oxygen, with the result that the floor coatings eventually become discolored and must be stripped and replaced for the sake of appearance.

About the end of World War II, complex polymers, some of which had been developed in efforts to produce substitutes for natural rubber, began to be substituted in emulsion form for the waxes. The most important of these products coming into early use were the styrenes and polystyrenes. The major disadvantage of these products was the tendency toward hardness, brittleness, and powdering. To distinguish them from the water emulsion waxes these products were referred to as "finishes," "polymers" or "polishes." Because of their hardness, the styrenes did not lend themselves to buffing and consequently were referred to as "non-buffable floor finish."

Later entries in the field of finishes have been the acrylics or polyacrylics and the polyethylenes. These materials are not as hard and brittle as the styrenes and possess greater toughness; they are also buffable to some degree. The acrylics generally have the disadvantage of being very difficult to remove with floor strippers. Often, combinations of the styrenes and acrylics, perhaps along with microcrystalline waxes of petroleum derivation were formulated in an effort to combine the advantage of each of the individual components. Certain types of molecular combinations are referred to as co-polymers. The trend in formulations has been away from the use of styrenes and to a greater use of the polyethylenes and the acrylics. Under average commercial conditions, the choice between the water emulsion waxes and the polymers or synthetic finishes usually favors the latter. Because the polymers form continuous surface films upon drying, rather than layers of discrete particles (as in the case of the waxes) there is less tendency toward slipperiness, the surfaces are more water resistant, and less buffing and repair are needed. An exception to this rule may be found in the case where floors are unavoidably exposed to unusually large amounts of sand and grit; under those conditions, it may be easier and more economical to buff the water emulsion wax frequently rather than to restore a polymer subjected to excessive scratching.

In the instance of terrazzo floors or of highly porous resilient floors, such as old asphalt, it is sometimes desirable to use a water-emulsion sealer. These sealers, whose solids are usually predominantly acrylics, have a greater ability to bridge floor porosity than do the finishes themselves; however they usually will not withstand the friction of traffic well and thus are used only as a base for the coats of finish which actually receive the wear. Some persons responsible for floor maintenance prefer to avoid the sealers because most of them are more difficult to strip than are the finishes; these persons simply use additional coats of the floor finish to obtain desired appearance.

In the application of water-emulsion sealers and finishes, one floor finish should never be applied over another unless it has been previously ascertained that the products are compatible. Thus, the usual first step in floor finishing is to strip the floor completely. This is accomplished with the chemical stripper used in concentrations recommended by the manufacturer, with the aid of a floor machine with suitable stripping pads and a wet vacuum and other necessary tools for picking up the emulsified floor finish and soil. If there is a heavy buildup of old finish on the floor, it is quite probable that recommended concentrations of stripper will not remove the entire coating in one application. The temptation to increase concentration beyond that recommended by the manufacturer in an effort to remove all old finish in one operation should be resisted, for to do so may result in "burning" or damaging the floor surface by excessive removal of plasticizers and pigments. The safe procedure is to re-strip those limited areas where the first application may not complete the job. Thorough rinsing of the floor to remove all old finish and stripper solution is essential to a good job of refinishing. Unless this step is properly carried out, the new finish may not bond properly or it may have a dull appearance. There seems to be no need for the "neutralization" of floors with vinegar or other acids if the floor is completely rinsed.

Generally the fastest and most satisfactory way of applying floor finish is with a mop. However, it is often desirable to use a 16-ounce rather than a larger size mop, and most workers prefer thin cotton strands (string mops) rather than heavy cord or synthetic fiber mops. The application of finish in thin coats is most important to avoid the hazards of tackiness, dullness, and slow drying.

Under reasonable conditions of temperature and humidity, a good polymer floor finish should dry in approximately 30 to 45 minutes. It should be completely dry to the palm of the hand before the second coat is applied. While either the first or second coat should be applied up to (but not on) the baseboard, all other coats should be kept about one nine-inch floor tile width away from the floor edge, since the absence of traffic in that area may otherwise permit the development of buildup. Unless the floor is excessively porous or the total solids in the finish fall below the desirable level of fourteen to eighteen percent, two or three coats of finish will normally give an excellent base.

## Maintenance of Floor Finish

In the past, the primary method of floor finish maintenance has been regular cleaning, periodic buffing, patching of heavily trafficked areas by blending in a partial wax coat with the mop, and total stripping and refinishing of the floor when indicated by the development of mottled appearance or the buildup of finish in the more lightly used areas.

Today, more and more custodial managers are coming to depend upon spray-buffing as the major approach to floor maintenance. Briefly defined, spray-buffing is a technique wherein the floor is first cleaned and then those surface portions damaged by traffic are restored by the application of a light spray of a dilute solution of the original floor finish (or a compatible substitute) and the buffing of this spray with a pad before it has dried. This technique will result in partial emulsification of the very top layer of finish, the repair of scratches and scuffs, and the removal of black marks, as well as the production of a tough surface with very high gloss. A necessary final step in the spray-buffing is dust mopping the treated area to remove residues of old finish which have been loosened by the spray-buff pad and left as a coarse residue on the surface.

The advantages and economies of spray-buffing have been well documented. Proper indoctrination and training of personnel is essential to convince the custodian of these advantages and to assure the greatest benefits from the process. It must be emphasized that spray-buffing is not a technique for coating large areas of a floor surface which actually needs the application of a full layer of finish, nor should it be depended upon to clean a dirty floor. The operator must understand the precautions necessary to avoid soiling walls and furniture with the spray and to prevent build-up in areas where the floor receives a little traffic; he must apply spray only where it is needed. He must be provided with the proper equipment —most importantly, good open mesh spray-buff pads, a good machine and suitable pad-driving block, and a convenient spray device. The proper use of the spray-buff technique can extend the time between floor stripping many fold; this in itself becomes a useful fact in the "sale" of this technique to workers. The advantage to management, of course, is the economy in time—as well as the less important economy in materials—which results.

## PROCEDURES FOR TERRAZZO FLOORS

While the use of water-emulsion sealers and finishes, along with the spray-buff technique, is generally applicable to maintaining terrazzo floors as well as resilient floors, conditions sometimes dictate other approaches. In any case, if a terrazzo floor is to be treated with any kind of finish, it is especially important that walk-off matting be provided at entryways where water might otherwise be tracked in and where slippery floors might result. If this precaution is taken, slip hazards seem to be encountered only under exceptional conditions.

The National Terrazzo and Mosaic Tile Association advocates that the use of floor finishes and sealers on terrazzo is not necessary. Yet, in practice, it is generally impossible to maintain uniformly good appearance on terrazzo floors without the use of either finishes or sealers. If the procedures already described for the use of the water-emulsion products are not to be followed, there remains a choice among three other frequently used systems of maintenance. In the past, many terrazzo floors were maintained by the application of a special soap, often based on linseed or similar oils, which left a hard water residue that was then buffed into the surface of the terrazzo. Today few authorities consider this a satisfactory approach.

The most popular type of sealer among terrazzo floor contractors is a solvent sealer based on synthetic waxes which do not discolor with oxidation and sunlight as do the carnauba and other natural waxes. This type of sealer is very easy to use and, since it can be stripped at least off the surface of the terrazzo, can seldom result in any serious difficulty. For large areas of terrazzo exposed to heavy traffic where the emphasis must be on economy in time more than on extremely high levels of appearance, this approach may be the choice; in fact, quite good appearance can be maintained with regular cleaning and buffing of the surface.

Other solvent sealers are available to provide surface coats which are relatively color stable and which can be patched with reasonable success as wear occurs. Generally these sealers are used without additional floor finish, and when traffic aisles occur which can no longer be satisfactorily spot-finished, the sealer must be removed with some type of organic solvent, sometimes a methylene chloride stripper. Stripping this type of product is a rather complicated operation requiring well-trained and supervised personnel; thus, this is probably the most expensive procedure for the maintenance of a high level appearance on terrazzo.

## RESINOUS POURED FLOORS

Several organic binders have been found suitable for the production of floors similar in many characteristics to classic terrazzo or, with the substitution of granite chips for marble chips, to granolithic floors. Since these binders are much less porous than Portland cement, many of the binder manufacturers and floor contractors advocate that they can be used without surface finishes. In practice, continued high-quality appearance seems to require the use of sealers or sealer and finish combinations. Under some conditions, a significant amount of powdering has been experienced with some of these floor types. Marble chips themselves are characterized by variable but significant amounts of porosity, and sealing may be desirable simply for protection of this aggregate, although discoloration or staining of the chips may not be noticeable if chips of mixed color are used.

One of the important advantages of the resinous poured floors, whether used with or without aggregates such as marble chips, is the reduction in weight both from the lower specific gravity of the organic binder and from the fact that these toppings can be poured much thinner than can Portland cement or classic terrazzo because of flexibility and other physical characteristics. In addition to these common characteristics of the resinous binders, there are numerous significant variations from one to another. These include variations in resistance to physical and chemical damage as well as variations in problems of application and maintenance. Careful consideration should be given to the relative advantages of individual binders before a decision is made on an installation of this type. After that, it is extremely important that the flooring contractor be experienced and qualified in the type of installation, since otherwise unsatisfactory results are practically a certainty.

It is the contention of some architects and installers of the resinous floors that long-range economies are achieved through the avoidance of floor finishes and the ultimate application of a new glaze or surface coat of the binder material itself when undesirable traffic patterns appear in the surface. It is not always made clear to the user of the flooring that many problems may be involved in this resurfacing. For instance, volatile solvents used in the binder which dissipate during the curing stage may be highly toxic, so that repair may require either evacuation of the building or rather complicated arrangements for safe ventilation. It may be necessary to restrict traffic from the areas under repair for a period of 48 hours or even more.

### CONCRETE FLOORS

Almost any non-acidic cleaning agent can be used on concrete floors without fear of damage to the flooring surface. The major problem in concrete maintenance stems from the fact that Portland cement is highly porous, so that grease and soils are readily absorbed; it is also subject to powdering and to the action of acids. Most of these problems can be eliminated through the proper use of sealers.

The only types of sealers which are suitable for use on new concrete are penerating seals which seal the pores, but which do not create surface films that may be raised by the hydrostatic pressure of uncombined water molecules in the fresh concrete. The most popular formulations of such sealers include the rubber and acrylic lattices and the silicate materials. Generally, these penetrating sealers contribute to the curing and hardening of the cement, but they do not provide significant protection against actual surface wear.

The most commonly used surface sealers for concrete are the phenolics, epoxies and polyurethanes. The phenolics are the oldest formulation of this type of sealer and have the disadvantage of considerable initial color and of darkening on exposure to oxidation and the effect of ultraviolet rays. The epoxies and polyurethanes are much more color stable, as well as tougher and more resistant to wear, than the phenolics. They—and in particular the epoxies—have the disadvantage of being much more difficult to remove than the phenolics, should that be necessary.

When a surface seal is to be used on concrete, consideration should be given to neutralization of excessive free alkali in the concrete surface. Unless this is done, there is a danger that poor bonding of the sealer will result and that the investment of time and sealer cost will be lost. This may not be necessary if the concrete is several years old and if it has been subjected to extensive washing, which may have removed the excess free alkali. As a precaution, it would probably be desirable to put down a test area of the sealer if it is planned to omit the neutralizing step.

At times a floor surface is too dense or slick to permit the seal to adhere. This situation can be improved with an etching rinse consisting of a mild acid solution. Always work with test areas in such cases.

In the application of a surface sealer, it is important that the first coat be thin enough to allow good penetration of the pores in the concrete. This can usually be provided for by cutting the first coat about one part of solvent to three or four parts of the sealer; the solvent used for this purpose should be the one recommended by the seal manufacturer. After the first coat has dried thoroughly, one or two successive coats are normally used to protect the floor against traffic. Under warehouse or manufacturing conditions, floor finishes, such as those used on resilent floors, are seldom employed. In some situations, their use is practical if a high level of appearance is desired, since they will protect the base coats of sealer and reduce the frequency of resealing.

### HARDWOOD FLOORS

Maple, oak and other hardwood floors are often used for their resiliency and for beautiful appearance. The best protection of such floors normally involves the use of phenolic or urethane sealers in solvent systems. If high slip resistance is required, as for a gymnasium floor, surface finishes are seldom used on the sealers. In other applications, either solvent waxes or water emulsion systems can be employed, since the floor is protected against moisture by the sealer itself.

### CERAMIC FLOORS

Ceramic and quarry tile floors are most commonly installed with Portland cement grout. While the tile itself is non-porous and highly resistant to the penetration of any type of soil, the grout areas present the same problems as those encountered with Portland cement floors. Floors made up of small non-glazed ceramic tiles can be successfully maintained with water emulsion sealers and finish. This approach is seldom successful with quarry tile because of the fact that such floors are normally installed where they are rather continuously exposed to moisture or to very heavy traffic. The most desirable approach in this case is to use organic binders in place of Portland cement for installation of the tile and then to avoid any surface coating. In some instances, good results have been obtained by using a penetrating sealer to close the pores of the grout, but at the same time buffing the floor vigorously to remove any sealer from the tile surface itself.

## A Note on Selection

A great many problems in floor care can be avoided through proper selection of easily maintainable types.

Sometimes the surface is the problem. Manufacturers of resilient floors are now preparing tiles that imitate travertine marble, and as mentioned above create an insoluble cleaning problem unless the cracks are filled, just as with natural "rotten marble." Embossed tiles are also very difficult to maintain; they cannot be spray-buffed well, and the raised portions cause any finish—and the tile as well—to wear and discolor more quickly.

Coloration is a constant problem—colors that are too light or too dark show marks and scuffing too easily in the first case, and show soil too easily in the latter.

The service operator is faced with many problems in cleaning, as is management in trying to provide the means or technique to solve them. Let's at least require our architects and interior decorators to avoid as many as we can!

# 39

# Solving Floor Care Problems

BELOW are a number of the problems, with their prevention and solution, often associated with floor care. Obviously, it cannot be a complete list, as there are literally countless problems that one encounters in floor maintenance. A special case, for example, would be the staining of resilient floors from the black solvent tracked off a wooden block floor (this can be controlled by sealing the approach area of the wood block floor with polyurethane seal, using plenty of matting and keeping it policed, and stripping the resilient floor more frequently than would normally be the case).

Problems listed in one category may have application in another. For example, conditions relating to the finishing of resilient floors will also be applicable to the finishing of terrazzo and other types of floors. The causes of a powdering floor finish applied on terrazzo will typically be the same as for a finish applied on a resilient floor.

## Resilient Floors

### INDENTATIONS

Permanent depression of the surface may be caused by one of a number of agents:

1. Furniture and equipment. This can be alleviated by moving the furniture or equipment an inch or so occasionally, such as when the floor is being stripped; or by using plates or gliders to spread the weight over a larger area of the floor.
2. Movement of furniture and equipment can cause grooves if the proper equipment for moving is not utilized. Desk-moving and file-moving equipment, for example, should be on hand in most organizations.
3. Women's shoe heels of the stiletto type can cause severe damage to floors. Some organizations have actually prohibited the use of such heels, while others suggest the use of larger heels or even provide heel caps which tend to enlarge their size.
4. The type of resilient material is an important factor. For example, asphalt tile has a much poorer recovery rate from indentation than does vinyl tile.
5. Resilient tiles are softened or even dissolved by solvents and oil; thus the oily type of dust mop treatment should be used sparingly so as not to soften the floor and cause numerous indentations.

### DAMAGED TILES

A great deal of damage to resilient floor tiles is caused by chemical agents, such as strong caustic or alkaline solutions or by brittle materials such as varnish or shellac. The use of any of these materials should be avoided, since they will cause the tiles to curl or pucker, crack, shrink, or round off at the corners. Neutral synthetic detergents should be used for general cleaning, and when stripping is required, specially formulated chemicals properly diluted should be used. Where floors have been damaged to a considerable degree, it may be necessary to replace them or to consider covering the area with carpet or other material.

### LOOSE TILES

The principal cause of loosened tiles is first their warping or curling through the use of excessive water and/or improper chemicals as described above, then the progressive deterioration of the mastic by additional water running under the tile. Floors should not be flooded for general repetitive cleaning, and the solution should be removed as quickly as possible. Where a floor shows some signs of damage, the life of the floor can be extended through the use of wet vacuums, which will remove a good deal more of the water than a mop and wringer.

### COLORS BLEED

The use of excessive oils or solvents, as in some types of dust mop treatments, soften the tile to such an extent that the colors run; this is especially prevalent in asphalt and rubber tiles. This condition may also be caused when, through error, a solvent-type wax is used (this type of wax, normally yellow in color, and of viscous or pasty consistency, should be reserved for use on wooden or cork floors).

### EXCESSIVE POROSITY

The floor surface may be damaged miscroscopically by the use of improper chemicals, such as harsh alkalies, or even normal detergents which are used in concentrations which are too high. The result is a "drinking up" of the floor wax or finish, so that it becomes difficult to achieve good gloss or surface protection. A similar effect is created when floor pads, either steel wool or synthetic, are used which are too abrasive. In this case, it is desirable to use a water-emulsion sealer to fill the pores after each stripping.

### TACKY SURFACE

This condition is caused by the improper selection or application of the wax or finish. Poor-quality waxes may be tacky because of the raw materials in the formula. When wax films are applied that are too thick, they may become tacky. A residual soap film, remaining because of improper rinsing, may also feel tacky. Improper cleaning may leave chemicals, dirt, and other foreign materials which can cause stickiness.

### SLIPPERINESS

So called "fast floors" may be caused by the use of an improper wax formulation or the application of overly-thick films of wax. A fine film of paper or chemical dust can also make a floor slippery, as can the presence of water or other liquids, either in droplet or film form. Proper protection of these areas with matting, and good cleaning techniques, will eliminate these latter problems.

### FUZZY SURFACE

Tiny particles of paper dust, fabric lint, chemicals, or other foreign materials may sometimes cling to a floor and present a fuzzy or wooly appearance and feel. This can be caused by a finish that is too tacky or soft, which is solved by the proper selection and application of the finish. Another cause is static electricity, and sometimes this may be cured by a change in the humidity, or the application of an anti-static agent to the cleaning solution; if these fail, then the area will simply need much more regular cleaning than before, preferably with the use of a wet vacuum or automatic scrubbing machine.

### DARKENING

Darkening of the floors is usually caused by a build-up of successive layers of wax or finish, mixed with soil; this would be especially prevalent around baseboards and under furniture. It is the result of improper floor care techniques, principally involving the placing of wax and finish in areas where it will not be worn off. A general darkening over the entire floor area may be caused by a redistribution of soil where the spray-buffing technique is mistakenly used for cleaning purposes, or where the resin-emulsion system is used; the dinginess is simply caused by successive layers of dirt and finish.

### BASEBOARDS SOILED

The soiling of baseboards is a reflection of poor mopping and waxing techniques. The untrained operator splashes water, wax, and chemicals on the baseboard, where they build up into an unsightly condition. Prevention through training is the best approach to this problem, since correction requires a good deal of labor through the use of a baseboard-scrubbing attachment to a floor machine, or a good deal of hand work.

### BLANCHING

The over-use of stripping solution can actually leach the color out of resilient tile, leaving a bleached or whitened appearance. This might be especially noticeable along baseboards, perhaps where a custodian has used stripper in excessively high dilution—or even straight from a container—in an attempt to dissolve a heavy build-up (the proper approach is to make successive scrubbings with stripper of the proper dilution). A general washed-out appearance can be caused by excessive application of detergents on the basis of frequency, concentration, or formulation.

### TRAFFIC LANES

Patterns worn in the floor by foot traffic may be caused by inferior finishes, which powder under pressure; or by improperly-maintained floors where dry buffing or spray-buffing is not provided to heal damaged surface areas. Where spray-buffing is not used, patch-waxing of the worn area may be successful if the material is applied in thin coats and properly feathered along the edges, and later dry buffed.

### FADING

Color fading may be caused by excessive sunlight (which might be controlled with shades or blinds) or by the use of strong cleaners, the same cause of bleeding and streaking.

### POWDERING

Excessive powdering represents a breakdown of the surface finish, and this can best be eliminated by the proper selection of a quality finish. Low temperature and humidity can aggravate this condition. We must not forget the possibility that the powdering may be tracked in, and that the granular powdering naturally caused by spray-buffing must be removed by dust mopping. Powdering may also be caused by the application of finish on a floor that has not been properly cleaned or rinsed, or by putting a hard finish on top of one that is soft (such as putting a polymer finish over a natural wax).

### WATER SPOTTING

White splotches caused by water typically occur on floors which have been finished with an inferior type material. The problem is greatly reduced when higher-quality finishes are used, and when the floors receive regular attention by buffing.

### DARK SPOTTING

The presence of dark circular splotches may be caused by dripping wax from a bucket or mop during the application; or by spraying wax in droplets that are too coarse or that are permitted to dry before being reached with a machine pad in the spray-buffing technique.

### DULLNESS

A flat, dull appearance may be caused by a porous floor absorbing the finish into the structure of the tile and leaving none on top to form a gloss, in which case a water emulsion sealer or additional coats of finish are required. If the floor has not been properly cleaned and rinsed, or if a poor grade of finish has been purchased, dullness can also occur.

#### SPLOTCHINESS

A floor showing irregular areas of varying reflectivity is generally caused by an improper stripping technique, where either all the old material was not removed, or some was allowed to redeposit. The reflection of light from the surface with the old material beneath it would be different, naturally, from a surface completely stripped and refinished.

#### SWIRL MARKS

Swirl marks may be caused by poor troweling of the concrete sub-floor, or by over-application of the mastic used for cementing the tile to the concrete. These two problems cannot be solved without replacing the floor, or covering it with another surface such as carpet. Swirl marks may appear in the floor when the finish is of inferior quality and does not level properly, or when it is applied too thickly.

#### WORN TILES

Resilient surfaces may be completely worn through, and even the concrete sub-floor dished out, where inadequate attention has been given to controlling the soil and providing a protective surface. Such wear areas typically occur at the bottom of a stair well, where a great deal of force is applied and also a turning motion is given, thus grinding the soil into the floor. Such wear areas should be protected with matting where possible, and given special attention in terms of cleaning and refinishing. Where necessary, the concrete floor should be repaired and new tiles put in.

## Terrazzo Floors

#### "BLOOMING"

As the cement grout ages, some dust will be formed in the crystallization and wearing process that percolates up through the pores of the floor, ending up as a white powder on the surface. Floors should be properly sealed with a membrane or polyurethane seal to prevent this occurrence.

### STAINING

Concrete grout is very porous, and acts almost like a sponge. The same sealing procedure that prevents blooming from the bottom up will prevent staining from the top down.

### PITTING

Pits in the terrazzo surface may be caused by the loosening of the marble chips or mechanical damage to the grout. In either case, the proper sealing of the floor will toughen its surface and tend to limit this damage. Where the damage has occured, the floor can be repaired by "densifying," or the holes filled in with clear epoxy resin. Specialists should be called on for such work.

### YELLOWING

An amber or yellow cast to the floor may indicate the use of an improper type of seal; for example, the phenolic seals are not transparent as are the polyurethane seals, and darken the floor. This darkness may also be caused by excessive coats of finish, the use of waxes of an inferior grade, or by improper cleaning.

### BLACKENED STRIPS

Black streaks alongside the metal divider strips often indicate that the strips are at a higher level than the surrounding terrazzo; the proper action here is a re-grinding of the strips to a flush condition by the terrazzo contractor.

### RUST SPOTS

Ferrous metals should not be placed in contact with terrazzo, as the presence of moisture causes rusting which can badly stain the grout. Steel wool is one of the worst offenders, as the fine bits of metal can get caught in cracks; steel wool pads should never be used on terrazzo floors. During construction, any nails, screws, or other metal objects should be removed daily to prevent not only rust stains but mechanical floor damage as well.

## Ceramic Floors

### WEAR AREAS

Where ceramic floors are sealed or finished, wear will take place on the square or rectangular surface of the tiles, but not in the rounded area of the tile or the grout which lies beneath, thus causing regular but unsightly wear patterns. When sealing ceramic floors, care must be taken to rub the seal into or off of the surface of the tile, concentrating it in the grout; and where finish will be applied to ceramic surfaces, it must receive regular attention so that wear spots do not become apparent.

### TILE LOOSENING

Grouting between the tiles can become damaged by alkaline detergents, or by imbedded soap, thus only neutral synthetic cleaners should be used in proper dilution. If the tile does become loosened, however, a repair or replacement will be required.

## Concrete Floors

### OIL SPOTS

Because of the porous structure of concrete, oils and other materials easily penetrate the surface and can actually stain the concrete through its entire thickness, and even a surface removal of the stain will not prevent the stain from reappearing as it re-enters the surface area through capillary action. Just as for terrazzo, which after all is made of marble chips and a cement matrix, the best action is prevention by sealing.

### CRACKING

There are many causes of cracks in concrete, dealing with the design of the floor and the stresses to which it is exposed. The important thing to remember is that once a crack has appeared, it should be patched as quickly as possible with any one of a number of commercial crack-filling agents, since a damaged floor is very apt to become still further damaged since it is no longer a continuous surface.

#### TRAFFIC PATTERNS

Traffic patterns may appear in concrete as a result of the surface being abraded to form a slightly dished-out condition. Matting should be used where soil is introduced to such areas, and patch-sealing should be performed in these wear areas more frequently. Finally, such areas should be given special attention with both dry and wet cleaning.

### Marble Floors

#### SLIPPERINESS

Marble, being a highly polished stone, is an inherently slippery material. Its slipperiness can be reduced by applying finishes, rather than leaving the floor unprotected! Of course, the causes of slipperiness to finished floors, as described above, also apply here.

#### SOIL RETENTION

Travertine, or "rotten marble" is an extremely difficult floor to care for. The numerous cracks and crevices trap soil of all kinds which becomes almost impossible to remove. A preventive approach is to fill all the pores with a clear epoxy resin. If this is not done, the use of a wet vacuum will help to remove a good deal of the soil, but the floor will never actually take on a clean appearance. Travertine is an exceedingly poor choice for a floor surface, although it is favored by some architects because of its unusual appearance.

### Wood Floors

#### WARPING

Warped, cupped, or twisted wood floors are typically caused by the over-use of water. If any water at all is to be used on wood floors, the wood should be properly sealed, and the water used sparingly and removed as quickly as possible. It is preferable to set up a floor care technique that avoids the use of water altogether through the use of treated dusting mops, solvent cleaners, solvent sealers, and the like.

### PEELING

When a varnish or seal is applied to floors, it may peel if the coating is too thick, or if the floor has not been roughened to permit proper adhesion. When applying sealers, it is best to apply a dilute penetrating seal at first, which improves the adhesion characteristics of the surface sealer which then follows after suitable drying time.

### RAISED GRAIN

As in warping, this condition is caused by failure to protect the floor through sealing, and by the overuse of water.

### BRITTLE CORK

Cork, being a form of wood, is susceptible to the same damage by overuse of water and improper detergents as is a planked floor. The maintenance of a cork floor should be conducted in a similar fashion to a wood gymnasium floor, for example.

# 40

# Carpet Selection and Care

## Do You Want Carpeting?

CAREFUL thought should be given to the actual desirability of carpeting an area, as a prelude to carpet selection. Much too frequently, this decision is related to cost alone, and even then the cost basis is often not an objective one.

Some of the important considerations in deciding to carpet or not to carpet are acoustics, aesthetics, bacterial control, comfort and prestige. It is sometimes desirable to use carpeting when there is a shortage of men available for custodial work, so that the cleaning job is made somewhat easier for women.

While carpeting may eliminate slip hazards associated with some floor finishes, proper installation is necessary to insure against tripping hazards. Excessive pile depth may present difficulties for enfeebled or elderly users; also, deep pile or an excessively spongy base may interfere with the movement of wheel chairs and carts.

## CARPET ECONOMICS

Naturally, those associations and industries having vested interests in each type of flooring material attempt to present the maintenance cost for their product in the most favorable way. This explains how parallel articles in the same issue of a given publication can show economic advantages for carpeting versus resilient tile, and cost savings for resilient tile versus carpeting! Similarly biased figures can be found for installation and maintenance costs of wooden floors, terrazzo and ceramic tile.

One of the most widely circulated and quoted comparisons, favorable to carpeting, has been criticized both by building owners and managers and by contract cleaners for exaggerating the probable useful life of carpeting and minimizing the useful life of resilient tile. Certainly a much more objective study on the comparative economics of these two types of flooring is that published by the industrial research unit of a leading university. However, this study appears to have serious shortcomings due to limited familiarity with housekeeping procedures. The statement is made that the study group did not observe the spray-buffing method of tile maintenance. Those experienced in this method know that it has wide commercial application, particularly where high levels of appearance are required, almost always with significant economies. Further, this study can be questioned on the basis of assigned capital equipment cost. The expenses for floor machines is projected on the assumption that full-shift usage will be made of each item—an unrealistic assumption, since this could not be achieved with economical assignment of worker time. Also, no wet pick-up vacuum is provided for, although this would certainly be needed in periodic stripping or heavy scrubbing.

Our own conclusions as to comparative cost of amortizing and maintaining carpeted floors versus resilient tile floors—based both on our own studies and on the reports of objective sources—is that they are far closer than partisans from either field concede. This is in the case where original installations of the different types of flooring are under consideration. If the comparison is being made strictly on the basis of maintenance cost, without consideration of installation and amortization, then the cost advantage will depend a great deal on the usage of the facility. Carpeting maintenance costs are generally advantageous in office areas where soiling and traffic are moderate. In a facility such as an airport, where traffic is extremely heavy, with extensive trackage of soil, studies have shown carpet maintenance costs to be substantially higher than those for terrazzo. One of the reasons for the higher carpeting maintenance cost is found in the present technological limitations on the productive capacity of equipment used for carpet care.

The foregoing discussion should lead to the conclusion that it is seldom desirable to decide upon the use of carpeting on the basis of economic factors alone. Further, before the investment is made, the prospective user of carpeting should read and listen to the available information about various types and should look at installations before deciding exactly what he wants.

### THE RIGHT CARPETING

Fiber type, weave, pile (height, weight and density), color and pattern all will affect the economics and maintainability of carpeting.

Rayon and cotton are not considered suitable fibers for commercial carpets. Of the remaining types available, nylon is probably the best for heavy commercial use, with wool being suitable for prestige applications. The other synthetics currently seem to be rated as follows:

Polyesters
Acrylics
Olefins
Modacrylics

For heavy wear, it appears that the Wilton and velvet weaves are most satisfactory. Axminster and knitted carpeting are generally somewhat inferior to the foregoing for commercial applications. High quality tufted carpeting may be satisfactory. For heavy commercial applications, continuous filament nylon woven with a tight loop, backed with sponge rubber and constructed so that moisture will not penetrate this rubber backing, seems to provide the best wear and permit easiest maintenance.

Generally speaking, the greater the pile weight, the better the wearing quality. This weight is generally stated in terms of numbers of ounces of pile yarn per square yard of carpeting. For good wear, nylon pile weight should equal or exceed 48 ounces. Pile height is the distance from the top of the backing to the top of the pile. In commercial carpeting, close weave is more important than pile height, but ordinarily a height of .25 inches would be desirable for nylon. It is important that cut staple fibers be avoided in nylon, since these are likely to pill or fuzz and almost certainly will give poor wear. Pick or row (referring to the number of loops per inch of carpet length) and pitch (referring to the number of pile ends in each 27 inches of width) are both measures of pile density.

Color and pattern are very significant in carpet maintenance. Extremely dark or extremely light colors, as might be anticipated, are difficult to maintain; desired general color effects can often be obtained through the use of tweedy mixtures, while retaining high resistance to evidence of soiling.

### CARPET PROTECTION

Probably the greatest hazard to carpeting is the grit which is allowed to become embedded in the pile and which then cuts the carpet fibers, resulting in their ultimate failure (or at least causing changes in light reflectance, with apparant color variations). A fundamental step in carpet care is to protect it from the trackage of soil through the use of walk-off matting and other devices for soil entrapment.

Carpeting should also be protected from the permanent depression of piling which can result from any furniture being allowed to remain in one position at all times. Relief from this problem can be obtained through the use of furniture leg cups or casters specially designed to transmit the load to the carpet backing or floor without allowing it to rest uniformly upon the pile itself. The furniture leg cups are designed with teeth which support the weight of the furniture and which also permit a major part of the piling to remain erect. The casters have a honeycombed surface which results in only a small proportion of the pile fibers being packed down.

### ROUTINE CARPET CARE

Since carpeting used in institutional and commercial applications is subjected to much heavier soiling than under household conditions, cleaning methods must be adjusted accordingly. The major destructive factor to carpeting is the gritty soil which penetrates the piling and abrades the fibers. Carpet sweepers and soft bristle brooms, often used in homes, have

A power sweeper may be used to remove surface soil in large carpeted areas under some conditions. (Tennant Company).

little value in the maintenance of carpeting under commercial conditions except for policing to remove trash. The major tool for cleaning carpeting is the vacuum cleaner.

Assuming that the carpet is of suitable color and texture, it is often possible to omit vacuuming for a day or so without this becoming evident except where coarse soil is present. This is one characteristic of carpeting which gives somewhat more flexibility in the maintenance program than is enjoyed with resilient or other smooth floor surfaces. Nevertheless, omission of daily vacuuming in traffic areas introduces the hazard of excessive damage to the carpet pile.

A distinction is commonly made between the daily vacuuming of trafficked areas and a less frequent—often weekly—vacuuming of non-trafficked areas. Some of those persons responsible for programming maintenance have set one time standard for spot vacuuming and another for overall vacuuming. Actually the major time difference lies in the percentage of the total area covered in the operation, although there is likely to be some factor of lost time resulting from travel about the area in the spotting operation, as compared to continuous vacuuming of the total area. The time requirement will also be affected both by the type of vacuum cleaner used and by the nature of the soil being removed. If trashy or linty soil is encountered, an upright with a driven brush will be much more efficient than will the simple wand-type vacuum; on the other hand, more difficulty is involved in reaching under low furniture with the upright type vacuum cleaner. A comparison of the overall results leads to the conclusion that the upright vacuum should be selected in most cases for the major daily cleaning operation, while the wand type vacuum, especially that adapted for wet pickup, may be very useful for policing heavily used areas during the day. In some instances, a carpet litter vacuum may be a needed policing tool where large amounts of litter accumulate, as perhaps in the corridor of a school building.

Whether the upright vacuum with driven brush or the wand type vacuum (either portable or central system) is used for daily cleaning of trafficked areas, it is essential to use a vacuum with driven brush for the less frequent overall cleaning of the area. In fact, most conditions dictate the use of the type of vacuum which might be considered a moderate duty pile lifter; this equipment actually has two motors, one driving the turbine to produce the vacuum, and the other driving the brush which raises the pile and loosens any embedded dirt and grit.

## SPOT CLEANING

There is a great difference in the extent to which greasy or wet soil will penetrate the various carpet fibers. Continuous filament nylon, which normally has a maximum moisture absorption of about six percent, is probably most resistant to actual staining, although various characteristics including its tendency to build up static electricity will sometimes cause the fiber to pick up surface soil. There is more opportunity for stains to penetrate wool fiber and the acrylics than there is with nylon.

Regardless of the carpet fiber, it is most important that any spills which could possibly spot the carpeting be picked up as quickly as possible. This need reflects the importance of good communications between the users of the building and supervisors or personnel in the housekeeping department. If accidental spills or soiling are reported at once, it is normally easy to determine the nature of the soil and consequently to select the proper cleaning method. On the other hand, when spots are allowed to remain, it is frequently a matter of guess work to determine their nature and to decide on a cleaning approach.

Most soils can be classified under two general headings: those which are water soluble and those which are oil soluble. This fact has resulted in the frequent use of the so-called "two-bottle spot cleaning kit." This is simply a convenient kit which can be prepared by the housekeeping department and which consists of one bottle of solution for removal of water-soluble soil and another for the removal of oily soil. Water-soluble soils can usually be removed by a diluted solution of neutral synthetic detergent with a slight amount (perhaps one teaspoon to a pint of solution) of white vinegar. The other unit in the kit consists of a dry cleaning solvent, which of course should not be highly flammable; nor should it consist of toxic solvents such as carbon tetrachloride.

Commercially assembled spot cleaning kits are available from several manufacturers, and these can be very useful. They normally contain a set of instructions showing which material should be used for removal of a given type of soil and just how it should be employed. In the use of these or any other spot cleaning materials, it is important that a preliminary test of each material be made on a swatch of the carpeting or in an area hidden from normal view to determine in advance that there is no tendency for the cleaning material to change the color of the carpet. Such a preliminary test will guard against serious discoloration of the carpeting, which often can be more noticeable than the spot being treated.

### DRY SHAMPOOS

Frequently the traffic flow in a carpeted area is such that limited portions of the carpeting becomes quickly soiled. Often these areas are too large for spot cleaning in the sense described above, but small enough that the housekeeping manager would like to avoid the usual "wet shampooing" approach. One approach to this situation is dry shampooing. Dry shampoo usually consists of an absorbent material, such as fuller's earth or a fine hardwood sawdust, impregnated with a volatile hydrocarbon solvent. Some of the first of these materials actually used carbon tetrachloride as the solvent and consequently were dangerous to the workers; pressure from health and industrial hygiene authorities has apparently resulted in those products now on the market being safe to the user. The method of application of these materials is to sprinkle them over the soiled areas and then brush them into the carpet piling, either manually or with specially designed electrically driven brushes. The dry shampoo is then allowed to remain on the carpeting until the solvent has volatilized. The evaporation of the solvent tends to draw up any soils which are soluble in it—probably by a wicking or capillary action—from the pile into the absorbent carrier of the dry shampoo. This dry material, along with the absorbed soil, is then picked up, and herein may lie a problem, since these materials are difficult to remove from the carpet pile. Generally a pile lifter, or at least an upright vacuum with a good motor-driven brush, is necessary for complete removal of the dry shampoo.

Certain formulations are available for "mop-on" cleaning of carpets. These may be wetting agents with emulsifiers and solvents which are effective in dispersion and suspension of soil. Normally these materials are diluted about one part to four or five parts of water and then used in the following manner. The cleaners are placed in a shallow open pan or basin of suitable size; then a small, short-strand mop (similar to a small standard dust mop but with shorter strands) is quickly dipped into the cleaner with the object of wetting only the ends of the strands. The mop is then shaken to remove excess solution and used with a brushing action to go over the soiled area of the carpeting. Very good results have been reported for this type of carpet maintenance under conditions such as those encountered in motels. Some of the manufacturers profess to include anti-soiling agents in their formulations; however it is important not to leave excessive cleaner residue in the carpet pile, for it apparently will accelerate the re-soiling of the carpeting if this is done. Some manufacturers advocate that the cleaner be used one day, and on the next day the same procedure be followed using clear water without cleaner.

### WET SHAMPOOING

For area rugs or carpets, there are advantages in removing them to a special location for shampooing. It then becomes possible to turn the rugs pile side down over a grating and to beat the back of the carpeting to loosen and remove any gritty material which is not taken out by the vacuum cleaner or pile lifter. Otherwise, the wet shampooing procedure is normally carried out in the usual manner, whether it is done in a special location or on the floor. Of course, it is not practicable to remove carpet where total areas are covered and where it is more or less permanently fixed to the floor.

While the procedure is referred to as wet shampooing, the actual objective is to shampoo the carpet piling with a detergent foam, so that the foam serves as the emulsifying and suspending agent for the soil. Until recent times, the equipment available has not permitted ideal control of the amount of moisture introduced into the carpeting. This equipment has normally consisted of a conventional single-disc floor machine, with a detergent solution tank mounted either on the handle or above the motor, a solution release control, and a shower feed brush through which the detergent solution was fed into the fibers of the brush. Originally, soft natural fibers, such as Tampico, were used for the brush fill. These have generally been replaced by nylon bristles, with better control of the actual bristle stiffness. In the procedure, a high foaming detergent, often with an organic solvent, is fed into the fibers of the brush so that enough of it is present to produce a high volume of foam when the machine is turned on and friction is produced between the brush fibers and the carpet pile. A properly trained operator can maintain a fairly consistent amount of wetting of the carpet pile by controlling the rate of feed and the intervals of feeding as the machine is moved from side to side on the carpeting. Generally carpet shampooing is a good team job; one operator will usually start picking up the foam with a wet vacuum cleaner after

The dry foam shampoo machine does a quick surface cleaning of carpet requiring little skill for the operator (Von Schrader Co.).

approximately twenty square feet of carpeting have been shampooed by the other operator. In this way, good operators can keep from wetting the carpet backing excessively. Over-wetting has many disadvantages, in addition to the fact that it requires excessive drying time: it can cause shrinkage of certain types of carpet backing even to the extent that the carpeting will pull away from the fastenings, or it can cause mildewing of the carpeting when poor drying conditions exist. Even with good technique and suitable temperature and humidity, this type of shampooing will normally require a drying time of four to twelve hours before the carpet can be subjected to traffic. Also, precautions should be taken to prevent rust spots when furniture with metal gliders is moved back onto the carpet, in case it is not completely dry.

A type of shampooing machine has been developed wherein the detergent is fed through special pumps so that foam is developed before it is fed into the brush of the machine. In this type of equipment, the brushes are usually mounted on a horizontal axis rather than on a vertical axis, as in the case of the conventional buffer-shampooer. The advantage of this type of equipment is that much less moisture is introduced into the carpet pile so that it is easier to train an operator and much more unlikely that the carpet will be over-wetted and damaged. Probably this type of equipment is not capable of cleaning the carpet pile as deeply as will the single disc floor machine type equipment, but it is so much easier to use that it is entirely practical to shampoo more frequently, which in the long run should result in better carpet care than with the other equipment.

In any shampooing operation, the initial step should be the use of a good pile lifter to remove all dry soil and to position the pile for best shampooing. After the foam has been picked up following the shampooing and the carpet has dried, a pile lifter should again be used to insure uniform lay of the pile and thus uniform light reflectance and appearance. Actually, with the foam type shampooer, many operators find it unnecessary to use a wet vacuum after the shampooing operation if this work is done frequently enough to prevent heavy soiling. In that case, the foam is simply allowed to dry on the surface of the pile and any dry soil thus left is then picked up with the pile lifter.

### SPECIAL TREATMENT OF CARPETING

Chemicals are now available for application to carpets and other textiles which significantly retard the penetration of soils. Some of these materials have proven highly effective and well worth the cost of purchase and application under certain conditions.

Anti-static agents which remain effective for reasonable periods of time (though the time will vary according to conditions of humidity and soiling) are now available. If static build-up is a significant problem, as it frequently is during winter in low humidity areas, the application of the materials may give adequate relief. A tendency to build up static is one of the major disadvantages of nylon carpeting. One application may not last the entire winter, the season when static is most noticed, but most reports indicate effectiveness for a period of two or three months.

At least one carpet manufacturer will supply carpeting containing conductive metal fibers woven in at proper intervals to insure against the build-up of static electricity. The addition of this conductive fiber does not cost much per yard, but the purchase of a full mill run may be necessary to obtain this feature.

The water-extraction system (mistakenly called "steam cleaning") uses equipment that injects warm water into the carpet, and removes it by vacuum in the same motion, using a wand attached by hose to a machine. It provides the best cleaning of all in-place systems, but is rather slow (and thus expensive), and requires a longer drying time for the carpet. Further, wool and the mastic used in glue-down carpets can be damaged if the water is too hot. For these reasons, the water-extraction system should be used only occasionally, such as after two or three dry-foam shampooings. For smaller quantities of carpet or rush jobs, contractors should be considered.

The water-extractor (mistakenly called "steam cleaner") does a very effective job in cleaning carpets, and with the proper attachment can also be used for upholstery cleaning (U.S. Floor Systems).

# 41

# Cleaning During Construction

THE natural tendency at a time of construction or renovation is to throw up one's hands and "give up" on housekeeping until things settle down and it can be handled in a normal, routine fashion. This is the same sort of attitude which is so commonly seen during periods of inclement weather, where snow, water, sand, or salt may be tracked onto the floors, making the preservation of the appearance of the floor a very difficult matter indeed.

But construction and renovation create conditions which require an actual *intensification* of sanitary maintenance activities, rather than a curtailment. The fact is, for new construction, finished surfaces must receive attention just as soon as they are in place; and for renovation or expansion, not only must the new surfaces be protected, but also the existing surfaces nearby must receive this protection as well.

At such times, then, we are faced with the very real problem of housekeeping becoming considerably more difficult—both from a psychological and practical performance standpoint—while at the same time it becomes more important. Even the fact that it is only a temporary condition brings problems of its own.

Let's discuss the necessity for this intensification of housekeeping, and some of the methods by which it can be accomplished.

## Construction Activities

During the rough construction or renovation period, where concrete is poured, steel beams put in place, bricks laid, and other rough construction work accomplished, we have no real interest in the housekeeping of that particular area for its own sake. It is up to the contractor to provide a level of housekeeping and orderliness within the construction area which will make his work efficient and safe. Nevertheless, we must protect adjacent or nearby areas from contamination by the soils and waste material produced in the construction.

There are several reasons for this:

1. Floors can become badly damaged by scratching or abrading by the tracking in of particles of concrete, metal chips, splinters, mud, etc. This is true not only of resilient floors, but also of floors normally thought to be quite durable, such as terrazzo or marble.
2. Air-borne waste particles and dust can damage mechanical systems, laboratory or office equipment, etc.
3. Comfort alone becomes an important factor. Tenants in a building will react strongly to unpleasant conditions caused by excessive dust and soil. Occupancy in a hotel will reflect directly the housekeeping level at such times.
4. Certainly in a hospital the patient must remain comfortable, but, even more important—and this is also true in school buildings—cross-infection control is extremely difficult when there is excessive dust and soil. Effective housekeeping at such times represents a direct involvement with the care of the patient or the provision of proper environment for the student, the occupant, or the tenant.

## The Principle of Isolation

The basic principle is one of isolation—that is, try to keep the renovation or construction work and its by-products completely isolated from that portion of the building which is not involved. Here are some effective steps to take in this direction:

1. Provide "air locks" in hallways, which will tend to confine the dust to the portion of the building under construction or renovation. These should be constructed of two sets of fabric, canvas, or plastic curtains, which must be pushed aside in order to walk through, but yet which will fall back into place. They should be the full length and width of the hallway opening. These curtains should be vacuumed regularly, but at least once per week. From a safety standpoint, it is desirable to have at least one portion of them transparent. Be sure that the area between the curtains is well lighted.
2. Place plenty of rubber mats and runners in connecting areas to entrap as much dirt and soil as possible, so as to keep this dust and soil from being ground into the older flooring and tracked throughout the building. An ideal location for such mats and runners is within the "air lock" area. These mats should be vacuumed daily and washed out at least once per week. The use of old carpeting is very helpful, and many organizations save discarded carpets to be able to use them for such a purpose.
3. Check the filters in the forced air system regularly so that they do not become "loaded" with construction dust. Keeping the filter system efficient will do much to cut down on air-borne dust.
4. Put up "off-limits" signs so as to prevent construction workers from tracking steel filings, stone chips, sand, dust and soil into areas not involved in their work.
5. Similarly, put up signs to prevent sight-seers from walking into the construction area and then tracking soil back into the original facility. In order to be able to enforce such a rule, and to satisfy a natural curiosity, a periodic tour of the new areas might be conducted for interested persons.
6. Equipment or machinery that is located near the construction area that is not regularly used should be covered with polyethylene sheeting or other protective material so as to keep out soil and dust.

## Continuing Maintenance

After all possible steps have been taken to isolate the soils created during construction from any existing facility, we must then look at the maintenance within the older areas, because despite the above precautions an additional soil load will undoubtedly be brought in. These steps must be taken:

1. The floors must be wet-cleaned more regularly, as dry mopping will not remove many of the heavier particles which become embedded in floors and their protective coating. The daily use of auto-scrubbing equipment is particularly desirable for this purpose in open areas.
2. Walls, trim and furniture should be vacuumed periodically to prevent the formation of permanent soils which occur when dust is permitted to form a hard material in association with water or other vapors, oils, etc.
3. Once the construction is complete, a thorough wall and trim washing may be required.
4. A protective coating of floor wax, finish, or seal should be applied to all floors before construction actually begins, so that as much of the wear as possible is confined to the protecting material, eliminating permanent damage to the floor surface itself. In order to prevent hard materials from being ground through the coating down into the floor causing permanent damage, the floor coating should be stripped and re-applied much more frequently than normal. When the project is finally complete, a terminal stripping and rewaxing of all affected floors is desirable.
5. The connecting areas between the existing facility and that under construction or renovation should be given especially careful attention, with daily wet cleaning being the rule.
6. Try to prevent the transfer of soils from one floor to another. This can be done by more frequent cleaning of stairwells and elevators, as well as entrance areas. Of particular value is the use of carpeting in elevators and entrance areas.

## Manpower Requirements

It is management's responsibility to provide the money necessary to be able to accomplish all the extra duties required at such times. This involves a recognition of the economic import of the situation in terms of damage to surfaces, and the effect on the public, tenant, guest, patient, and employee attitudes.

If the construction or renovation program is an extensive one and lasts over a good period of time, it will undoubtedly not be possible for the existing housekeeping staff to handle this extra work even if the original staffing were efficient. Here are some of the ways in which to provide the additional time required:

1. The use of overtime for existing employees. This is undesirable except for a short period of time, since persons who become used to overtime feel that something has been taken away when it is discontinued, and a severe morale problem ensues. Also in such cases, one must consider the fatigue factor.
2. Temporary employees may be hired either specifically for the extra jobs required, or else to permit a redistribution of work assignments.
3. Additional full-time permanent employees might be hired to handle the extra work-load, and these employees can become the cadre for the staffing of the new facility. This is very desirable from the standpoint of having a ready-made, properly trained work force which can correctly maintain the new facility from the day it is first occupied.
4. A very good source of help in such cases is the various contract cleaning establishments, which can contract to perform specific cleaning jobs at assigned frequencies, or to clean entire areas for a fixed period of time. In these cases, it is very important to specify materials, methods, frequencies, and supervision.

## Liaison with the Contractor

If the contractor (or the construction department where the work is done internally) is approached before the work begins, it is entirely possible to make agreements on methods that will be used in the construction that will considerably reduce the amount of soil created. Such methods can actually be written into the specifications and, in many cases, will not increase the cost appreciably. Where the cost is increased, the added amount must be considered in the light of its value in terms of preventive rather than corrective action. Some of these steps include:

1. Wet-grinding rather than dry-grinding can be specified for a number of operations, which will sharply limit the dust problem.
2. The laborers should be instructed regularly to clean up accumula-

tions of dirt and dust and to remove them from the building before they are spread around.

3. The actual removal of the waste, particularly plaster and concrete waste, should be done in such a way as to avoid clouds of dust. This would require the avoidance of dumping the materials from upper levels. Keeping such waste wetted down also helps.
4. The use of sweeping compound, treated dust mops and cloths, and vacuums in clean-up operations is desirable.
5. From another standpoint, management is certainly desirous of maintaining the health and safety of the construction workers, who should be required to wear hard hats, dust masks, gloves and goggles where necessary.

On the final building clean-up, the typical construction contract requires the contractor to finish floors with a protective coating. Generally, the material used is of a low grade and often may not even be compatible with materials which mangement may wish to use. In such cases, it is most desirble for the contract to indicate the use of specific materials and methods of application. Again, some organizations prefer that this work be done by their own people—it is good training for new workers, for example—and receive an allowance from the contractor who is thus relieved of this responsibility.

Military doctrine gives us a clue as to effective housekeeping action during construction and renovation: when the situation becomes difficult, attack and attack again. Do not permit a retreat—it can turn into a rout. Face the problem and fight it!

# 42

# Cross-infection Control

## What and Where?

THE term "cross-infection" has become uncomfortably familiar in hospitals in recent years. It refers to those infections which are actually acquired within the hospital by patients or others, as opposed to infections which may have existed in the particular individuals upon entering the hospital environment. It should be noted that cross-infections are *not* the exclusive property of hospitals, and that they may occur in any factory, office building or institution where there is close contact among people; in fact, the common cold is transmitted as a cross-infection. Thus, in any environment, reasonable sanitation precautions need to be observed to protect personnel involved. However, there are factors which invest the cross-infections in hospitals with far greater danger than those encountered elsewhere. First, the conditions which normally bring people to the hospital have developed strains which are resistant to antibiotics, so that it may not be readily possible to assist the already-weakened patient to fight the newly acquired infection.

While the germs ("pathogenic organisms" or "pathogens") causing cross-infections are introduced into the hospital environment by people and are most readily acquired by person-to-person contacts, it must be assumed that they can be shed from the original carrier into the air, onto surfaces or objects in the hospital, and then transmitted to the victim by these secondary agents. A leading authority in this field (Dr. Carl W. Walter) says "exemplary technic on the part of the professional staff is effective only in a hygienic environment."

Germs enter the body through inhalation, through the mouth and through lesions in the skin. Perhaps the most common means of infection, second to direct contact, is from air-borne organisms, and these are often present because of improper cleaning of floors and other surfaces.

The organism most notoriously associated with hospital cross-infections is Staphylococcus aureus, somewhat romantically known as "golden staph"; this organism can cause pneumonia, meningitis and infections of the ear, eye and throat, as well as many other illnesses. Other less highly publicized sources of cross-infection include Mycobacterium tuberculosis, and such organisms as Proteus vulgaris, causing kidney and other infections, and Pseudomonas aeruginosa, often causing pneumonia.

## Importance of Cleaning and Disinfection

Obviously it would be desirable to destroy as many as possible of the pathogens which potentially may cause cross-infection while they are present on surfaces and before they are transmitted to possible victims. This is especially true because of the resistance of these organisms to drugs, as already mentioned. Fortunately, there is no relationship between the drug resistance and resistance to disinfectants. It was once widely believed, even among physicians, that a thorough job of cleaning was sufficient to control germs on hospital surfaces. Certainly, many of the micro-organisms are associated with and carried by soiled particles. However, cleaning is not enough because it is never possible to remove all organisms due to the nature of the micro-organisms themselves and to that of surfaces on which they are deposited. A very large proportion of the pathogens under consideration fall within the classification of bacteria. It is generally assumed that the average bacterium is about 1/25,000 of an inch in diameter. Obviously, a particle of this size is not visible to the naked eye and is revealed only by a powerful microscope. Many pathogens are viruses, which are considerably smaller than bacteria and in fact are visible only by means of the electron microscope. It can be inferred that particles so tiny can readily be concealed and protected in minute faults in surfaces from which they will not necessarily be removed by cleaning operations. Thus it is important to use disinfectants to destroy as many as possible of those remaining.

In actual practice, the use of cleaner-disinfectants in a single-step operation designed both to remove soil and to disinfect surfaces permits the greatest operational efficiency. When the cleaning and disinfection operation is carried out daily, the soil loads developing in the interim are seldom so heavy as to interfere with the effectiveness of the chemical agents, as long as proper techniques are observed.

Admittedly, it has not been possible to formulate a cleaner-disinfectant providing adequate pathogenic destruction without some compromise in the cleaning ability which could otherwise be built into the product—a fact which also emphasizes the need for good techniques. It is essential that mopping or other cleaning solutions be changed frequently enough to prevent inactivation of the disinfectant by organic or inorganic matter picked up by the mop or other cleaning tool. It is impossible to set a general frequency for the changing of cleaning solutions which would be the right one for all situations because of variations in the water supply, the types and amounts of soil, and other influences. As a general rule, it may be said that mop water should be changed after mopping about 1500 square feet of floor surface which is only lightly or moderately soiled; the solution should never be allowed to become obviously dirty. No matter how little area has been covered, a bucket of cleaning solution should never be taken from an isolation area to any other area of the hospital.

Authorities have determined that for greatest assurance in cross-infection control in critical areas, such as the operating room and the delivery suite, the cleaner-disinfectant solution should be poured onto the floor without dripping the mop into the bucket, only using the mop to spread the solution and to apply pressure to scrub obvious accumulations of organic matter. A sufficient amount of solution should be put down to allow at least two full minutes of contact time, and the solution should then be removed from the floor with a wet vacuum. This procedure gives reasonable assurance of the emulsification and dispersion of organic matter and of sufficient time for the germicide to be effective in controlling pathogens. The relatively large quantity of solution being used minimizes the chance for the chemical to be inactivated by the excessive organic matter.

## Disinfectant Fogging or Misting

Numerous studies have been made of the fogging or misting technique and its effectiveness in disinfection of contaminated areas. Initially, it was hoped that this procedure could eliminate the need to wash walls and other surfaces after isolation or "dirty cases" and thus permit both earlier return of rooms to service and reductions in the manpower required for cleaning and disinfection. Authorities in the field of infection control have concluded after review of many applications of this procedure that the results may be quite variable, depending to a considerable extent upon the amount of disinfectant deposited on the surfaces; also there seem to be wide variances in the effect on air-borne organisms. Consequently most of those authorities have discouraged dependence on this technique, although some of them agree that the procedure has limited merit in providing a significant initial reduction of organisms present. Without any question, if the fogging technique is used, the decontamination must be completed by spraying or washing all surfaces with sufficient quantities of a cleaner-disinfectant solution to insure complete wetting and removal of significant soil accumulations.

In summary, the misting procedure appears to offer a valid substitute for the time-honored but inefficient and inconvenient practice of airing out an isolation room for 24 hours prior to its cleanup and return to service. Application of the technique under varying conditions with presumably competent bacteriological monitoring has shown reductions in the total counts of organisms of a magnitude exceeding 95% in some cases. If these results are accepted, it must be conceded that the technique is achieving at least as much as the airing-out process in respect to gross reduction of organisms and that it provides a significant safety advantage as a preliminary to the terminal cleanup by properly trained, equipped and garbed housekeeping workers. Presumably all items already in the room during the isolation case would be left there during the misting procedure, but trash and linens would be transferred to plastic bags for later removal after the exterior surfaces of those bags had been substantially disinfected during the misting. The actual details of the procedure should be reviewed by the Infections Committee, and housekeeping personnel should then adhere to all specified steps.

## Formulations of Cleaner-Disinfectants

No chemical disinfectant formulation which is safe in use both for personnel and for surfaces involved can be expected to kill 100% of all the pathogens exposed to it. Many disinfectants are highly selective and consequently do not destroy some of the most significant germs. Thus, it is important to choose one that does the job as completely as possible under practical use conditions. At the same time, the disinfectant should carry a minimum hazard of irritation or sensitization to personnel or damage to structural surfaces.

The most popular type of cleaner-disinfectant for hospital use is the synthetic phenolic. As the name implies, these products are chemically related to phenol or carbolic acid, but the molecule has been modified by introduction of chlorine or oxygen or even benzene rings at various points in the structure of the basic molecule. The result of this modification is both greater disinfecting ability and reduced corrosiveness of the molecule. The individually modified phenolics have been found to differ, one from another, in their effectiveness on various micro-organisms. Consequently, the trend in formulation, in order to provide broad spectrum control, is to include more than one of the synthetics.

Other materials which are used in the formulation include certain alcohols, wetting agents and synthetic cleaners, and sometimes soaps. The presence of soap generally tends to reduce the likelihood of skin sensitivity on the part of using personnel, but it creates films with hard waters which may insulate the surfaces of conductive floors and thus create static and electrical problems. Hence, products based on synthetics rather than soaps are usually selected.

Well-formulated phenolics will readily destroy Staphylococcus aureus—the organism which first caused great alarm over cross-infections when they were found to develop resistance to the antibiotics. They will also destroy the principal causative organisms of tuburculosis and other difficult-to-kill pathogens such as the gram-negative Pseudomonas aeruginosa and Proteus vulgaris.

In the selection of a phenolic disinfectant (as in the case of any other type) primary consideration of claims should be based on the A.O.A.C.

Use-Dilution Test. This should indicate the amount required for control of various organisms. Comparative evaluation of products with respect to cost can be made intelligently by relating the number of ounces per gallon necessary to provide control, along with the cost per gallon of the concentrate.

Quaternary ammonium chlorides can be formulated to kill a wide range of pathogens, but some of these products have not proved effective on the Bacillus tuberculosis, and many are not effective on Pseudomonas aeruginosa. Further, the formulations of synthetic phenolics seem to yield better cleaning results than those of the quaternary ammonium chlorides especially with oily soils.

Another popular type of germicidal-detergent is the iodophor. Generally these should not be classified as cleaner-disinfectants because of the inadequacy of the detergent ingredients to cope with heavy organic soil; this difficulty arises in part from the fact that, for germicidal activity, the iodophors must be in acid solution (preferably below 6.0 pH) which fact alone reduces their ability to remove organic soils. Because of the acidity of iodophors, they are actually better in removal of mineral soils than are most neutral or alkaline detergents; but soils of that type tend to raise the pH of the solution (that is, reduce acidity) and thus detract from the germicidal activity. Another disadvantage of the iodophors is that, despite the use of wetting agents to "complex" the iodine and hold it in solution, there is a tendency for them to stain porous surfaces. As for germicidal potential, they are comparable to the phenolics.

At the present state of the art of disinfectant formulation, no classes of products beyond the above three seem to merit serious consideration for cross-infection control in hospitals. Some of those offered must be classed primarily as bacteriostats, rather than germicides, and are considered inadequate by hospital bacteriologists, since the primary aim is to kill as many of the pathogens as possible rather than merely to suspend their growth. The organo-metallic compounds fall in this class. Other materials are so toxic or so corrosive as to be considered unsafe for use; these include the mercurics and formalin.

In summary, good formulations of synthetic phenolics appear to be the best choice for use in general hospitals or those specializing in treatment in tubercular cases. For schools and institutions which are the habitat of persons in normal health and which do not by design collect large numbers of persons suffering from communicable diseases or infections, less stringent control is required; hence, cleaner-disinfectants such as the quaternary ammonium chlorides should prove entirely acceptable, as long as they give adequate results in cleaning. One of the purposes of using cleaner-disinfectants is to control odors through the destruction of microorganisms which might break down any organic residues to produce offensive aromatic materials—an objective which the quaternary ammonium chlorides will accomplish while giving the added advantage of effectiveness in tying up existing aromatics.

## Isolation and Critical Areas in a Hospital

Hospitals will generally have specific rooms or wards which are designated for the isolation of patients suffering from highly communicable or infectious diseases. In some instances, the patient found after admission to be suffering from such a condition may be isolated in the area to which already admitted, rather than moved to a designated isolation area. The objective of course is to help prevent the spread of the organisms of the infected patient.

Another type of isolation patient is the one suffering from extensive burns or a comparable condition where it is desirable to limit the number of people contacting that patient and thus limit the possibility of entry of infectious organisms into damaged tissue. It is analogous to "protective custody."

Definite controls for isolation areas should be established by the Infections Committee. Restrictions should exist on the number of persons entering the areas and all such persons should be required to use protective devices such as gowns, masks, caps and shoe covers, which should be properly discarded upon departure from the room.

During the stay of the isolation patient, the amount of project cleaning must be limited, but certain daily procedures will need to go on. It is best for the fewest possible number of people to enter the isolation area; hence it is often desirable to assign one person to do all cleaning tasks, rather than to divide the work between a maid and custodian as is often done in normal patient areas. Standard procedures must exist for the complete cleaning of the isolation area once the patient or patients have departed.

The operating (OR), delivery and maternity (OB), and the emergency (ER) suites are generally considered to be critical areas. The infection resistance of occupants in these artas may be lowered by the shock of surgical procedures or childbirth and extra hazards exist when surgical incisions or wounds break the skin barrier.

In some hospitals, all cleaning in these critical areas is performed by personnel in the nursing services department. More commonly, building surfaces are cleaned by housekeeping personnel, while the equipment used by doctors and nurses is often left to the responsibility of nursing service personnel. In a rather large percentage of hospitals, agreements are reached between housekeeping and nursing services designating which equipment is cleaned by the one and which by the other.

Obviously, where there may be a division of responsibility, there should be close coordination between the supervisors of the departments concerned and—where housekeeping cleans specialized equipment—on the methods and frequency of cleaning. Generally it seems most advisable that the housekeeping department not be given responsibility for cleaning x-ray equipment, sterile storage cabinets (except when emptied by nursing service), suction units or stands carrying anesthesia equipment or bottles of anesthetic gas; however it is quite appropriate for housekeeping to clean the bases or legs of such equipment in conjunction with the cleaning of floor surfaces.

## Conductive Floors

Whenever explosive anesthetic gas may be used in OR, OB or emergency suites, special floors providing controlled ranges of conductivity are a legal and practical necessity to guard against the possible explosion of these gases. For every type of explosive gas, there is a range of mixtures of gas and air which will be explosive. Since critical concentrations may quite likely be reached in operating rooms and similar areas, and since the mixture might be set off by an electrical discharge of any kind, precautions in the form of conductive floors are required. These floors are designed to provide a controlled amount of resistance to the passage of an electrical current or static discharge. At the lower limit of resistance, the passage of a current directly to the floor from electrically-wired equipment must not occur; at the upper limit of resistance, there must not be a sufficient difference in potential between a surface which tends to collect electrons by static buildup and one which does not, to cause electrons to be discharged through the air to produce an arc or spark. While we speak of "conductivity of the floor" the controls actually established are in terms of ohms of electrical resistance. Under Code 56 of the National Fire Protection Association, Section 252, the average of readings taken in five locations shall fall between 25,000 and 1,000,000 ohms, while extremes are listed for individual readings.

There are three common types of conductive floors, including terrazzo, mosaic tile and vinyl tile. All of these achieve their conductivity through impregnation of proper amounts of certain conductive materials, and the floor must be laid in such a way as to be grounded to the framework and foundation of the building. With all types of conductive floors, it is important that no cleaner or finish used on the floor surface shall interfere with conductivity. Certain sealers are used by the contractor laying conductive terrazzo floors for the initial curing or sealing; also, there are some manufacturers who sell sealers and finishes alleged to be satisfactory for maintenance use on the surface. Most manufacturers of sealers and finishes do not recommend that their products be used on conductive floors, and most authorities on housekeeping are similarly wary of the application of *any* coatings once the floors have been put into use.

Only a few years ago, soap-base cleaners were the only ones available, and problems were frequently experienced with the development of hard-water films on conductive floors, tending to insulate the surfaces and interfere with conductivity. Present-day synthetic cleaner-disinfectants seldom form hard-water films and thus, with proper techniques, should not produce problems on conductive floors. If there has been a loss of conductivity due to past practices, the first attempt to solve the problem should be a thorough cleaning with a chemical designed to remove any existing film. A good synthetic-base wax stripper will often serve this purpose. There are terrazzo cleaners based on chelating agents which are safe and often effective in scum removal. The use of acids should be avoided on any of these floors, especially on terrazzo and grouted mosaic tile, because of the probable deleterious effect of the acid on the Portland cement binder.

Conductivity can sometimes be improved in conductive terrazzo floors through the use of a standard calcium chloride solution in accordance with the procedure published by the National Terrazzo and Mosaic Tile Association. There is the possibility that past cleaning techniques may actually have removed the carbon black or other conductive component originally incorporated in the surfacing material. If this has happened, no cleaning procedure, however thorough, will permanently restore the conductivity.

If none of the suggested methods for restoring a defective conductive floor proves successful, advice should be sought from the original flooring contractor (if known) or from some other flooring authority. In the case of terrazzo, it is sometimes necessary that the floor be reground to remove that portion of the topping wherein the conductivity has been destroyed. This is an expensive procedure which should be performed only by experts.

## Carpeting in Hospitals

As of this writing, there has not been a definitive answer to questions regarding the microbiological safety of carpeting in patient rooms and critical areas of hospital environments. Several articles have appeared on this subject, some of which have purported to show that the use of carpeting would present no infection hazard. In some instances the research was badly flawed because sampling or other procedures were carried out by personnel not technically qualified to perform this work with the degree of dependability demanded by the project. In other instances, the design of the experimental work or the purpose for which it was established did not result in answers to this basic question. There is doubt that adequate techniques have been developed to make possible a definitive answer to the question of microbiological safety of carpeting. It should be noted that the significant questions of safety do not relate to public areas, but rather to patient rooms and critical areas likely to be frequented by persons harboring and shedding highly contagious and often antibiotic-resistant organisms, further, there is great difficulty in controlling odors in carpeted areas, such as those involving incontinent patients.

For use in the non-critical areas of hospitals, it would seem wise to select carpeting of the most easily maintained type. This is likely to be a continuous-loop, closely-woven pile, with sponge rubber backing so applied or of such construction that it does not permit moisture to permeate into the backing itself. To avoid the possibility that particles of fiber lint may become airborne and serve as vehicles for distribution of micro-organisms, it appears best to use continuous filament pile, rather than spun staple fibers.

Few of those persons whose opinions on carpet maintenance in the hospital seem worthy of consideration have concluded that there is any particular advantage in the use of disinfectants applied by spray or otherwise to the carpeting. One study compared the results in terms of airborne organisms which were recovered when carpets were vacuumed daily versus those obtained when vacuuming was done on alternate days or every third day. That investigator concluded that his data indicated vacuuming on alternate days provided optimum bacterial control results for the labor investment and indeed appeared to give conditions as good as those when vacuuming was done daily. (From the standpoint of appearance and carpet wear, daily vacuuming of trafficked areas is still needed in most instances.) In that particular study it was concluded that general shampooing of the carpeting would be needed at about six-month intervals, with more frequent repetition of this procedure in heavily trafficked areas.

## Conclusions

There are differences among authorities as to the degree of importance of environmental sanitation in cross-infection control. But there is no doubt that it plays a part. Housekeeping personnel are members of the infection-control team, and their role deserves to be carefully planned, implemented and supervised.

# 43 Common Mistakes in Cleaning

IN our consulting work in housekeeping management, we have come across a number of oft-repeated mistakes made in cleaning. Although in consulting work we find the approach of positive improvement through "working smarter" to achieve the greatest benefits, yet we also find that a review of common mistakes can be both interesting and instructive.

Naturally, other mistakes are being made, but these are the ones most often seen; and they are also the ones contributing to the greatest loss in time and quality.

## Dust Mopping

1. A number of operational errors occur in dust mopping, the basic dry floor-care job. The purpose of the dust mop (be it treated cotton, impregnated paper, or the like) is to catch and hold soil on its porous and sticky surface. A good rule to follow in dust mopping is to keep the mop on the surface to be cleaned until it is necessary to vacuum or remove the

mop head, and to lift it as little as necessary to avoid dropping some of the dust from the mop surface and causing it to become airborne.

A basic principle that should be followed in performing manual work of any kind is to utilize as few muscles as possible, and to favor the use of the larger rather than the smaller muscles.

Combining the principles mentioned above, dust mops are best used in open areas, such as corridors, by moving the mop forward simply with a walking motion, with the handle of the mop stick held on or near the hip by one hand.

The common mistake made here is to swing the mop from side to side, using many small muscles of the arm and torso, which fatigues the worker while throwing dust from the floor and from the mop into the air.

2. The dust mop is one of six basic long-handled manual tools; each is to be used with its own proper motion for best results. Whenever a dust mop is used in the same way as one of these other tools, poor results will be achieved.

The swinging motion mentioned above, which uses so much effort and creates so much dust, is the characteristic motion for the *wet mop.*

A very common error is to utilize the dust mop as if it were a *push broom,* with a sweeping, then lifting motion, ending with dropping the dust mop back to the floor near the beginning point. This lifting motion tends to throw a good deal of soil into the air, just like the swinging motion, but merely in a different direction.

Another long-handled manual tool is the *corn broom,* where working sideways, or at right angles, to the work simplifies the motion, but this merely makes dust mopping more difficult.

Another such tool is the *deck brush,* where the proper motion is a back-and-forth scrubbing motion using a good deal of pressure. When this technique is used with a dust mop, too much effort is used for the results achieved. Normally, the weight of the dust mop itself provides sufficient pressure, and one pass over the surface to be cleaned is sufficient.

Finally, the *floor squeegee* is best used by tapping it at the end of the forward stroke so that drops of water are not left in the dried area on the back swing. When the dust mop is handled in this way, the tapping motion does indeed release a good deal of the dust that has accumulated on the mop, but a great deal of this will go into the air, later depositing a film of dust on numerous horizontal surfaces, which will then require more cleaning than normal.

In congested areas, a swivel-type dust mop is best used, permitting the mop to slide around furniture legs and other obstructions with a simple twist of the mop stick. The size would range from about fifteen to twenty-four inches in length, depending on the congestion of the furniture, and the skill and strength of the worker. With good training, the worker can use the dust mop with one hand, leaving the other hand free for moving furniture and other obstructions. The use of dust mops in open areas is described above.

3. Dust is held to the surface of the dust mop by the nature of the fibers; this natural retention is improved by the addition of an impregnating material. This impregnation is well controlled when disposable paper is used, or sometimes when a rental mop system is used. In the latter case, the impregnating material is charged into the washing machine at the final rinse cycle (this is the same material and method used for impregnating cotton matting).

In larger housekeeping departments, it is quite feasible to operate a washer-extractor so that this impregnation can be done on the premises; but in smaller operations, where the rental system is not used, the problems in mop treatment can be avoided by having all mops treated by one trained custodian at the end of his work shift, so that the fibers are properly impregnated and ready for use the next day.

Where each worker is permitted to treat his own mop, we often find two common mistakes being made: over-treating and under-treating of the mop head.

Over-treating will put so much liquid material into the mop head that it is not all absorbed, so that in use a film of treatment is actually deposited on the floor; this can cause slipping hazards, as well as a softening of the floor through chemical attack. The appearance of numerous indentations in the floor caused by furniture legs may indicate that the tiling has been softened by such chemical action.

Under-treatment means that the mop head will not hold the dry soil properly, and that a good deal of it will become airborne in one of a number of ways, so that the dust settles back down both on the floor which the worker was supposed to have cleaned, as well as on all other nearby surfaces.

4. There is, of course, a limit to the amount of soil which a dust mop can hold in its fibers and on its surface. On very heavily soiled floors,

other types of cleaning—such as automatic scrubbing, or wet mopping, or dry vacuuming—should be considered so that the dry mopping does not cause an immediate "loading" of the dust mop, so that it is only efficient for a short while. As the dust mop begins to overload, it develops a slick, hard leading edge. It is true that the usefulness of the dust mop can be extended by exposing new surfaces, but it is a very common mistake to over-use dust mops far beyond their ability to retain soil. Naturally, the result is somewhat similar to wet mopping a floor with dirty water, since it causes a redistribution of soil.

In many facilities, such as hospitals, and in other areas where especially high quality is desired, such as building lobbies, dust mops are collected, laundered, and retreated daily. This is a very desirable practice. With the disposable paper dry mops, these can be discarded at will as they become filled.

The day of the dust box, into which mops were shaken, seems to be gone; but an even more valuable method of removing soil to extend the usefulness of the dust mop, is by dry vacuuming, either with a tank-type vacuum or with a central vacuum system. Some of the central systems have built-in floor slots specifically for the cleaning of dust mops.

5. When dust mops are mishandled, they bump and rub furniture legs and baseboards, causing unsightly build-ups of soil that are later difficult to remove. In some organizations, everything from the floor level up is stained up to a height of about four inches.

This problem can be resolved, of course, through a simple training program.

6. Even if the dust mop is properly treated, the proper motion is used, and the worker well trained, productivity can be sharply limited, and the worker greatly fatigued, by providing a handle that is simply too short. Standard mop sticks—whether they be used for dust mops, wet mops, squeegees or the like—are only fifty-four inches long. Studies show that a mop stick when standing on the floor should reach to eye level, in order to be used without the worker bending over. When purchasing manual equipment that uses wooden or metal "sticks," it is desirable to purchase a stick that is at least five feet long, and in many cases six feet long. Where the handle is too long, it can be cut down to size.

7. All dust mopping, of course, is a corrective operation to remove soil from floors once it has entered a building. A good deal of floor work,

both dust mopping and other types, can be avoided through the use of matting used as a preventive device to trap soil at entrances and other locations before it can be tracked throughout a building.

As wage rates go higher and higher, the use of matting, both the rental type and the owned type, becomes more and more worthwhile.

A common mistake is made in the use of matting, however: unless it is policed regularly, it can become so filled with the very soil that one is attempting to trap, that this soil is picked up and transferred to other areas just as if the mat were not there.

Depending on the weather and the type of soil, the matting needs vacuuming, shaking, or otherwise cleaning, periodically—sometimes as much as every hour. Where the matting becomes so wet that it can no longer be cleaned, it should be replaced with dry matting. This attention must not be considered as time lost, because the more policing time given, the more soil will be eliminated from adjacent floor areas.

## Wet Mopping

1. Probably more soil is removed from buildings every day through wet mopping than through any other cleaning technique. Yet, a number of common mistakes are made in this procedure that outwardly looks so deceptively simple.

The error most common to wet mopping—and of all mistakes made in housekeeping, probably the most common and one of the most significant —is, very simply, failure to change the mopping water frequently enough. Cleaning personnel will use mop water longer than desirable because they do not want to take the time or effort necessary to dump out the soiled water and prepare a new detergent solution; this is one reason why the location and equipping of custodial closets at strategic points is so important. Some workers will use mopping water too long through a misdirected attempt to save money by not wasting detergent.

A detergent solution can contain only a given amount of soil; the detergent molecules become "used up" in their envelopment of soil particles through the chemical phenomenon called micelles. The results of over-use of detergent are unhappy: the soil is redistributed, often with the display of the very swirl marks that put it down, to completely negate the efforts of the custodian. In terms of cross-infection, the mopping can create conditions worse than existed before the job was done!

2. Just as failure to change mopping water represents an under-use of detergent, it is also quite common to over-use detergent. Untrained custodians feel that the more detergent that is used, the easier the job will be on them, expecting the additional chemical to do more of the work. It is too bad this is not the case, since it would vastly simplify the economics of cleaning. In actuality, an over-use of detergent can cause so much suds that either an additional rinsing operation is necessary, or a sticky, unsightly film is left to attract and hold soil. Even where suds are not formed in excess, such a film may be deposited.

Consumption of detergents should be controlled through dispensing pumps, measuring cups, or proportioning devices, not so much to save the cost of the product—which although of importance, is a relatively minor item—but rather to assure that efficient cleaning is performed.

3. One of the most difficult of all custodial jobs is the cleaning of baseboards. Special brushes, floor machine attachments, and even special machines have been designed for this purpose. Yet, most of the difficult part of baseboard cleaning can be avoided altogether through proper mopping techniques.

The untrained, unskilled worker falls into the correct pattern of wet mopping—moving backward while swinging the mop in a "figure S" movement—very easily; but if he is adjacent to a wall, at the end of each stroke his mop will come into contact with the baseboard or the lower part of the wall. Since the mop will be charged with a good deal of soiled water, this pounding action deposits the soiled water onto the baseboard. It has the same affect as if one would slap a wet, dirty rag against a wall. Frequent repetitions of this mopping technique create a soiled area that is very unsightly and difficult to clean. The problem is compounded further when wax is applied in the same manner; the finish deposited on the baseboard simply seals in the soil beneath.

All this can be avoided through a simple technique known as "striping" or "lining" the baseboard. The mop is pulled along parallel to the baseboard, right alongside it, but preferably not touching, so that we have a mopped area that extends from the baseboard usually about eight or ten inches. Then, when the "figure S" pattern is started, the worker stays far enough away from the wall so that the ends of the strokes merely overlap the mopped area parallel to the wall, with the mop never actually touching the baseboard.

4. When wringing a mop, it is unavoidable that some of the soil will

be left in the yarn, trapped by the fibers, especially near the head of the mop. But we can improve the effectiveness of the mopping procedure by reducing to a minimum the retained soil and water, so that on the next application to the floor we will not be putting both the soil and water back down.

With an inefficient wringer or inefficient wringing action, a good deal of the weight that is being lifted in the mop is never disposed of, and this becomes useless effort. Of course, when the mop remains relatively soiled as it goes back to the floor, some of the soil in the mop is transferred elsewhere.

The three principal types of mop "wringers" are the roller, side-pressure, and downward-pressure, in increasing order of effectiveness. With the roller wringer, it is difficult to squeeze near the head of the mop, and a good deal of physical effort is required with relatively poor results. The side-squeezer wringer is more effective, but has the disadvantage of spraying water on the floor at times. The downward-pressure wringer is by far the most effective type to use, also requiring the least effort.

While speaking of mop wringing, it is worth mentioning that mops should be washed out in a washing machine or mop-washing device on occasion; this removes most soil and bacteria and, combined with the hanging of the mop so that it can receive sufficient airing, avoids problems with odors due to souring. The fan-type mop—which has a band of cut tape sewn near the bottom of the strands—washes well without knotting, and also performs well in use, as it lays flat on the floor without balling up. Some housekeepers will bleach their mops during laundering for better appearance.

5. Mop handles can be dangerous weapons—when left projecting from a bucket. A number of accidents have been caused by such placement of mop handles, where people have walked into them and received serious facial or eye injuries. The same problem can occur when mops are carried carelessly. When mops are left in buckets, they should be kept next to walls or corners, and leaned against the wall if they are not self-supporting; when carrying mops, the stick should be held upright (one convenient method is to put the head in an empty bucket, with the bucket being held in one hand and the stick in the other).

Mops can also cause damage to surfaces by mishandling, the most common example of which is pulling the mop stick straight up from a roll type wringer in a low-ceiling area with resulting damage to the ceiling.

6. Assuming that "if a little is good, a lot is better," custodians will use too much water in the mopping process. The flooding of floors, particularly if the area stays completely wet for some time, and this is repeated regularly, can cause severe damage to resilient floors; eventually the tiles will curl at the corners and edges, due to the loosening of the mastic at the joints; this curling accelerates the loosening action, so that in time the tiles are completely lifted from the floor. This same action may be observed when water damage occurs through pipe leakage or from rain blown in through open windows.

## Spray-Buffing

1. The spray-buffing technique is one of the most productive floor care systems developed in recent times. Many of the organizations which have tried this technique, unhappily, have discarded it following apparently unfavorable results. Usually this is not because the organization did not have suitable surfaces for this technique, but because one or more common errors associated with it had been made. As a result, many housekeeping departments which should be taking advantage of the spray-buffing technique are not doing so.

The most common error in the use of the spray-buffing technique is considering it as a *cleaning* method; thus, it is used on floors which are too heavily soiled; in this case, the floor often looks worse than before any action was taken at all!

In spray-buffing, a thick, loosely woven nylon or animal hair pad buffs a floor which has been wetted with a dilute polymer finish containing a small detergent component. This action tends to remove the abraded finish, impregnating it into the pad. Also removed are rubber heel burns, scuff marks, and a small amount of soil; the floor is left with a thin film of new finish. Where the floors are not overly soiled, the results become increasingly attractive, both from the standpoint of appearance and cost.

2. It is a mistake to use any floor pad which may be handy for spray-buffing. Pads which are too coarse will scratch the floor; pads which are too fine will load too quickly. And pads which are too thin will curl or fold.

Most pad manufacturers now provide products specifically made for the spray-buffing technique; color-coding helps to identify them.

3. If the pads are not given regular attention after use, the material

removed from the floor can harden and becomes difficult to remove. A good deal of money has been wasted by throwing away pads which could have been restored to usable condition. Often, a small section of pad (such as the material which comes from the center hole) can be used to brush off dryer soil. Pads can also be rinsed out after use. Where the build-up is heavy or hardened, a soaking in a stripping solution is helpful. For severe conditions, the pad can be run through a washing machine, or even steam-cleaned.

Once the pad has been torn, it can either be trimmed down for use on a smaller machine, or cut up to be used as scouring pads by hand, or tied to the heel of a wet mop for spot cleaning.

4. A number of types of sprayers are available for applying finishes while spray-buffing; these can amount to anything from a small plastic finger-bottle costing less than a dollar, to a relatively expensive pressurized or electrical atomizing system.

Some of these systems clog very easily, so that if the spray nozzle is not flushed out after each day of use, the finish hardens in the orifice, either making the sprayer very difficult to clean or causing it to be thrown away.

There are a few spray devices which do not clog after use; but where the slightest trouble is encountered, it should be made standard practice to clean the spray unit by spraying clear water through it following each day of use. Unfortunately, it's a common mistake for spray-buffing to be put in general use for a few days, then for custodians to revert to a former method of floor maintenance because all the sprayers have become clogged.

5. As the custodian moves his floor machine forward into the wetted area while spray-buffing, he continues to spray a few feet in advance of his machine. But he can make the mistake of spraying too far in front of the machine—which is easily done, particularly if the spray head has been adjusted to coarse droplets—in which case the solution may actually dry on the floor before the machine reaches it. Even though the machine then goes over the area later, it still leaves a pattern of dark circles on the floor, which require a good deal of effort for removal.

Care should be taken in the adjustment of the spray nozzle, and the length of the spray pattern.

6. While speaking of the spray pattern, some workers will make the error of spraying walls, furniture, the floor machine, and even the worker

himself! Actually, keeping the spray directed two or three feet away from walls or furniture is best, since the damp pad will move a quantity of the material to the sides, but properly in smaller quantities, since these areas are less heavily trafficked.

7. As the pad rotates on a dampened floor surface, it will pick up a mixture of soils, including old wax material. A certain amount of this soil works its way up into the pad (hence the desirability of a thick pad), but a good deal, unavoidably, stays on the surface. After a time, the pad becomes "loaded"—actually the surface becomes slickened with the soil. Before this slick surface develops to any extent, the pad should be turned over or a fresh pad obtained. If the pad is used too long, very little cleaning action will take place, and the appearance and condition of the floor will be harmed rather than helped.

8. There is a final step to the spray-buffing technique that may be forgotten, with unhappy results: a dust mopping to remove the loosened finish and soil, to prevent its being tracked over the floor.

## Wax Stripping

1. Just as in wet mopping, one can use too much detergent for the job. But while in wet mopping the over-use of detergent merely takes more time, in wax stripping it can actually cause damage to the floor. Most strippers are alkaline in nature, which gives them the ability to penetrate the finish or wax surface, and to break it up for emulsification and suspension in the solution. Too much of this caustic can actually leach the color from resilient tile—a process also known as "burning."

In order to remove a heavy build-up adjacent to baseboards, custodians have applied overly-concentrated stripping solutions, or even poured on or mopped on undiluted stripper; perhaps you have seen corridors that have been blanched or discolored down both sides, irrevocably damaging the floor, by just such a mistake in cleaning technique.

2. Again as in wet mopping, putting down too much water and leaving it down too long will permit the solution to work its way between the tile joints, and then under the tiles, causing them to loosen and curl, thus damaging the floors permanently as above, but in another way.

3. After the stripping solution has been mopped on the floor and scrubbed, the resulting slurry of solution and old finish is removed preferably with a wet vacuum, or with a mop. If the removal were not rapid

enough, by the simple process of evaporation, the floor would dry out again—and this leads to probably the most common mistake made in wax stripping. If the floor is permitted to redry before the old material is removed, then the finish is redeposited on the floor, and forms a new film, which is just about as difficult to remove as if the floor had never been stripped at all. If removal is not attempted, this redeposited film leaves an unsightly blotch on the floor, which is then covered up with new wax, to create a very poor looking floor indeed.

4. Somewhat similarly, blotches may be left by failure to strip completely, which most often occurs on a floor with a considerable build-up of numerous coats of finish, probably as a result of previous poor floor care techniques. Naturally, this is most often to be expected near baseboards and under and around furniture. With heavy buildups, it may not actually be possible to penetrate all the layers of finish with a single stripping, and successive applications of stripping solutions, scrubbings, and removals may be necessary in these built-up areas. It's always a good practice to examine the floor from a sharp angle to see if all the old wax has indeed been removed. Where only a few small spots remain, these can be removed individually without the need for an entire restripping of the floor (it's good to bear in mind the minimizing of the amount of water used on the floor).

5. When the floor has been completely stripped of old wax, it will still be covered with a fine film of detergent; this should be rinsed well with clear water, which would then be removed by vacuum (again preferably) or mop, so as to provide a truly clean floor surface to receive the new finish.

Where the mistake is made of rewaxing the floor which has not been rinsed after stripping, the detergent membrane can actually soften the finish from beneath, creating a tacky or slippery floor, as well as one which is dull in appearance. As usual, the error leads to more work than would have been done otherwise.

6. It is good practice to clean baseboards while stripping floors. For one thing, baseboard cleaning unavoidably causes solution to run on to the adjacent floor area, which would damage a finished floor. Then, the buildup of soil and finish on baseboards proceeds along with that of the adjacent floor area. It is not only expedient but simple to clean both areas simultaneously with a baseboard scrubbing machine or baseboard devices on regular floor machines.

Actually, the baseboard should be considered a part of the floor, since it is made of similar material, that has simply been curved up the wall a distance of a few inches! And, custodians should not consider a floor properly stripped where the baseboard remains soiled.

## Waxing

1. "If a little bit is good, then a lot is better"—this fallacious reasoning is applied to many aspects of custodial maintenance, and contributes to one of the most common problems in waxing. Custodians, by laying down thick coats of wax, feel that they are saving themselves work because of the additional protection to the floor, and believe they will not have to refinish the floor as frequently.

Unfortunately, thick coats of wax cause many problems, and generally leads to a great deal more labor than the application of two or three thin coats; in addition, it is almost impossible to get a high-quality appearance without the use of thin coats.

Thick films have the tendency to be much softer than thin films, and are thus more susceptible to scuffing and smearing. This softness also contributes to the retention of a good deal of soil, causing a dingy if not downright dirty appearance. Finally, the thicker film is apt to be a good deal slipperier than the thinner film.

The benefits of the spray-buffing system, which provides an almost microscopic film thickness, well demonstrates the values of the thinner coats of wax or finish.

2. The unsightly darkening of floors adjacent to walls and furniture, caused by a build-up of successive layers of wax and soil is not as infrequently seen as it should be. It is a mistake to assume this condition is caused by improper cleaning or stripping techniques; rather, it is caused by improper application of wax or finish. Because of the way the human body is arranged, and the fact that people don't like to rub up against walls when they're walking, footsteps are rarely made closer than six inches to a wall or other vertical obstruction. The principal reasons for waxing a floor are protection of the floor surface, and appearance; and both of these objectives can be achieved with a single coat of wax in untrafficked areas. Further, buffing (wet or dry) tends to spread the wax from the center of the floor to the sides, especially the softer waxes but also, to a lesser extent, the harder finishes.

After a floor has been stripped, only a single coat of wax or finish should be applied adjacent to a wall. Many custodians prefer to make this the second coat, therefore covering up the edge of the first coat that might begin several inches away from the wall. It's often helpful to follow the line of a resilient tile joint in applying wax away from the wall, which would put the coat about nine inches away from the wall.

3. In an effort to improve quality or cost, purchasers of custodial materials will change finishes and waxes from time to time. But this change should not be made without testing the new material with respect to its compatibility with materials which are already on the floor, if they are not to be stripped first. For example, where harder materials are put over softer materials, such as applying a polymer finish over a carnauba wax, cracking and powdering of the top finish would be likely.

Since floor finish materials are rather complex, containing natural waxes, synthetic resins, plastic polymers, emulsifiers, levelers, etc., the testing of samples befores the purchase and application of large quantities can save a great deal of time and expense. Where there is any doubt, the new material should be applied following a complete stripping of the old.

4. A number of improperly maintained floors take on a very dingy cast, appearing grey, brownish, or sometimes taking on a mottled white cast. Although this can be due to failure to mop or sweep the floors regularly, very often the soil is actually sandwiched between two layers of wax, when wax is put down on a floor that has not been properly cleaned. After this has been done, dust mopping or wet mopping will not remove the trapped soil, and the floor will stay in very poor condition until a complete stripping is performed.

5. Time is often a problem in the refinishing of floors, especially in those organizations which operate on more than one shift, or where floor finishing project teams have not been set up, and each area-assigned worker is responsible for his own project work near the end of his shift. This time pressure may result in the application of a second coat of wax over a first coat which has not properly dried. Drying time is affected by a number of factors: the humidity, temperature, air movement, the thickness of the wax film, the nature of the sub-surface, formulation of the finish, etc. Normally, a twenty- to thirty-minute drying time is necessary if the material has been properly applied, and the first film must be at least dry to the touch.

Where the first coat has not dried properly, the application of the second coat will tend to re-emulsify the base film. At the least, the result would be a dingy looking surface. In addition, the reemulsification would not be uniform, and a mottled or splotchy appearance might result, perhaps along with powdering, bubbling, etc.

6. The mishandling of mops probably creates more work for custodians than any other improper technique. Just as in wet mopping mentioned previously, the striking of baseboards with wax will deposit material there, which causes an unsightly appearance as well as a difficult removal problem later on. The same thing can happen to furniture legs, file cabinets, doors and door jambs, and the like. The training of custodians and maids in proper mop-handling technique is an indispensible part of a good housekeeping program.

7. In applying the wax film, a clean cotton mop is the best applicator with which to cover fairly large areas (some skilled custodians prefer the "string' type of cotton mop, which has finer strands than the usual cotton mop; those who prefer lambs wool applicators, however, are not apt to get both good coverage and results).

To produce the proper thin film, the mop must be wrung fairly well, and moved in a "figure S" pattern, with the worker stepping backwards with each swing. He must avoid rubbing the mop in a back-and-forth motion, since this causes a bubbling of the wax because of the emulsifiers which are part of its formulation. When the bubbles break, the wax dries with thousands of miniature craters in its surface, which collect soil and present a very poor appearance.

8. Housekeeping managers appreciate an interest in the conservation of materials, but there is an important exception to this. Wax is an emulsion; a dispersion of minute droplets of material in the water medium. It is this emulsion that gives waxes and finishes their characteristic milky color (of course, milk itself is an emulsion). Emulsions are rather sensitive, and bacterial action, for example, can tend to "break" the emulsion by causing the globules to clump together and fall out of suspension.

Once a wax has been poured out of its original container, it should be considered as lost, since the pouring of a quantity back into the container may introduce bacteria-laden soil, which can spoil the entire drum or can. This spoilage may not be immediately noticeable, and a good deal of effort will be wasted in applying this faulty material to a floor, only to find out that it gives a very poor appearance.

Thus, when pouring out these emulsions for finishing or waxing, it is best to be conservative, obtaining a little less than might be thought needed. It's always easy to pour out more, but any material remaining should go down the drain.

## Buffing

1. The first step in dry buffing is to prepare the floor by removing the loose surface soil with a dust mop. Of course, if the floor is quite soiled, a damp-mopping or wet-mopping would be necessary.

Where buffing is performed without this prior cleaning, the action of the floor machine will cause a scattering of the soil not only over the floor, but up into the air, from which it settles back down onto the floor as well as all other surfaces, such as furniture, ledges, equipment, and the like. Dust mopping is a very easy and quick job, and can save the expenditure of a good deal of effort later on.

2. The single-disk floor machine is by far the most common type of floor care equipment in use. The rotation of the brush will tend to move the machine to the right or left depending on whether the handle is raised or lowered. Workers must be trained in the "feel" of the machines in open areas before they are permitted to use them in congested areas, to avoid bumping—which causes both damage and marking—of furniture, walls, doors, and other objects.

3. Where an operator has not learned the proper handling of the equipment, he is apt to "fight" the machine rather than control it, and he can become fatigued very quickly before the performance of a reasonable amount of work. Proper training permits even smaller women to handle such machines with relative ease. Further, the development of vibrating and oscillating-type equipment also tends to limit this particular type of problem.

4. Synthetic floor pads are now being manufactured in four basic grades: Stripping, scrubbing, spray-buffing, and dry buffing. It is a rather common mistake to choose the wrong pad for the dry-buffing technique. Naturally, the pad that is too aggressive will cut through the finish, and sometimes through the floor as well!

Fortunately, a number of manufacturers are beginning to color-code pads to indicate their use, and this greatly simplifies training custodial workers in proper pad selection.

In addition to these synthetic pads, other materials are available for dry buffing. These include steel wool (also available in various grades) and only the finer grades should be used for buffing; of course, steel wool should not be used on terrazzo because the small metal chips which might remain imbedded in the floor can cause serious rust stains; and terry-cloth pads and lambs wool pads (which are used for ultra-sheen polishing).

## Restroom Care

1. Probably the most frequent omission in restroom cleaning, and one which is often discussed, is the cleaning of foreign matter hidden under the lips of toilet bowls and urinals. This is the source of a number of serious odor problems in restrooms, since bacteria multiply very quickly in such an environment, bacteria of course being the principal source of non-transient restroom odors. The custodial worker should be equipped with a small mirror, either of the pocket or dentist's type, for inspecting these hard-to-get-at surfaces. Normally an acid-type descaler, applied with a bowl mop, is required to remove these hidden soils and stains.
2. Acid descalers do such a good job on soil removal that it becomes a common error to use it too frequently. Normally, descaling should be performed at the frequency of from once per week to once per month, depending on the hardness of the water (and thus the rate of scale deposit) as well as the amount of use the fixtures are given. More frequent flushing means more scale deposit.

   It is not rare to find custodians using acid descalers on a daily basis, yet these materials can damage restroom fixtures and surrounding surfaces, particularly metals, and also represent a safety hazard in its application since the acids are corrosive to the skin and especially hazardous for the eyes.

   Descaling, then, should be set up on a calendar performance basis, and in some cases performed by special projects workers who are properly trained and equipped to do nothing but this in all areas.
3. "Tunnel vision" is a defect we all seem to have; it is not so much physical as it is psychological. That is, we tend to see things at eye level or below, since we often walk or stand with eyes downcast, but rarely do we notice things above our level. Thus, a restroom that might otherwise be perfectly clean might well have overhead areas that require a good deal more attention.

Examples of such areas are ceiling diffusers, ventilation grilles, piping, light fixtures, and duct surfaces.

With a little attention, this problem is easily remedied, since overhead areas normally require only periodic (such as monthly) attention. As with many other aspects of housekeeping in all types of areas, in restroom cleaning we may be performing daily repetitive jobs adequately, but fail to give proper attention to those tasks which are performed infrequently. Good housekeeping requires attention to both!

4. Perhaps you have walked into a restroom and found a crumpled-up paper towel balanced neatly on the top of a torpedo-type waste receptacle. Or worse, the receptacle surrounded by litter which has been thrown on the floor. And yet, on inspection, one finds that the receptacle is far from being full.

   A still closer inspection reveals the cause: the push plate on the receptacle has not been properly cleaned or policed. A visitor washing his hands and then drying them simply does not wish to resoil his hands on a dirty push plate—it seems to be an invitation for him to avoid using the receptacle.

   Thus we find that much of the litter in wash rooms is caused by failure to provide proper cleaning and policing, and it is hardly all due to public apathy.

5. The custodian opens the door to a restroom stall, cleans the fixture, the walls and floor, and then closes the door; this results in the common error of failure to clean the inside surface of the stall door, and the wall area which it hides. This can only be done, of course, by the custodian entering the stall, closing the door, and then cleaning it. It should be borne in mind that the visitor normally has nothing to look at but that soiled back surface of the door!

6. The only positive approach to restroom deodorization is thorough cleaning and disinfection, thus removing and killing odor-producing bacteria.

   The substitute of odor-masking and perfuming devices such as drip fluids, deodorant blocks, wicks, and aerosols can be helpful in certain difficult cases where ventilation is not adequate, or where the public demands it merely through habit, yet the use of these devices—they should not be used nearly so frequently as they are found—should not represent an excuse to avoid good restroom cleaning practices.

## Room Care

1. Perhaps you have heard stories, or had experiences, where an office worker will deliberately place a bit of refuse under his desk, then keep a record on his calendar of how long it remains there until it is removed by a maid or custodian. This points out one of the rather common mistakes in the care of such areas as offices, classrooms, patient rooms, guest rooms, and the like: the failure to clean beneath furniture.

   It must be remembered that when a person sits down, he is able to see a good deal more of the floor under the furniture than when he is standing.

   Cleaning personnel must be trained to clean underneath desks, tables, beds, and other furniture. This would include removing litter, dry mopping smooth floors (and wet mopping and buffing periodically), vacuuming and spotting carpeted floors.
2. Just as failure to clean *beneath* surfaces is a problem, so is a failure to clean *above* them. Thus we find collections of dust on the tops of lamps, over doors, on the tops of tall furniture such as cabinets, on high ledges, and the like.

   We have witnessed on a number of occasions hospital operating rooms, whose sanitation requirements are extreme, where the operating lamp, directly over the patient, has its upper surface covered with a film of soil, while everything else in the room may be spotlessly clean!
3. If one area were to be selected as that which is most often missed in room cleaning, it would be the hinge area of the door frame at the entry to the room. Where this surface is omitted in the cleaning schedule, it is for one of two reasons. In one case, the door may normally stay closed, with the hinge area being covered by the door itself, and thus not seen either by the worker or his supervisor. Interestingly, the hinge part of the door frame is invisible even when one opens and closes the door; it can only be seen when the door remains open.

The other possibility is a jurisdictional problem: where one worker is assigned to clean the room, and another to clean the corridor, the door frame becomes a "gray area," lying neither in the room nor in the corridor, and each worker leaves it to the other. As in military operations, the border areas may be very important but receive little attention unless specific instructions are prepared. It is usually best to assign the door frame to the person cleaning the room itself.

4. Waste baskets are often left uncleaned, with numerous unhappy results ranging from an unpleasant odor to the attraction of vermin.

   This cleaning omission may occur for failure to schedule the periodic washing of the receptacles, since the person emptying the receptacle normally does not have the cleaning solution, sponge, and cloth necessary to do this job. An excellent way to wash waste baskets is with a mobile double-tank cart with drainage rack, which can move from one room area to the next, the baskets being cleaned on a project basis by two persons, one collecting and returning the baskets, the other washing them at the cart. Of course, waste baskets should be standardized so that it is not necessary to return any one basket to its original location.

   Another approach to this problem, this time from a preventive rather than corrective basis, is the use of plastic liners. In quantity purchase, these cost little more than a cent each, and might last a week or more depending on the waste that has been put into the basket. They are particularly helpful where workers put wet waste—such as coffee—in waste baskets.

5. It is not uncommon for hand prints on walls, doors, partition glass, and other vertical surfaces to be ignored in the daily cleaning routine. Where "spotting" is done daily, it becomes a simple task to keep them removed, but where they are permitted to accumulate, the net result is often a complete wall washing or even painting job.

   Further, human psychology being what it is, uncorrected problems of this type can amount to invitations for more of the same! Just as people do not hesitate to throw litter on top of other litter, but would hesitate to "throw the first stone," so they would not mind putting hands or feet on walls which are already marked and not receiving attention, but would hesitate to mark a clean area.

6. Just as in rest room care, we continue to be faced with "tunnel vision," and tend to ignore areas above the six-foot level. So, it is not uncommon to see rooms that are well cleaned in other respects, having dirty ceilings, grilles and louvers laden with soil, and pipes and ducts covered with films of dust.

   Fortunately, the correction of these overhead problems is a simple matter, often most easily handled with a tank-type or pack-type dry vacuum equipped with an extension wand for reaching these overhead areas.

7. Finally, a common mistake in room cleaning that is very difficult to cope with insofar as floor care is concerned, is one which is caused by management's failure to require the occupants of the room to keep areas clear. Floors which are cluttered with boxes full of business forms, old files, and the like, cannot be properly cleaned because no liquid can be used on these floors for fear of damaging the contents of the containers; it would even be difficult to dry-buff or spray-buff such an area; this leaves only dust mopping which, in time, will leave the floor in an unsatisfactory condition.

   The same is true of cluttered furniture, where workers will leave tables and desks covered with papers and supplies. It is normally best to have a ruling that the custodial personnel will clean only exposed areas of such furniture, since the re-stacking of papers and other materials may lead to a loss of time or information that could be rather costly. But even this is not wholly satisfactory, since the desk user, on clearing his desk, finds "islands" of uncleaned areas, which he may feel reflects on the custodial department rather than on himself!

   Where management can do this, other than in executive areas, it is desirable to have personnel responsible for the cleaning of their own desk tops whenever there are any papers or other material on the surface; but where the user completely clears the surface (except for telephone, dictating machine, and such) this becomes a signal for custodial personnel to dust, and even wash or polish the desk if needed.

   Many room areas provide this same problem on vertical surfaces as the horizontal surfaces just discussed. That is, walls and partitions may be cluttered with calendars, notices, cartoons, and lists which makes their spotting and cleaning very difficult. And the removal of these materials may actually take away part of the painted surface. Management does well to limit the nature of wall-hung items.

## Housekeeping Equipment

1. The technological advance that is most often sought by cleaning management is the development of floor machines, vacuums, and other motorized equipment that can be operated with a light-weight battery. Meanwhile, we are required to use equipment with electric cords, and this leads us to a common mistake in the use of equipment: simply, failure to clean the cord.

   The electric wire—whether it be for a vacuum cleaner or for a floor machine—is naturally pulled across soiled floors, and thus becomes well soiled itself. When the cord is moved about it transfers soil to furniture legs, door frames, clean floors, and other surfaces.

   Workers should be trained, when putting the equipment away at the end of each use, to clean the cord while winding it onto the proper part of the machine. If this cleaning is performed after each use, it can be simply done with a cloth or sponge dampened in detergent solution. If it is done infrequently, however, it will be necessary to use stronger cleaning means, such as scouring powder or paste, or a warm stripping solution where some of the soil consists of wax or finish.

2. The equipment itself suffers from this lack of cleaning attention, and although there is not the problem of soil transfer so much as is experienced with the electric cord, dirty equipment does provide a poor appearance and, in the case of wet vacuums, may lead to foul odors.

   If equipment cleaning is done after each use, as with cleaning the electric wire, the time requirement is rather slight. Workers should be instructed that no equipment should be put away until it has been properly cleaned, and of course the short time required for this should be allowed.

   Equipment cleaning should involve manual equipment as well as power equipment; buckets, wringers, mop handles, and other such items should be cleaned after use.

Where such a program is first started, and a considerable build-up of soil must be removed first before beginning a daily cleaning program, the use of soak-tanks charged with a warm stripping solution, or the use of a steam-cleaning machine, can be very helpful.

3. It is natural to judge workers by the tools that they keep and the work station where they prepare for their jobs.

   It is often possible to judge the morale of the custodial workers—and of the entire department as well, as a matter of fact—as well as to be able to judge the effectiveness of the supervision—by simply examining custodial closets to see whether they have been properly organized and kept clean.

   The walls should be washed, the floors scrubbed, the door louvers vacuumed. The room should be kept lighted. Tools with handles should be hung in a vertical position, while chemicals and small supplies should be neatly stored on shelves. Empty boxes and litter should be removed.

   A simple example of care of tools in the custodial closet is the washing of mops; where wet mops are merely thrown into a corner while still containing soil, they may "sour," the result of the multiplication of millions of bacteria. If the mops are washed out after each use—and of course it is desirable to have removable mop heads rather than the old "stick" type—and hung up to dry, this problem can be avoided. In some places where the appearance of the mop is important, such as hospitals, it is even desirable to bleach them so that they will have a whiter appearance.

   Equipment that is kept clean, stored in a closet that is also kept clean and is well organized, will help develop pride of performance, and the public image of the custodial department.

The foregoing list of common housekeeping errors, hopefully, will help us discover areas of potential improvement, which can then be approached positively; we sincerely hope that this list will not be used as a vehicle for wholesale criticism.

APPENDIX A

# Typical Rules for Custodians

1. Each custodian must be regular in attendance and must not remain away from work without permission from his supervisor unless illness or emergency makes it impossible for him to report to work.
2. The custodian must be ready to begin work promptly at starting time and must not leave his work area until quitting time.
3. The custodian should report any suspicious stranger to his supervisor.
4. Each custodian is responsible for his tools and equipment and must either return them in clean and proper order to the storage area or turn them over to the next shift worker.
5. Each custodian in regular contact with the public must wear the required clothing or uniform while on duty. His appearance must always be neat.
6. Custodians must be courteous to the public, to fellow workers, to callers, and to company officials with whom they come in contact.
7. Custodians must not disturb papers, drawings, or other documents which have been left on tables or desks, nor must they open drawers, cabinets, etc., unless specifically told to do so by the supervisor.
8. Custodians must keep their own lockers clean and share in keeping the locker area clean and neat.
9. The custodian should economize in the use of electricity by turning off lights not needed unless the supervisor instructs otherwise.
10. Smoking is prohibited in restricted areas.

11. If a custodian discovers a fire, he is to turn in the alarm immediately.
12. Custodians may be disciplined for repeated failure to pay just debts.
13. Gambling is not permitted in the buildings or on the grounds.
14. The use of liquor or intoxicants is not permitted and will be met with disciplinary action.
15. Custodians must report any accident to the supervisor.
16. The custodian should report to his supervisor any repairs needed for equipment, furniture, light fixtures, wiring, tools, and other items.
17. Each custodian will see that the windows in his area are closed when leaving.
18. The custodian must turn in to his supervisor any lost articles that are found.
19. Custodians are expected to follow the prescribed safety rules.

APPENDIX B

# Sanitation Test Example

| TRUE FALSE TEST | *True* | *False* |
|---|---|---|
| 1. A waxed floor is always more slippery than an unwaxed floor. | ( ) | ( ) |
| 2. Lifting the handle of a floor machine while it is running causes it to move to your right. | ( ) | ( ) |
| 3. A wooden floor that is warped and buckled has probably been cleaned with too much water. | ( ) | ( ) |
| 4. The proper way to dust mop a corridor is to swing the dust mop from side to side in a figure "8" pattern. | ( ) | ( ) |
| 5. Dirty baseboards are often caused by poor handling of mops. | ( ) | ( ) |
| 6. The reason for a dust-mop treatment is to make the dust mop slide easier on the floor. | ( ) | ( ) |
| 7. The proper way to use sweeping compound is to sprinkle it on the entire area to be swept, then sweep it up one section at a time. | ( ) | ( ) |
| 8. The proper way to get dust off an uneven surface is to blow it off, using the exhaust from a vacuum, and then sweep it up. | ( ) | ( ) |
| 9. Wax should always be applied from wall to wall. | ( ) | ( ) |

10. Putting too much water on the floor can cause asphalt tile to curl up at the edges. ( ) ( )
11. Quarry tile is probably the easiest type of floor to keep clean. ( ) ( )
12. Solvent is a good material to use on asphalt tile floors. ( ) ( )
13. The use of the wrong type of cleaner can cause asphalt tile floors to crack. ( ) ( )
14. Thin coats of wax do a better job than thick coats. ( ) ( )
15. A mirror is a good tool for checking rest-room sanitation. ( ) ( )
16. Pouring wax back into its drum can cause the entire drum to spoil. ( ) ( )
17. An acid-type cleaner should be used on rest-room fixtures daily. ( ) ( )
18. If you wait a few minutes after applying a cleaning solution before beginning scrubbing, you can let the chemical cleaner do part of the work for you. ( ) ( )
19. It is not necessary to wait until water-emulsion waxes are completely dry before beginning buffing. ( ) ( )
20. A dusty or dirty light fixture gives out less light than a clean one. ( ) ( )

In the above tests, only the following numbers are true: 3, 5, 10, 11, 13, 14, 15, 16, 18 and 20.

## PRACTICAL PERFORMANCE TEST

Workers must demonstrate the proper care and use of the following:

a. Push broom (storage, reversing the block, combing, proper handling).
b. Dust mop (storage, treating the dust mop, proper method of use).
c. Wet mop and wringer (storage, selection of size, trimming the strands, method of use, and washing the mop and wringer).

APPENDIX C

# Container Dimensions

Container dimensions vary from manufacturer to manufacturer. The outside dimensions indicated here may be considered as average or typical figures.

TIGHT-HEAD STEEL DRUMS

For Liquids

| *Gallons Capacity* | *Diameter in Inches* | *Height in Inches* |
|---|---|---|
| 65 | 23 | 40 |
| 55 | 23½ | 35 |
| 35 | 18½ | 34 |
| 30 | 19 | 29 |
| 20 | 16 | 28 |
| 15 | 16 | 21 |
| | PAILS | |
| | For Liquids, Pastes, and Powders | |
| 12 | 15 | 20 |
| 10 | 13 | 21½ |
| 7 | 12 | 17½ |
| 6 | 12 | 15½ |
| 5 | 12 | 13 |

## OPEN-HEAD STEEL DRUMS

For Pastes and Jellies. Also for Granular Materials When Liners are Used

| *Gallons Capacity* | *Diameter in Inches* | *Height in Inches* |
|---|---|---|
| 55 | 24 | 34½ |
| 30 | 19½ | 29 |
| 20 | 19½ | 20 |
| 15 | 16½ | 21½ |

## FIBER DRUMS

For Powders. Also for Liquids When Liners or Inserts are Used.

| *Gallons Capacity* | *Approximate Pounds Capacity* | *Diameter* | *Height* |
|---|---|---|---|
| 67 | 450-500 | 23½ | 38½ |
| 55 | 350-400 | 22 | 36 |
| 41 | 250-300 | 20½ | 31½ |
| 32 | 200-250 | 19 | 28½ |
| 26 | 150-200 | 18 | 29½ |
| 20 | 125-150 | 16 | 25 |
| 13 | 75-100 | 14½ | 21 |
| 7 | 35-50 | 11½ | 18 |

APPENDIX D

# Stain Removal for Carpets

The removal of stains from carpets should be done carefully and cautiously. Try to clean a small inconspicuous "test" spot before beginning the stain removal of a larger area. By all means try to determine the cause of the stain before attempting removal. Supervise the work carefully. Many variables (surfaces, dilutions, application, supervision, etc.) offset the results obtainable.

In general, work from the outside to the inside of the stain area to prevent spreading and formation of rings. For fresh stains, blot up as much of the staining material as possible before beginning the stain removal.

1. ACIDS: Use ordinary household ammonia to neutralize the acid but allow the ammonia to stay on only a moment. Rinse thoroughly with a cloth and cold water.

2. CANDLE WAX: Remove as much as possible with a dull knife. Cover with a clean, dry blotter and then press with a warm iron. Sponge with Chlorothene (toxic chemical—handle carefully, do not inhale fumes, provide adequate ventilation).

3. CANDY: Scrape off with a dull knife. Apply Chlorothene. Sponge with hot water. (For chocolate, use lukewarm water, then follow with pepsin powder rubbed in, then sponged out after 30 minutes.)

4. CHEWING GUM: Scrape off with a dull knife. Apply Chlorothene.

5. COFFEE: Lather on a paste of egg yolk and warm water. Add a few drops of alcohol for an old stain. Rinse with clear water. (If creamed coffee, treat for cream.)

6. CREAM, ICE CREAM, OR MILK: Place a pan under the rug at point of stain. Pour on hot water, allowing it to soak for a while. Remove traces with hot water and cloth. Finish with Chlorothene.

7. ENAMEL, LACQUER, PAINT, VARNISH: Apply turpentine, then wash with strong, lukewarm soapsuds. Rinse with clear water and let dry. For a dry stain, apply half benzene-half denatured alcohol, working out the softened paint with a dull knife. Repeat as necessary for complete removal.

8. GREASE, OIL, OR TAR: Apply Chlorothene with a cloth, using a circular motion and working from the edge of the stain to the center to avoid causing a ring.

9. INK: Place a pan under the rug at point of stain. Pour on milk, allowing it to soak for a while. Use Chlorothene to remove grease left by milk.

10. MUD: Allow to dry thoroughly, then brush off and vacuum.

11. SHELLAC: Place paper under rug to protect floor, then rub with alcohol.

12. SOOT: Cover with dry salt. Then, sweep both up together and vacuum.

13. TEA: If clear tea, use mild soap and clear water. If it contains cream, use Chlorothene.

APPENDIX E

# Stain Removal for Floors

Use all chemical agents with extra care to prevent fire, skin irritation or injury, and damage to surfaces. Try to destain a small inconspicuous "test spot" before beginning a larger operation. Many variables (surfaces, dilutions, application, supervision, etc.) affect the results obtainable.

A. BLOOD

1. Asphalt Tile, Linoleum, Vinyl, Wood: Rub with cloth dampened in clear cold water. If stain persists, dampen cloth with ammonia.
2. Concrete, Marble, Oxychloride Cement, Terrazzo: Rub with cloth dampened in clear cold water. Bleach with peroxide if stain persists.

B. CHEWING GUM

1. Concrete, Linoleum, Marble, Oxychloride Cement, Terrazzo, Wood: Remove gum with putty knife. Apply alcohol, rub with clean cloth.
2. Asphalt tile: Remove gum with putty knife. (The use of dry ice will make the gum easier to remove.) Do not use alcohol on asphalt tile.
3. Vinyl: Remove as much as possible with putty knife. Rub with # 0 steel wool dipped in all-purpose synthetic detergent solution.

C. GREASE OR OIL

1. Asphalt Tile, Linoleum: Scrub with warm all-purpose synthetic detergent solution. Rinse with clear water.
2. Concrete: Pour alcohol on spot. Rub on spot. Rub with clean cloth.
3. Marble, Terrazzo, Oxychloride Cement: Pour solvent on spot, cover with fuller's earth and let stand for several hours. Repeat if necessary.
4. Vinyl, Wood: Pour kerosene on spot. Permit to soak for a short time. Wipe dry with a clean cloth. Wash with all-purpose synthetic detergent solution, rinse dry.

D. INK

1. Asphalt Tile, Concrete, Linoleum, Marble, Oxychloride Cement, Terrazzo: Use warm all-purpose synthetic detergent solution. If stain persists, mix 2 tablespoonfuls sodium perborate in pint of hot water. Mix whiting to form paste. Apply to spot and let dry.
2. Wood: Apply solution 1 part oxalic acid crystals to 9 parts of warm water. Permit to stand until dry. Mop with clear water.

E. IODINE OR MERCUROCHROME

1. Asphalt Tile: Warm neutral soap solution.
2. Concrete, Linoleum, Wood: Apply alcohol and rub with clean cloth.
3. Marble, Oxychloride Cement, Terrazzo: Apply alcohol and cover with fuller's earth.
4. Vinyl: Wash with all-purpose synthetic detergent, rinse, then dry. If stain persists, scrub with scrubbing powder and warm water.

F. PAINT

1. Asphalt Tile: Rub with steel wool and all-purpose synthetic detergent solution. If area is large, use steel wool on buffing machine.
2. Concrete: Scrub with one pound trisodium phosphate in one gallon of hot water; rinse with clear water.
3. Linoleum: Rub with # 0 steel wool dipped in turpentine. Wash with all-purpose synthetic detergent solution and rinse.
4. Marble, Oxychloride Cement, Terrazzo: Rub with # 0 steel wool dipped in turpentine.
5. Vinyl: Rub with # 0 steel wool dipped in kerosene.
6. Wood: Use oxalic acid solution or one pound trisodium phosphate in 1 gallon of warm water.

## G. RUST

1. Asphalt Tile: Rub with # 0 steel wool and all-purpose synthetic detergent solution.
2. Concrete: Dissolve 1 part sodium citrate in 6 parts water. Mix with equal parts of glycerin. Make a paste with whiting and apply to stain. For bad stains, wash with sodium citrate solution 1 to 6 parts, add pad soaked in sodium hydrosulfite for 10 to 15 minutes. Wash thoroughly with water.
3. Linoleum, Vinyl: Apply solution 1 part oxalic acid to 9 parts warm water. Let dry. Rinse thoroughly with clear water.
4. Marble, Oxychloride Cement, Terrazzo: HORIZONTAL surfaces; to $\frac{3}{4}$ gallon water add 1.9 pounds sodium citrate and 1 pound sodium hydrosulfite. Add enough water to make a gallon solution. Cover stain with solution and let stand $\frac{1}{2}$ hour. Absorb with cloth by rubbing. Rinse with clear water. VERTICAL surfaces; make paste with whiting and 3 ounces sodium citrate and 3 ounces sodium hydrosulfite. Apply with putty knife and allow to remain 1 hour. Wash with sodium citrate.
5. Wood: Wash with all-purpose synthetic detergent. Rub with # 0 steel wool if necessary.

## H. SOLE AND HEEL MARKS

1. Asphalt Tile, Linoleum, Wood. Rub with # 0 steel wool or wash with all-purpose synthetic detergent solution.
2. Concrete, Marble, Oxychloride Cement, Terrazzo: Wash with all-purpose synthetic detergent solution and rinse.
3. Vinyl: Rub with # 0 steel wool dipped in all-purpose synthetic detergent solution.

## I. TAR

1. Asphalt Tile: Remove surplus with putty knife. Do not put kerosene on asphalt tile. Wash with warm, all-purpose synthetic detergent solution.
2. Concrete, Linoleum, Vinyl, Wood: Remove tar with putty knife. Soak with kerosene. Rub with clean cloth. Wash with all-purpose synthetic detergent solution.
3. Marble, Oxychloride Cement, Terrazzo: Remove surplus with putty knife. Soak with alcohol and cover with fuller's earth.

APPENDIX F

# Standard Times

These standard times are estimates for an average situation. The figures must be considered as guides only, because of the many variables affecting the performance time of housekeeping jobs. Some of these variables are:

1. The worker
   a. Ability to work with others
   b. Age
   c. Attitude
   d. Experience
   e. Fatigue
   f. Health and limitations
   g. Intelligence and literacy
   h. Manual dexterity
   i. Morale
   j. Motivation and incentive
   k. Physical strength and build
   l. Quality of supervision
   m. Training
2. The surface to be cleaned
   a. Condition of sub-surface (level, smooth, etc.)
   b. Condition of surface (smoothness, porosity, etc.)

c. Degree of soil
d. Total area involved
e. Traffic
f. Type of soil

3. Work conditions
   a. Amount of obstruction
   b. Attitude of other employees
   c. Comfort level
   d. Desired quality of result
   e. Distance to custodial facilities
   f. Evaporation rate
   g. Illumination intensity
   h. Method of work
   i. Presence of others
   j. Quality of materials
   k. Type, size, and conditions of equipment being used

The time standards are developed on the assumption that all factors indicated above are at an average level, a condition which, of course, would be almost impossible to achieve in practice. As your housekeeping program develops, the time requirements for various jobs will be lessened. It would be necessary to take time studies to establish rates for a specific situation. Specific time studies may be found to vary as much as 50% with respect to a given estimate. Where figures are given in ranges, the variability is even more pronounced.

## Comprehensive Jobs

(These average time estimates include set-up and put-away time, some non-productive time, and assume a reasonable mechanization and training level.)

## Washrooms, Locker Rooms and Related Areas

WORK LOAD PER 8-HOUR DAY

| | |
|---|---|
| Swing & locker rooms (incl. damp mopping) | 15,000 - 18,000 sq. ft. |
| Swing & locker rooms (no damp mopping) | 17,000 - 20,000 sq. ft. |
| Toilet rooms in office areas | 120 fixtures (basins, commodes, urinals) |
| Toilet rooms in plant areas | 107 fixtures |
| Toilet rooms (alternate estimate) | 3,000 - 4,000 sq. ft. |

## General Cleaning Operations

WORK LOAD PER 8-HOUR DAY

| | |
|---|---|
| Elevators, freight: no damp mopping | 24 elevators |
| Elevators, passenger: including damp mopping | 24 elevators |
| Overhead office areas: dusting and vacuuming | 7,000 - 10,000 sq. ft. |
| Stairs: sweep treads, dust handrails | 60, 12-ft. flights |
| Stairs: mop and rinse | 45, 12-ft. flights |
| Storage & supply areas: sweep floors, dust horiz. | 40 - 60,000 sq. ft. |
| Unobstructed areas: manual sweeping | 60 - 80,000 sq. ft. |
| Unobstructed areas: power sweeping | 400 - 600,000 sq. ft. |

## Policing Operations

WORK LOAD PER 8-HOUR DAY

| | |
|---|---|
| Lobbies and corridors | 200,000 sq. ft. |
| Stairs | 180 flights |
| Swing & locker rooms | 45,000 sq. ft. |
| Toilet rooms in office areas | 360 fixtures |
| Toilet rooms in plant areas | 320 fixtures |

## General Cleaning Operations

TIME PER OPERATION

| | | | |
|---|---|---|---|
| Chair, cafe, wash | 36 sec. | Dispenser, towel, refill | 90 sec. |
| Classroom (40 desks), dust | 5 min. | Dust executive office | 3 min. |
| Desk or table, clean & wax | 10 min. | Gen'l ofc. dusting 1,000 sq. ft. | 12 min. |
| Desk or table, strip & rewax | 15 min. | | |
| Desk or table, wash | 6 min. | Partition, wax or polish, 1 ft. | 90 sec. |
| Desk top, glass, wash | 115 sec. | | |
| Dispenser, napkin, refill | 90 sec. | Table, cafe, damp wipe | 90 sec. |
| Dispenser, soap, refill | 60 sec. | Trash can, wash | 4 min. |

Vacuum & wash drop light 3.3 min.
Vacuum & wash fluor. light 5.9 min.
Wash receptacle, large, empty 30 sec.
Wash receptacle, small, empty 15 sec.
Windows, washing, typical 40 - 60 windows

## Floors

MINUTES PER 1,000 SQUARE FEET

| | *Degree of Obstruction* | | | |
|---|---|---|---|---|
| *Maintenance Operation* | *None* | *Slight* | *Medium* | *Heavy* |
| Auto scrubber-vac, single pass | 16–20 | 18–24 | — | — |
| Buff, 16″ single-disc machine | 23 | 32 | 38 | 42 |
| Buff, 19″ single-disc machine | 15 | 25 | 30 | 35 |
| Carpet shampoo, Dry Foam | 65 | – | 80 | – |
| (Same, but before and after) | 110 | – | 125 | – |
| Carpet shampoo, water extraction | 400 | – | – | – |
| Damp mop | 14 | 17 | 20 | 28 |
| Dust mop | 7 | 9 | 11 | 13 |
| Hose and squeegee | 20 | 25 | 36 | 43 |
| Rewax (apply 1 coat wax only) | 16 | 19 | 22 | 27 |
| Scrub, manual | 75 | 105 | 120 | 135 |
| Scrub, 16″ single-disc machine | 50 | 60 | 85 | 95 |
| Scrub, 19″ single-disc machine | 25 | 30 | 40 | 45 |
| Spray-buff, 19" machine | 30 | 40 | 50 | – |
| Strip once and rewax | 110 | 140 | – | – |
| Sweep, administrative areas | 9 | 11 | 13 | 16 |
| Sweep, plant areas | 11 | 14 | 18 | 22 |
| Vacuum carpets | 20 | 23 | 29 | 35 |
| Vacuum, dry | 15 | 19 | 23 | 28 |
| Vacuum, wet | 29 | 33 | 37 | 45 |
| Wet mop and rinse | 32 | 36 | 40 | 48 |

## Furnishings

DUSTING, SECONDS PER ITEM OF AVERAGE SIZE.
(DOUBLE THESE FIGURES FOR DAMP CLEANING.)

| Item | Seconds | Item | Seconds |
|---|---|---|---|
| Air conditioner, unit | 90 | Lamp, wall fluor. | 8 |
| Ash tray | 15 | Lamp, desk fluor. | 18 |
| Book case, 36″ x 40″ | 35 | Lamp, with shade | 35 |
| Cabinet, 3′ x 6′ | 108 | Mural, 3′ x 5′ | 45 |
| Calculator, covered | 8 | Pencil sharpener | 15 |
| Chair, large | 43 | Pictures, framed | 15 |
| Chair, medium | 35 | Rack, 6′ coat and hat | 90 |
| Chair, small | 22 | Radiator, open | 180 |
| Cigarette stand | 25 | Radiator, enclosed | 30 |
| Clock, desk | 8 | Sand urn | 60 |
| Clock, wall | 20 | Spittoon | 180 |
| Coat tree | 15 | Table, large | 40 |
| Desk, large | 28 | Table, medium | 35 |
| Desk, medium | 23 | Table, small | 22 |
| Desk, small | 18 | Telephone | 9 |
| Desk trays | 8 | Tel. switchboard | 110 |
| Dictator, covered | 8 | Typewriter, covered | 7 |
| Door, flush, dust | 25 | Vending machine | 60 |
| Door, glassed, dust | 40 | Venetian blind | 210 |
| File cabinet | 25 | Sofa or divan | 150 |
| Fire extinguisher | 16 | Waste basket | 16 |
| Glass part. dust sq. ft. | 1.2 | Wall, dust per sq. ft. | 2.1 |
| Window ledge per ft. | 2 | | |

## Walls, Ceilings and Glass

CLEANING, SECONDS PER ITEM.

| Item | Seconds | Item | Seconds |
|---|---|---|---|
| Door, wash both sides | 150 | Glass part., opaque, wash, per sq. ft. | 3 |
| Glass part., clear, wash, per sq. ft. | 8 | Wall, painted, wash, per sq. ft. | 9 |

| | | | |
|---|---|---|---|
| Wall, marble, wash, per sq. ft. | 5.5 | Wall, vacuum, per sq. ft. | 4.7 |
| Wall, tile, wash per sq. ft. | 9 | Windows, wash, per sq. ft. | 7.5 |

## Rest Rooms

CLEANING, SECONDS PER ITEM.

| | | | |
|---|---|---|---|
| Basin, incl. soap disp. | 120 | Mirror, average | 30 |
| Bradley basins, semicircl. | 180 | Mirror, large | 60 |
| Bradley basins, circular | 300 | Receptacle, paper towel | 10 |
| Dispenser, napkin | 13 | Shelving, per sq. ft. | 12 |
| Dispenser, paper towel | 7 | Toilet, incl. partition | 180 |
| Door, spot clean | 50 | Urinal | 120 |
| Drinking fount. | 110 | Wainscot, per 10″ | 3 |
| Fixtures, destain | 180 | | |